Springer Tracts in Civil Engineering

Springer Tracts in Civil Engineering (STCE) publishes the latest developments in Civil Engineering - quickly, informally and in top quality. The series scope includes monographs, professional books, graduate textbooks and edited volumes, as well as outstanding PhD theses. Its goal is to cover all the main branches of civil engineering, both theoretical and applied, including:

- Construction and Structural Mechanics
- Building Materials
- Concrete, Steel and Timber Structures
- Geotechnical Engineering
- Earthquake Engineering
- Coastal Engineering; Ocean and Offshore Engineering
- Hydraulics, Hydrology and Water Resources Engineering
- Environmental Engineering and Sustainability
- Structural Health and Monitoring
- Surveying and Geographical Information Systems
- Heating, Ventilation and Air Conditioning (HVAC)
- Transportation and Traffic
- Risk Analysis
- Safety and Security

Indexed by Scopus

To submit a proposal or request further information, please contact:

Pierpaolo Riva at Pierpaolo.Riva@springer.com (Europe and Americas)
Wayne Hu at wayne.hu@springer.com (China)

Vahid Vakiloroaya

HVAC and Building Services Rules of Thumb

A Quick Reference Guide

Springer

Vahid Vakiloroaya
Construction Manager
Pacman Air Pty Ltd
Sydney, NSW, Australia

ISSN 2366-259X ISSN 2366-2603 (electronic)
Springer Tracts in Civil Engineering
ISBN 978-3-032-10284-3 ISBN 978-3-032-10285-0 (eBook)
https://doi.org/10.1007/978-3-032-10285-0

This Springer imprint is published by the registered company Springer Nature Switzerland AG
The registered company address is: Gewerbestrasse 11, 6330 Cham, Switzerland

Preface

In modern construction and facility management, heating, ventilation, and air conditioning (HVAC) and building services play a pivotal role in creating safe, comfortable, and energy-efficient environments. From residential homes to sprawling industrial complexes, these systems form the backbone of operational functionality, influencing air quality, thermal comfort, and sustainability outcomes. Yet, navigating the complexities of HVAC and building services can be daunting for both seasoned professionals and newcomers to the field.

This book, *HVAC and Building Services Rules of Thumb: A Quick Reference Guide*, is designed to serve as an accessible yet comprehensive guide. It distills key principles, practical calculations, and actionable insights into a format that allows for quick referencing without compromising on depth. Whether you're an engineer, technician, architect, or a facilities manager, this book provides essential knowledge at your fingertips.

Why Rules of Thumb?

Rules of thumb serve as simplified guidelines derived from industry best practices and historical data. While they may not replace detailed design and engineering processes, they offer a starting point for decision-making. These benchmarks help professionals make initial assessments, troubleshoot issues, or validate complex calculations with a high degree of confidence. However, the results of *Rules of Thumb* shall not be considered as final design nor it shall be used for construction phase. The rules of thumb and guidelines presented in this book are meant for early design and educational purposes only. Users must validate and adapt them to their specific projects, local codes, and standards. The author holds no responsibility for decisions based solely on these guidelines.

What's Inside?

Each chapter of this book tackles a specific aspect of HVAC and building services, starting with foundational principles and advancing to specialized topics. Topics like energy efficiency, load estimation, HVAC system sizing and selection, equipment sizing, air and water distribution system's calculation, and fire safety are paired with practical examples and tips that can be directly applied to real-world scenarios. You'll also find practical examples in some chapters showing how rules of thumb can be used at the early stages of projects.

The book's format is intentionally structured to facilitate quick reference. Tables, charts, and simplified diagrams supplement textual content, while each section concludes with a summary of key takeaways. This layout ensures that even under time-sensitive conditions, the critical information is readily accessible.

Who Should Read This Book?

This book is tailored for:

- **HVAC Designers and Engineers**: Looking for quick access to validated principles and formulas
- **Facility Managers**: Seeking to understand system performance and maintenance requirements
- **Architects and Builders**: Collaborating on integrated design projects requiring HVAC considerations
- **Students and Trainees**: Gaining foundational knowledge in HVAC and building services

Bridging Theory and Practice

Understanding HVAC systems requires more than theoretical knowledge; it demands a hands-on perspective that only real-world applications can provide. This book strikes a balance by emphasizing practical strategies while reinforcing them with theoretical underpinnings. From estimating cooling and heating loads to designing ductwork for a space, this guide equips readers with the tools they need to navigate diverse challenges confidently.

In a world increasingly focused on energy conservation and sustainable development, the role of HVAC and building services has never been more critical. This book aims to empower professionals to design, implement, and maintain systems that align with both functional and environmental goals.

As you turn the pages, you'll uncover insights that simplify complexity and illuminate best practices. Whether you're in the early stages of a project or troubleshooting an existing system, this guide promises to be an indispensable companion on your professional journey.

Important Consideration

The rules of thumb provided in this book are intended solely for use during the early design stages of HVAC projects. They are not a substitute for detailed engineering calculations, professional judgment, or adherence to applicable codes and standards. The final design and its implementation are the responsibility of the user, who must verify all parameters and conditions specific to their project. The author assumes no liability for decisions or outcomes based on these guidelines.

A Note of Gratitude

No work is truly a solo endeavor, and this book is no exception. I would like to express my heartfelt gratitude to my colleagues in both academia and industry who have shared their knowledge, experiences, and insights with me over the years. Your guidance, support, and collaboration have been instrumental in shaping this book and enriching my understanding of this dynamic field.

Sydney, NSW, Australia Vahid Vakiloroaya

Competing Interests The author has no competing interests to declare that are relevant to the content of this manuscript.

Contents

About the Author

Vahid Vakiloroaya, PhD (Mechanical Engineering), is a distinguished mechanical engineer with over 22 years of dedicated experience in the HVAC and building services industry. He holds a Doctorate in Mechanical Engineering and is widely recognized as a thought leader and innovator in the field.

Throughout his career, Dr. Vakiloroaya has contributed extensively to the advancement of HVAC systems through cutting-edge research, innovative design methodologies, and the development of sustainable building solutions. His work spans residential, commercial, and industrial applications, with a focus on improving energy efficiency, indoor environmental quality, and system performance.

Dr. Vakiloroaya has authored numerous peer-reviewed publications and technical papers, making significant contributions to the body of knowledge in HVAC design, thermal systems optimization, and energy management. His expertise is frequently sought by industry professionals and academic institutions alike.

Beyond his technical contributions, Dr. Vakiloroaya is passionate about education and knowledge sharing. His extensive practical experience, combined with his academic background, allows him to bridge the gap between theory and application, making complex engineering concepts accessible and actionable for practitioners.

This book reflects Dr. Vakiloroaya's commitment to empowering engineers with practical insights, proven methodologies, and reliable rules of thumb for designing efficient and effective HVAC systems.

Chapter 1
Use of Rules of Thumb

1.1 Overview

Rules of thumb in HVAC engineering are simplified guidelines or approximate methods used to make quick decisions during the design, selection, and evaluation of HVAC systems. These are particularly useful in the early stages of a project or when detailed calculations are impractical due to time constraints or lack of data. While not a substitute for precise engineering analysis, they provide valuable starting points for professionals to estimate system requirements and costs.

1.2 When to Use Rules of Thumb

Early Design Stage

- During the conceptual or preliminary design phase, rules of thumb provide quick approximations that help engineers evaluate design feasibility and define broad system requirements.
- They assist in developing initial cost estimates and defining the scope of work before detailed calculations are performed.

Quick Assessments

- Rules of thumb can be used to rapidly verify if preliminary system sizes, capacities, or configurations are within an acceptable range.
- They are particularly valuable during informal discussions with stakeholders or when immediate decisions are required.

Budgetary Analysis

- They help in estimating equipment sizes and associated costs when detailed information is unavailable.

V. Vakiloroaya, *HVAC and Building Services Rules of Thumb*, Springer Tracts in Civil Engineering, https://doi.org/10.1007/978-3-032-10285-0_1

- This is useful for providing clients with a high-level understanding of project scope.

Comparative Analysis
- During the evaluation of multiple options, rules of thumb allow for quick comparisons to identify viable alternatives.
- They enable engineers to narrow down choices before diving into detailed modelling.

1.3 When Not to Use Rules of Thumb

Detailed Design Phase
- Rules of thumb lack the precision required for final HVAC design. As the project progresses, detailed calculations based on actual building loads, occupancy patterns, and climatic data must be conducted.

Complex or Specialized Systems
- For unique applications, such as laboratory ventilation, clean rooms, or high-performance buildings, the assumptions behind rules of thumb may not apply.
- These systems require precise modelling and validation to ensure compliance with stringent performance criteria.

Compliance and Certification
- Building codes, energy standards, and green certifications often require rigorous documentation and precise calculations, which rules of thumb cannot provide.
- Oversimplified design using rules of thumb may lead to non-compliance or suboptimal system performance.

System Optimization
- Rules of thumb do not account for factors like energy efficiency, lifecycle costs, or integration with other building systems.
- Detailed analysis is essential to achieve an optimal design tailored to specific project requirements.

1.4 Practicality of Rules of Thumb in Early Design Stages

Rules of thumb offer a straightforward way to make informed decisions without the need for extensive calculations or simulation tools. This can save time in the early stages of a project, where detailed inputs may not yet be available.

Facilitating Communication

- They provide a common language for discussions among engineers, architects, and clients. This helps ensure everyone understands key design parameters and project constraints.

Guiding Initial Decisions

- Using rules of thumb, engineers can establish baseline values for equipment sizes, airflow rates, and system configurations. These serve as starting points for more detailed design iterations.

Cost and Resource Efficiency

- Early-stage decisions based on rules of thumb can help allocate resources more effectively, focusing detailed calculations on the most critical areas.

Risk Reduction

- By identifying potential oversizing or undersizing issues early, rules of thumb act as a sanity check to avoid significant errors later in the design process.

Real-World Practicality

- Many rules of thumb are derived from years of industry experience and often align closely with practical outcomes, making them reliable for initial assessments.

Applicability for Quick Verifications

- In situations where design changes are proposed late in the process or during construction, rules of thumb provide a fast way to assess feasibility without starting detailed calculations from scratch.

A Word of Caution

While rules of thumb are invaluable tools, they are not substitutes for detailed engineering design. They should always be validated with precise calculations and cross-checked against project-specific data to ensure accuracy and compliance. Using them responsibly ensures their value as a complement to, rather than a replacement for, rigorous design methodologies.

Chapter 2
HVAC Briefing Procedures

2.1 Overview

Heating, Ventilation, and Air Conditioning (HVAC) systems, along with associated building services, are fundamental components in modern structures. These systems ensure comfort, safety, and efficiency in diverse environments, from homes and offices to hospitals and industrial plants. In this chapter, we will explore the history, purpose, and scope of HVAC and building services, setting the stage for the detailed discussions in subsequent chapters.

2.2 The History and Evolution of HVAC System

The concept of controlling indoor environments dates back centuries, with ancient civilizations relying on rudimentary techniques. For example, the Romans used hypocaust systems to provide underfloor heating in bathhouses and villas. Meanwhile, in ancient Iran, passive cooling methods and wind towers were implemented to combat extreme temperatures.

Modern HVAC systems began to take shape in the late nineteenth and early twentieth centuries with the advent of mechanical ventilation, refrigeration, and central heating. Key milestones include:

- **1902**: Willis Carrier's invention of mechanical air conditioning.
- **1930s**: The development of ducted air systems for homes and commercial spaces.
- **1970s**: A shift toward energy-efficient designs following the energy crisis.
- **2000s and Beyond**: Integration of smart technology, automation, and sustainability initiatives.

V. Vakiloroaya, *HVAC and Building Services Rules of Thumb*, Springer Tracts in Civil Engineering, https://doi.org/10.1007/978-3-032-10285-0_2

Today's HVAC systems are increasingly sophisticated, combining advanced controls, eco-friendly refrigerants, and energy-efficient designs to meet the demands of modern construction.

2.3 The Purpose of HVAC and Building Services

The primary goals of HVAC and building services include:

Thermal Comfort: Maintaining a comfortable temperature within buildings, regardless of external conditions.
Air Quality: Ensuring clean, filtered air to prevent health issues and improve productivity.
Energy Efficiency: Optimizing system performance to reduce energy consumption and operational costs.
System Reliability: Ensuring uninterrupted functionality through robust design and proactive maintenance.
Safety and Compliance: Meeting building codes, safety standards, and environmental regulations.

Thermal Comfort
Following parameters identify thermal comfort for a building.

- **Temperature Ranges:**
  - Maintain indoor temperatures between **20–22 °C** (winter) and **23–25 °C** (summer).
- **Relative Humidity:**
  - Keep indoor humidity levels between **40% and 60%** to ensure comfort and minimize mold growth.
- **Air Movement:**
  - Air velocities should not exceed **0.15 m/s** in occupied zones to avoid drafts.
- **Radiant Heat:**
  - Avoid large temperature differences between surfaces and air; ensure wall, floor, and ceiling temperatures are within **2–3 °C** of room air temperature.
- **Fresh Air Supply:**
  - Provide a minimum outdoor air ventilation rate as per standards (e.g., ASHRAE 62.1 or local regulations). Typically, **8–10 L/s per person** is recommended.

Suitable indoor temperature for different type of buildings are shown in below table:

Building type	Suitable indoor temperature (°C)	Suitable relative humidity (%)
Residential (living areas)	20–24	30–50
Residential (bedrooms)	18–22	30–50
Office buildings	21–25	30–60
Retail spaces	20–24	30–60
Hospitals (patient rooms)	22–24	40–60
Hospitals (operating rooms)	20–23	50–60
Schools (classrooms)	20–24	30–50
Libraries	20–23	40–50
Museums	20–22	40–55
Hotels	20–24	30–50
Laboratories	18–22	30–50
Data centres	18–27	40–60
Industrial (light)	16–20	30–50
Industrial (heavy)	15–18	30–50
Gyms/fitness centres	20–22	40–50
Theatres	21–24	30–60
Restaurants	20–24	30–60
Warehouses (general)	16–20	30–50
Warehouses (cold storage)	−18 to 5	N/A
Greenhouses	18–24	50–80

2.4 How HVAC Systems Ensure Thermal Comfort

- **Temperature Control:**
  - HVAC systems use thermostats and advanced sensors to maintain consistent indoor temperatures.
  - Variable Air Volume (VAV) systems and zoning improve localized control of temperature.
- **Humidity Regulation:**
  - Dehumidifiers or humidifiers integrated with HVAC systems ensure optimal indoor humidity levels.
  - Cooling coils in air-handling units (AHUs) remove excess moisture during cooling.
- **Air Distribution:**
  - Properly designed ductwork and diffusers ensure even air distribution, avoiding hot or cold spots.
  - Adjustable air diffusers and dampers allow fine-tuning of airflows in different zones.

- **Fresh Air and Ventilation:**
  - Mechanical ventilation systems introduce fresh air, dilute indoor pollutants, and maintain oxygen levels.
  - Heat recovery ventilators (HRVs) and energy recovery ventilators (ERVs) improve ventilation efficiency.
- **Radiant Systems:**
  - Radiant heating and cooling panels maintain a uniform surface temperature, reducing radiant asymmetry.
- **Advanced Controls:**
  - Smart HVAC systems equipped with occupancy sensors, CO_2 detectors, and programmable settings adapt to changing conditions in real time.
- **Acoustic Comfort:**
  - Proper insulation and low-noise equipment contribute to thermal comfort by reducing HVAC system noise.
- **Integration with Building Design:**
  - Passive strategies like shading, high-performance glazing, and proper insulation complement HVAC efforts to achieve thermal comfort.

2.5 Scope of Building Services

Building services encompass more than just HVAC systems. They include:

- **Electrical Systems**: Power distribution, lighting, and backup systems.
- **Plumbing Systems**: Water supply, drainage, and sanitary fittings.
- **Fire Safety Systems**: Sprinklers, alarms, and evacuation protocols.
- **Building Management Systems (BMS)**: Centralized control of HVAC, lighting, and other systems.
- **Elevators and Escalators**: Ensuring safe and efficient vertical transportation.

HVAC often intersects with these services, requiring an integrated approach to design and maintenance.

2.6 Components of an HVAC System

Understanding HVAC begins with familiarising oneself with its key components:

- **Heating Systems**:
  - **Furnaces**: Use combustion to generate heat, typically powered by gas or oil.
  - **Boilers**: Heat water to produce steam or hot water for space heating.
  - **Heat Pumps**: Transfer heat from one location to another, often used in mild climates.

- **Ventilation Systems**:
  - **Natural Ventilation**: Relies on windows, vents, and architectural design to circulate air.
  - **Mechanical Ventilation**: Uses fans and ductwork to ensure consistent airflow.
  - **Exhaust Systems**: Remove stale air, odours, and contaminants from spaces.
- **Air Conditioning Systems**:
  - **Chillers**: Provide cooling by removing heat from water or air.
  - **Split Systems**: Compact units commonly used in residential and small commercial settings.
  - **VRF/VRV Systems**: Variable Refrigerant Flow/Variable Refrigerant Volume systems for large, multi-zone applications.
- **Control Systems**:
  - Thermostats, sensors, and building management software regulate temperature, airflow, and system performance.

2.7 Energy Efficiency in HVAC Systems

With the increasing emphasis on sustainability, energy efficiency is a cornerstone of modern HVAC design. Strategies to achieve energy-efficient systems include:

High-Efficiency Equipment: Using energy-efficient chillers, boilers, and heat pumps.
Variable Speed Drives (VSDs): Adjusting motor speeds to match load requirements, reducing energy consumption.
Zoning: Dividing buildings into zones with independent temperature controls to optimize comfort and efficiency.
Energy Recovery Systems: Capturing waste energy from exhaust air or fluids and reusing it.
Renewable Integration: Incorporating solar panels, geothermal systems, or wind power into HVAC operations.
Building Automation Systems (BAS): Using advanced controls to monitor and optimize HVAC performance in real time.

2.8 Challenges in HVAC and Building Services

Despite advancements, the field faces several challenges:

- **Energy Consumption**: HVAC systems account for a significant portion of a building's energy use. Balancing comfort with efficiency remains a priority.
- **Climate Change**: Rising global temperatures demand systems that can adapt to extreme weather conditions.

- **Indoor Air Quality (IAQ)**: As buildings become more airtight for energy efficiency, maintaining air quality becomes increasingly complex.
- **Integration**: Coordinating HVAC with other building services requires seamless collaboration between disciplines.

2.9 HVAC Briefing Procedures: Early Design Stage

The early design stage of an HVAC system, often referred to as the briefing stage, is a critical phase where the groundwork for the system's success is established. This stage involves gathering key information, aligning stakeholder expectations, and identifying project-specific requirements. A well-executed briefing procedure ensures that the HVAC design aligns with the overall goals of the building project.

Key Steps in HVAC Briefing Procedures

Stakeholder Consultation

Engage with architects, engineers, building owners, and end-users to understand the functional and operational requirements of the building. Key discussions should cover:

- Building usage (e.g., residential, commercial, industrial).
- Occupancy patterns and density.
- Special requirements, such as temperature-sensitive equipment or cleanroom standards.

Site Analysis

- Conduct a thorough evaluation of the project site to identify environmental and contextual factors, including:
- Local climate conditions.
- Site orientation and shading.
- Availability of utilities such as electricity, water, and natural gas.

Establishing Design Criteria

Define the key design criteria based on project goals and compliance requirements, such as:

- Indoor temperature and humidity ranges.
- Ventilation rates as per ASHRAE or local standards.
- Energy efficiency targets and sustainability certifications

Preliminary Load Calculations

- Perform initial heating and cooling load calculations to estimate system capacity requirements. This step ensures the system is neither undersized nor oversized, both of which can lead to inefficiencies and higher costs.

Budget and Timeline Considerations

Develop a preliminary budget and timeline for the HVAC system. This includes:

- Initial cost estimations for equipment and installation.

- Long-term operational and maintenance costs.
- Integration with overall project schedules.

Compliance and Standards Review

- Ensure that all proposed designs align with relevant and local building codes, industry standards, and regulations. Addressing compliance early minimises the risk of costly redesigns or project delays.

Documentation and Communication

Compile findings and design criteria into a clear and comprehensive briefing document. Share this document with all stakeholders to ensure alignment and obtain necessary approvals before advancing to detailed design phases.

Benefits of a Well-Executed Briefing

- **Clarity**: Ensures all stakeholders have a shared understanding of project goals and requirements.
- **Efficiency**: Reduces the likelihood of costly changes during later stages of design or construction.
- **Optimization**: Provides a solid foundation for creating an energy-efficient and cost-effective HVAC system.
- **Compliance**: Addresses regulatory requirements early in the process, avoiding potential legal and operational issues.

2.10 Drawings

In the HVAC and building industry, various types of drawings are used throughout different phases of a project, each serving a specific purpose. Followings represent breakdown of the differences between tender drawings, shop drawings, and construction drawings:

Tender Drawings

- **Purpose**: These drawings are used during the bidding process for a construction project. They provide a general overview of the design intent and specifications.
- **Content**:
  - Show the overall design, layout, and spatial relationships of the building and its systems.
  - Include basic dimensions, materials, and design concepts but may lack detailed information regarding the installation.
  - Serve as a basis for contractors to prepare accurate bids and proposals.
- **Focus**: Primarily concerned with aesthetics and building layout rather than specific construction details.

Shop Drawings

- **Purpose**: These detailed drawings are prepared by the subcontractors or manufacturers after a contract is awarded. They are used to illustrate how specific components will be fabricated and installed.
- **Content**:
  - Provide precise details about specific HVAC equipment, ductwork, piping, and systems.
  - Include dimensions, material specifications, assembly instructions, and installation methods.
  - Often show modifications to the original design as approved by the relevant authorities or based on site conditions.
- **Focus**: Concentrated on the production and installation aspects of specific systems, ensuring that all components fit properly as per the construction documents.

Construction Drawings

- **Purpose**: These drawings are finalized and used for the actual construction phase. They guide contractors in the physical building of the project.
- **Content**:
  - Include detailed plans, elevations, sections, and specifics of the HVAC installation.
  - Show dimensions, materials, and methods for construction, as well as notes about codes and standards to be followed.
  - May incorporate revisions based on shop drawings or change orders.
- **Focus**: Serve as the authoritative source of information for on-site construction, ensuring all aspects are executed according to the design and specifications.

Summary of Differences

Feature	Tender drawings	Shop drawings	Construction drawings
Stage in project	Pre-construction (Bidding phase)	Post-bid (Pre-construction phase)	Construction phase
Detail level	General overview with basic details	Highly detailed for fabrication	Detailed for construction
Purpose	To obtain bids and proposals	To illustrate fabrication and installation	To guide actual construction
Prepared by	Architects/Engineers	Subcontractors/ Manufacturers	Architects/Engineers
Focus areas	Aesthetics, layout	Equipment, installation specifics	Compliance with design and specifications

Chapter 3
HVAC Design Map

3.1 Overview

Designing, calculating, constructing, and commissioning HVAC systems and building services require a structured, step-by-step approach to ensure efficiency, functionality, and compliance with regulations. This chapter signifies an "HVAC Design Map" outlining the key stages in the process.

3.2 Design Stages

Preliminary Planning

- **Understand Project Requirements:** Review architectural plans, building function, occupancy, and owner goals.
- **Evaluate Climatic Conditions:** Analyse local weather data to determine heating and cooling needs.
- **Establish Design Criteria:** Define thermal comfort parameters, indoor air quality (IAQ) standards, noise levels, and energy efficiency goals.
- **Preliminary Load Estimation:** Use rules of thumb or simplified tools to approximate heating and cooling loads.

System Selection

- **Choose HVAC System Type:** Based on building type, budget, energy efficiency goals, and space availability (e.g., VAV, chilled beams, VRF systems).
- **Evaluate Options:** Compare initial costs, operating costs, and maintenance requirements.
- **Consult Standards:** Ensure compliance with relevant standards (e.g., ASHRAE, CIBSE, or local codes).

V. Vakiloroaya, *HVAC and Building Services Rules of Thumb*, Springer Tracts in Civil Engineering, https://doi.org/10.1007/978-3-032-10285-0_3

Detailed Load Calculations

- **Perform Heat Load Calculations:** Account for internal and external heat gains using software tools or manual methods.
- **Consider Ventilation Loads:** Calculate outdoor air requirements based on occupancy and standards like ASHRAE 62.1.
- **Factor in Diversity:** Adjust calculations for simultaneous operation of systems.

Equipment Sizing and Selection

- **HVAC Equipment:** Size chillers, boilers, air handling units, and terminal devices.
- **Ductwork and Piping:** Determine sizes for air ducts and water pipes, ensuring adequate flow and pressure.
- **Ancillary Components:** Select pumps, fans, filters, dampers, and control systems.
- **Energy Efficiency:** Prioritize high-efficiency equipment and systems.

System Design and Layout

- **Prepare Drawings:** Develop ductwork, piping, and wiring layouts.
- **Coordinate with Other Disciplines:** Resolve clashes with architectural, structural, and other building services.
- **Accommodate Accessibility:** Ensure maintenance access to critical components.
- **Optimize Zoning:** Divide spaces into zones for effective temperature control.

Regulatory Compliance and Approvals

- **Submit for Approvals:** Present designs to relevant authorities for permits.
- **Address Feedback:** Revise designs as per regulatory requirements.

Construction Phase

- **Procurement:** Order materials and equipment based on finalized designs.
- **Installation:** Supervise the installation of HVAC components.
- **Quality Control:** Ensure installations match design specifications.

Testing and Commissioning

- **System Testing:** Verify system performance under varying loads.
- **Air and Water Balancing:** Ensure even distribution of air and water flow.
- **Fine-Tuning Controls:** Optimize control systems for efficiency and comfort.
- **Documentation:** Provide as-built drawings and operation manuals.

Handover and Maintenance

- **Handover to Facility Management:** Train staff on system operation and maintenance.
- **Scheduled Maintenance:** Develop a maintenance plan to ensure longevity and efficiency.

Benefits of an HVAC Design Map

- **Clarity:** Provides a roadmap for stakeholders, ensuring all aspects are addressed.
- **Coordination:** Enhances collaboration between disciplines.
- **Efficiency:** Streamlines the design, construction, and commissioning process.

3.3 Questionary List for Designer

This checklist ensures a comprehensive review of the HVAC design and helps engineers identify potential gaps, ensuring the design is ready for real-world implementation.

Design Compliance and Objectives

1. Have I met all local, national, and international building codes and standards (e.g., ASHRAE, CIBSE, NCC, NFPA, etc.)?
2. Does the design meet client requirements and project specifications?
3. Have all relevant environmental sustainability and energy efficiency standards been addressed?
4. Are all system capacities (cooling, heating, ventilation, and dehumidification) aligned with the calculated loads?

HVAC System Performance

5. Have the cooling and heating loads been accurately calculated, considering internal and external heat gains?
6. Are the selected HVAC systems appropriate for the building type, usage, and climate?
7. Does the system provide sufficient ventilation to meet indoor air quality (IAQ) requirements?
8. Are thermal comfort parameters (e.g., temperature, humidity, air velocity) within acceptable ranges?

Equipment Selection

9. Have I selected the most efficient and cost-effective equipment (e.g., chillers, boilers, heat pumps, AHUs, etc.) for the application?
10. Are the capacities of major equipment (e.g., chillers, cooling towers, fans, pumps) adequate and not oversized or undersized?
11. Is redundancy required for critical equipment, and has it been included in the design?
12. Have noise and vibration control measures been incorporated for equipment?
13. Are the selected filters appropriate for the required air quality levels?

Ductwork and Air Distribution

14. Are the duct sizes optimized for airflow and pressure drop?

15. Have I avoided excessive bends and fittings to minimize pressure losses?
16. Is there adequate space for duct insulation?
17. Have I ensured proper placement of diffusers, grilles, and dampers for uniform air distribution?

Piping and Water Distribution

18. Are pipe sizes optimized for flow rates and pressure drops?
19. Have I included provisions for thermal expansion, such as expansion tanks and loops?
20. Are valves, pumps, and heat exchangers properly located and sized?
21. Have I specified pipe insulation to prevent heat loss/gain and condensation?

Controls and Automation

22. Have I integrated Building Management System (BMS) or standalone controllers for efficient operation?
23. Are sensors and control points (temperature, pressure, humidity) appropriately located?
24. Have I ensured compatibility of all control systems with equipment?
25. Have I included provisions for energy monitoring and fault detection?

Space Requirements

26. Have I allocated sufficient space in mechanical rooms for installation, maintenance, and operation?
27. Are clearances around equipment adequate for service access?
28. Is there enough space for duct and pipe routing, including insulation and supports?

Energy Efficiency and Sustainability

29. Have I incorporated energy recovery devices where feasible (e.g., heat recovery wheels, economizers)?
30. Is the HVAC system designed for the lowest possible energy consumption while maintaining performance?
31. Have I considered renewable energy integration (e.g., solar, geothermal)?
32. Have I included high-efficiency motors, variable frequency drives (VFDs), and low-energy equipment?

Safety and Risk

33. Are fire dampers, smoke dampers, and stair pressurization systems included where required?
34. Have I ensured proper exhaust ventilation for hazardous areas (e.g., kitchens, labs)?
35. Have I addressed all health and safety concerns, including Legionella prevention in water-cooled systems?
36. Are emergency power provisions included for critical HVAC equipment?

Construction Feasibility

37. Have I coordinated with structural, electrical, and architectural teams to avoid clashes?
38. Is the design constructible with available materials and technology?
39. Are equipment sizes and weights suitable for transportation and installation on-site?
40. Have I provided detailed specifications, layouts, and installation instructions?

Cost Considerations

41. Is the design cost-effective without compromising performance?
42. Have lifecycle costs been evaluated, including operating and maintenance costs?
43. Are alternate materials or systems available to optimize costs?

Documentation and Deliverables

44. Have I included a complete set of drawings, including plans, sections, and details?
45. Are equipment schedules, specifications, and datasheets complete?
46. Have I provided clear instructions for balancing and commissioning?
47. Are as-built drawing requirements defined for post-construction?

Testing, Commissioning, and Handover

48. Are all testing and commissioning procedures defined and documented?
49. Have I provided operation and maintenance manuals for the client?
50. Are training sessions scheduled for the client's staff on system operation?

Checklist Examples for Specific Applications

- **Data Centres**: Is redundancy (N + 1 or 2 N) included for cooling systems?
- **Hospitals**: Are isolation rooms and clean zones designed with proper pressure gradients?
- **Commercial Buildings**: Are zoning and VAV systems optimized for varying occupancy levels?
- **Industrial Facilities**: Have I addressed specific exhaust or process ventilation needs?

3.4 Guide to Fire Engineering Knowledge for HVAC Engineers

Fire engineering is an essential aspect of HVAC design to ensure occupant safety, property protection, and compliance with building codes. HVAC engineers must consider fire safety principles, equipment, and systems integration to achieve a fire-resilient design. Below is a detailed guide with rules of thumb for key fire engineering considerations in HVAC systems.

- **Fire Dynamics**:
  - Understand fire behaviour, heat release rates, smoke movement, and the impact on HVAC systems.
  - Ensure HVAC designs minimize smoke and fire spread.
- **Building Codes and Standards**:
  - **NFPA Standards**: Key references include NFPA 90A (air-conditioning systems), NFPA 92 (smoke management systems), and NFPA 101 (life safety code).
  - **Local Codes**: Comply with national/local fire safety regulations (e.g., IBC, BS 9999).
- **Passive Fire Protection**:
  - Design ductwork to maintain compartmentalization using fire-resistant materials and dampers.
- **Active Fire Protection**:
  - Integrate systems like sprinklers, smoke control, and fire detection within HVAC designs.

HVAC Fire Engineering Rules of Thumb

Fire Dampers

- **Purpose**: Prevent fire and smoke spread through ducts penetrating fire-rated walls and floors.
- **Installation**:
  - Install dampers at fire-rated barriers, ensuring compliance with a **2-h fire resistance rating** for most applications.
  - Use combination fire/smoke dampers where both fire and smoke control are required.
- **Rule of Thumb**:
  - Size dampers to match duct cross-sections, maintaining airflow while meeting fire-resistance requirements.

Smoke Dampers

- **Purpose**: Control smoke movement and prevent its spread through HVAC systems.
- **Rule of Thumb**:
  - Position dampers within smoke zones to confine smoke.
  - Operate dampers based on signals from smoke detectors.

Duct Insulation

- **Fire-Resistant Insulation**:
- Use insulation rated for **1–2 h** of fire resistance on ducts passing through fire-rated walls or spaces.
- **Smoke Control Ducts**:
  - Use non-combustible materials (e.g., mineral wool).

Smoke Extraction Systems

- **Purpose**: Remove smoke from fire zones to aid evacuation and firefighting.
- **Design Criteria**:
  - Maintain **smoke extraction rates** between **6 and 12 ACH (air changes per hour)** for spaces like lobbies, stairwells, and atria.
  - Smoke extraction fans should operate under fire conditions, typically up to **300 °C for 2 h**.
- **Rule of Thumb**:
  - Use mechanical extraction for large spaces or natural ventilation for smaller spaces.

Stairwell Pressurization

- **Purpose**: Prevent smoke ingress into evacuation routes.
- **Pressurization Levels**:
  - Maintain stairwell pressure between **25 and 50 Pa** to prevent smoke infiltration.
- **Fan Sizing**:
  - Select fans to achieve design pressures without exceeding **60 Pa** to allow easy door operation.
- **Rule of Thumb**:
  - For high-rise buildings, ensure pressure is uniform across the height of the stairwell.

Air Handling Units (AHUs) and Fire Modes

- **Shutdown**:
- AHUs should automatically shut down in fire mode to avoid circulating smoke.
- **Smoke Control**:
  - Dedicated smoke handling AHUs may be designed to operate during fire conditions.
- **Rule of Thumb**:
  - Provide isolation dampers to segregate fire zones during AHU shutdown.

Emergency Power

- **Purpose**: Ensure critical fire protection and smoke control systems operate during power outages.

- **Rule of Thumb**:
  - Connect smoke extraction fans, pressurization systems, and fire dampers to emergency power.

Sprinkler System Integration
- Avoid HVAC components obstructing sprinkler coverage.
- Coordinate AHU and ductwork placement to ensure proper fire suppression coverage.

Materials and Fire Resistance
- **Ductwork**:
- Use galvanized steel or fire-rated materials for ducts.
- Non-combustible duct linings, such as mineral wool, for fire and smoke resistance.
- **Sealants**:
  - Apply fire-rated sealants to duct penetrations in fire-rated walls.
- **Insulation**:
  - Use materials with a high fire-resistance rating (e.g., rock wool or ceramic fiber).

Smoke Control and Evacuation Strategies
- **Zoning**:
- Divide buildings into smoke zones to isolate fire incidents.
- **Natural Ventilation**:
  - Use strategically placed openings for smoke exhaust in small or low-rise buildings.
- **Mechanical Ventilation**:
  - Use fans to extract smoke in large or complex buildings.

Fire Engineering Design Checklist for HVAC
- **Duct and Damper Design**:
- Ensure proper placement of fire and smoke dampers.
- Use fire-resistant ducts where required.
- **Smoke Control Systems**:
  - Design compliant smoke extraction or containment systems.
  - Consider redundancy in smoke control fans.
- **Equipment Shutdown and Fire Modes**:
  - Automate AHU shutdown or operation in fire mode.
- **Pressurization Systems**:
  - Verify stairwell and escape route pressurization meets code requirements.
- **Material Selection**:
  - Use non-combustible and fire-rated materials.

- **Compliance**:
  - Validate design against NCC, NFPA, IBC, or local codes.
  - Coordinate with fire protection engineers.
- **Emergency Power**:
  - Ensure critical fire safety systems have backup power.

3.5 Guide to Electrical Engineering Knowledge for HVAC Engineers

Electrical engineering knowledge is crucial for HVAC engineers to design systems that operate safely, efficiently, and in compliance with codes. Electrical integration ensures the seamless functionality of HVAC components, from motors and controls to power supplies.

Fundamentals of Electrical Engineering in HVAC

- **Electrical Power Basics**:
  Understand single-phase and three-phase power systems for residential and commercial HVAC systems.
  Familiarize with terms like kilowatts (kW), kilovolt-amperes (kVA), and power factor to calculate loads.
- **Voltage Levels**:
  Common HVAC systems operate at 120 V, 208 V, 230 V, or 400 V, depending on system size and application.
- **Current Types**:
  Most HVAC systems use alternating current (AC); however, variable frequency drives (VFDs) may convert to DC for motor speed control.

Electrical Equipment in HVAC Systems

- **Motors**:
  Types: Induction motors, variable speed drives (VSDs), and electronically commutated motors (ECMs).
  Rule of Thumb: For large HVAC systems, use three-phase motors to improve efficiency and reduce current draw.
- **Transformers**:
  Step-up or step-down transformers are used to match HVAC equipment voltage to the power supply.
  Rule of Thumb: Size transformers at 125% of the expected load to allow for future expansion.
- **Variable Frequency Drives (VFDs)**:
  Control motor speed to improve energy efficiency.

Rule of Thumb: VFDs can reduce motor energy consumption by 20–50%, depending on load variations.

Electrical Rules of Thumb for HVAC Engineers

- **Wire Sizing**:
  Rule of Thumb: Size conductors to handle 125% of the full load current (FLA) for safety and compliance.
- **Voltage Drop**:
  Ensure voltage drop does not exceed 3% for branch circuits and 5% for feeders.
  Rule of Thumb: Use larger cables for circuits longer than 100 feet to maintain voltage levels.
- **Breaker Sizing**:
  Overcurrent protection should be 125% of the FLA for motors.
  Rule of Thumb: For a 10 HP motor, use a breaker rated for at least 30 amps.
- **Power Factor Correction**:
  Install capacitor banks to maintain a power factor of 0.95 or higher.
  Rule of Thumb: Correct power factor for systems with inductive loads (e.g., motors) to reduce energy costs.

Emergency Power Systems

- **Backup Generators**:
  Ensure critical HVAC systems, like stairwell pressurization fans, remain operational during power outages.
  Rule of Thumb: Sizing of emergency generators should account for the starting current of large motors (2–6 times the running current).
- **Uninterruptible Power Supply (UPS)**:
  Protect control systems and sensors during power fluctuations.
  Rule of Thumb: Select a UPS with a capacity 20–30% higher than the total load to ensure reliability.

Integration of Electrical and Control Systems

- **Building Management Systems (BMS)**:
  Use protocols like BACnet and Modbus to integrate HVAC controls into centralized systems.
  Rule of Thumb: Ensure communication devices are compatible with existing control networks.
- **Thermostats and Sensors**:
  Install temperature and pressure sensors at critical points.
  Rule of Thumb: Place sensors away from direct airflow to improve accuracy.
- **Control Panels**:
  Design panels to house breakers, relays, and contactors for easy access and maintenance.
  Rule of Thumb: Label all components for quick identification.

Material Selection and Design Considerations

- **Cables**:
  Use insulated conductors with temperature ratings suited to the installation environment.
- **Duct Insulation**:
  Ensure non-combustible insulation to meet fire safety standards.
  Rule of Thumb: For ducts passing through fire-rated walls, use 1–2 h fire-rated materials.

3.6 Ingress Protection (IP) Ratings

In the HVAC industry, **Ingress Protection (IP) ratings** such as IP54, IP56, etc., are used to classify and define the levels of protection provided by enclosures for equipment against the intrusion of solids (like dust) and liquids (like water). These ratings are defined by the **IEC 60529** standard and are critical for determining the suitability of equipment for specific environments.

Structure of IP Ratings

- **First Digit (Solids Protection)**: Indicates the level of protection against dust and solid objects.
- **Second Digit (Liquids Protection)**: Indicates the level of protection against water ingress.

Common IP Ratings in HVAC Applications

- **IP20**:
  - Protection: Basic protection against solid objects larger than 12.5 mm. No protection against water.
  - Use: Indoor equipment in clean, controlled environments (e.g., control panels in offices).
- **IP44**:
  - Protection: Protection against solid objects over 1 mm and splashing water from any direction.
  - Use: Fans, dampers, and equipment in semi-covered outdoor locations.
- **IP54**:
  - Protection: Limited protection against dust (dust ingress won't interfere with operation) and water splashes.
  - Use: HVAC components like outdoor units in moderately dusty or damp environments.

- **IP55**:
  - Protection: Protection against limited dust ingress and jets of water from any direction.
  - Use: Exhaust fans, motors, and other outdoor equipment exposed to rain or hose-down cleaning.
- **IP56**:
  - Protection: Protection against limited dust ingress and strong jets of water or heavy seas.
  - Use: Equipment in coastal areas, marine applications, or highly wet environments.
- **IP65**:
  - Protection: Dust-tight (no ingress of dust) and protection against water jets from any direction.
  - Use: Outdoor HVAC enclosures and rooftop units in dusty or wet areas.
- **IP66**:
  - Protection: Dust-tight and protection against powerful water jets or heavy seas.
  - Use: HVAC systems in harsh industrial or coastal conditions.
- **IP67**:
  - Protection: Dust-tight and immersion in water up to 1 m for 30 min.
  - Use: Specialized HVAC equipment in flood-prone or underwater applications.

Importance in HVAC

- **Equipment Durability**: Ensures that HVAC equipment can withstand environmental challenges like dust, moisture, or direct water exposure.
- **Safety Compliance**: Prevents hazards like electrical shorts caused by water ingress or dust accumulation.
- **Application-Specific Design**: Enables proper selection of equipment for specific environments, such as industrial plants, hospitals, or outdoor installations.
- **Reduced Maintenance**: Protects internal components, reducing the need for frequent cleaning or repair.

Example Applications

- **Rooftop Units (RTUs)**: Typically require IP54 or higher to handle outdoor exposure.
- **Cooling Towers**: May need IP55 or IP56 to withstand water spray and outdoor conditions.
- **Marine HVAC Systems**: Often require IP66 to IP68 for protection against saltwater and high humidity.
- **Cleanrooms**: Require dust-tight equipment (IP65 or higher) to maintain air quality standards.

Chapter 4
HVAC Project Management and Site Supervision

4.1 Overview

In the HVAC construction industry, the roles of a **Construction Manager**, **Project Manager**, and **Site Manager** often overlap but have distinct focuses. Project management primarily emphasizes planning, scheduling, budgeting, and overall project coordination, ensuring the design, procurement, and compliance aspects align with project goals. Site management, on the other hand, focuses on on-site activities, including workforce supervision, safety enforcement, quality control, and daily task execution. The overlap occurs as both roles require collaboration to address installation challenges, maintain timelines, and ensure the system is installed per design specifications, creating a need for close coordination between planning and on-site operations. However, the **Project Manager** typically retains overall authority and accountability for project completion and success. This chapter states a detailed breakdown of their tasks and responsibilities.

4.2 Construction Manager

- **Primary Focus:**
  - Overseeing the entire construction process, including scheduling, resource allocation, and adherence to codes and standards.

Key Responsibilities

- **Planning and Coordination**:
  - Develop detailed construction plans for HVAC systems.
  - Coordinate with architects, engineers, subcontractors, and vendors.

V. Vakiloroaya, *HVAC and Building Services Rules of Thumb*, Springer Tracts in Civil Engineering, https://doi.org/10.1007/978-3-032-10285-0_4

- **Budget Management**:
  - Ensure the project stays within budget.
  - Approve and monitor expenditures related to HVAC equipment and labour.
- **Compliance and Quality Assurance**:
  - Ensure adherence to industry standards, safety regulations, and building codes
  - Inspect the quality of HVAC installations.
- **Risk Management**:
  - Identify potential risks and implement mitigation strategies.
  - Oversee safety protocols on-site.
- **Communication**:
  - Act as a liaison between stakeholders (owners, designers, and contractors).
  - Provide regular updates to clients and higher management.

4.3 Project Manager

- **Primary Focus:**
  - Managing the HVAC project lifecycle, from initiation to completion, ensuring all objectives are met on time and within scope.

Key Responsibilities

- **Project Initiation**:
  - Define project scope, objectives, and deliverables for HVAC systems.
  - Develop a project charter and work breakdown structure (WBS).
- **Scheduling and Timeline Management**:
  - Create and monitor detailed project schedules using tools like MS Project or Primavera.
  - Coordinate milestones for material delivery, equipment installation, and commissioning.
- **Resource Allocation**:
  - Assign tasks to teams (designers, installers, and technicians).
  - Ensure availability of necessary tools, equipment, and materials.
- **Stakeholder Management**:
  - Communicate progress, challenges, and changes to stakeholders.
  - Handle change orders and approvals efficiently.
- **Documentation and Reporting**:
  - Maintain records of HVAC designs, contracts, and change orders.
  - Prepare progress reports for senior management and clients.

- **Commissioning and Handover**:
  - Oversee HVAC system testing, commissioning, and client training.
  - Ensure proper documentation and certifications are delivered at project closeout.

List of Questions for HVAC Project Management

Followings are the list of questions that an HVAC project manager should prepare and address throughout the entire project lifecycle, categorized by project phases. This list should serve as a guide to ensure comprehensive planning, execution, and completion of an HVAC project.

Planning Phase

- **Project Scope:**
  - What are the specific HVAC requirements for the project?
  - Are there any special considerations for the project (e.g., green building certification, redundancy)?
  - What are the project deadlines and milestones?
- **Budget:**
  - What is the allocated budget for HVAC systems?
  - Are there opportunities for cost-saving without compromising quality?
  - Are there contingencies in place for unexpected expenses?
- **Regulations and Standards:**
  - What are the local and national codes (e.g., ASHRAE, NCC, AS1668) applicable to the project?
  - Are there specific environmental or energy efficiency standards to meet?
- **Stakeholder Communication:**
  - Who are the key stakeholders, and what are their expectations?
  - How will regular updates be communicated to stakeholders?

Design Phase

- **System Design:**
  - What type of HVAC system is best suited for the building (e.g., VRF, Chilled Water, DX)?
  - Is the design load calculation complete and accurate?
  - Are the system's capacities adequate for future expansions?
- **Equipment Selection:**
  - What are the criteria for selecting HVAC equipment?
  - Are there constraints on size, weight, or noise levels for the equipment?
- **Coordination with Other Trades:**
  - How will the HVAC system integrate with electrical, plumbing, and structural designs?

  - Are there potential conflicts in routing ductwork and piping?

- **Energy Efficiency:**
  - Is the HVAC design optimized for energy efficiency?
  - Are renewable or alternative energy sources being considered?

Procurement Phase

- **Vendor and Supplier Selection:**
  - Are all vendors and suppliers prequalified?
  - Have warranties, delivery timelines, and service contracts been discussed?
- **Material and Equipment:**
  - Are the ordered equipment and materials compliant with specifications?
  - Is there a process for tracking and inspecting deliveries?

Construction Phase

- **Site Preparation:**
  - Is the site ready for HVAC installation (e.g., clearances, power supply)?
  - Are all safety measures in place?
- **Installation:**
  - Are installation procedures aligned with the design specifications?
  - How will quality control and inspections be conducted during installation?
- **Coordination:**
  - Are there conflicts between HVAC installation and other trades on-site?
  - How are changes to the scope or design being managed?

Testing and Commissioning Phase

- **Performance Testing:**
  - Are all components functioning as per design specifications?
  - Is the system meeting required airflow, temperature, and humidity levels?
- **Balancing:**
  - Has the system been properly balanced for airflow and water flow?
  - Have the pressure drops across dampers, filters, and coils been checked?
- **Controls:**
  - Are all control systems (e.g., BMS) calibrated and operational?
  - Are fail-safe and emergency systems tested?
- **Documentation:**
  - Are as-built drawings and O&M manuals complete and accurate?
  - Have all warranties and certifications been obtained?

Handover Phase

- **Client Training:**
  - Is the client trained on the operation and maintenance of the HVAC system?
  - Are troubleshooting guides and emergency procedures provided?
- **Handover:**
  - Are all snag lists and pending items resolved?
  - Has a final walkthrough with stakeholders been conducted?

Maintenance and Operations Phase

- **Post-Installation Support:**
  - Is there a maintenance schedule in place?
  - Who is responsible for post-installation service and repairs?
- **Performance Monitoring:**
  - Are energy consumption and system performance being monitored?
  - Is there a feedback loop to address client concerns?

General Questions Across Phases

- What are the project risks, and how are they mitigated?
- Are there lessons learned from previous projects that can be applied here?
- What are the key performance indicators (KPIs) for project success?
- Are all team members aware of their roles and responsibilities?

4.4 Site Manager

- **Primary Focus:**
  - Overseeing the physical implementation of HVAC systems on-site, ensuring daily operations run smoothly.

Key Responsibilities

- **On-Site Supervision**:
  - Direct and supervise HVAC installation teams, subcontractors, and technicians.
  - Monitor work progress and resolve on-site issues.
- **Safety and Compliance**:
  - Enforce adherence to safety standards (e.g., OSHA) and PPE requirements.
  - Ensure compliance with local building codes and project specifications.
- **Material and Equipment Management**:
  - Verify deliveries of HVAC materials and equipment.
  - Maintain inventory and prevent delays caused by shortages.

- **Quality Control**:
  - Inspect HVAC installations (ducts, piping, units) for alignment with design specifications.
  - Ensure proper sealing, insulation, and system integrity.
- **Daily Reporting**:
  - Update the project manager on daily activities, challenges, and progress.
  - Maintain site records, including attendance, equipment usage, and incidents.
- **Coordination with Trades**:
  - Work with electrical, plumbing, and structural teams to avoid conflicts.
  - Schedule and oversee testing and balancing (TAB) for HVAC systems.

List of Questions for an HVAC Site Manager
Below is a detailed list of questions an HVAC site manager must address at various stages of a project to ensure efficient site management, compliance, and successful execution:

Pre-construction Phase
- **Site Readiness:**
  - Is the site cleared and prepared for HVAC installation?
  - Are access points for equipment delivery and storage clearly identified?
  - Are temporary utilities (power, water, etc.) available on-site?
- **Documentation and Approvals:**
  - Are all permits, approvals, and risk assessments complete?
  - Are construction drawings, schedules, and specifications up to date?
- **Resources:**
  - Is the manpower allocated sufficient for the project timeline?
  - Are all required tools, equipment, and materials available and inspected?
- **Safety and Compliance:**
  - Are safety procedures and protocols established on-site?
  - Are all workers trained on safety standards, including PPE usage?
- **Coordination:**
  - Are roles and responsibilities for each trade clearly defined?
  - Have potential conflicts with other trades been addressed?

Installation Phase
- **Progress Tracking:**
  - Is the work proceeding according to the project schedule?
  - Are delays being documented and addressed?

- **Material Management:**
  - Are all materials and equipment inspected upon delivery for quality and compliance?
  - Are storage conditions appropriate for HVAC materials (e.g., ducts, pipes, insulation)?
- **Installation Quality:**
  - Is the installation following approved shop drawings and standards?
  - Are tolerances for ductwork, piping, and equipment alignment being maintained?
- **Site Coordination:**
  - Are clearances for duct and pipe routing maintained as per design?
  - Are changes in design or scope being communicated to the design team?
- **Safety:**
  - Are all workers adhering to safety guidelines during installation?
  - Are risks associated with heavy equipment and high-level installations mitigated?

Testing and Commissioning Phase

- **Pre-Testing:**
  - Are all components (ducts, pipes, coils, etc.) cleaned and free of debris?
  - Are pressure tests and leak tests conducted on ductwork and piping?
- **System Testing:**
  - Are airflows, water flows, and pressure drops within design parameters?
  - Are control systems, sensors, and actuators calibrated?
- **Commissioning:**
  - Are all HVAC systems tested under full and part-load conditions?
  - Are performance benchmarks achieved for temperature, humidity, and airflow?
- **Documentation:**
  - Are commissioning reports and test certificates completed and submitted?
  - Are as-built drawings being updated as required?

Post-installation and Handover Phase

- **Handover Preparation:**
  - Have all punch-list items been resolved?
  - Is the HVAC system operating as intended and verified by third-party testing?
- **Client Training:**
  - Is the client trained on system operation, controls, and maintenance?
  - Are O&M manuals and warranty documents handed over?

- **Site Clean-Up:**
  - Has all construction debris been removed from the site?
  - Are all access points and installed systems protected from potential damage?

Maintenance and Operational Phase

- **Post-Handover Support:**
  - Are service schedules provided to the client?
  - Are arrangements in place for preventive maintenance and troubleshooting?
- **Performance Monitoring:**
  - Is energy consumption being monitored and optimized?
  - Are complaints or issues from occupants being addressed promptly?

Comparison of Roles

Aspect	Construction manager	Project manager	Site manager
Focus	Overall project execution and compliance	Planning, scheduling, and meeting objectives	On-site supervision and implementation
Scope	Entire construction process	Specific HVAC project lifecycle	Daily site-level activities
Interaction level	High with all stakeholders	High with clients, engineers, and construction teams	High with workers and subcontractors
Key tools	Contracts, budgets, regulations	Schedules, reports, documentation	Checklists, drawings, safety protocols

Collaboration Between Roles

- The **Construction Manager** ensures overarching alignment with project goals and compliance.
- The **Project Manager** works closely with the Construction Manager to track progress and manage resources.
- The **Site Manager** reports to Project Manager to provide updates on field conditions and challenges.

4.5 Step-by-Step HVAC Commissioning Plan

Objective To ensure that all HVAC systems and components are installed, tested, and verified to operate as intended according to design specifications and owner requirements.

Pre-commissioning Phase

Planning and Preparation

- **Develop a Commissioning Plan**:
- Outline scope, objectives, responsibilities, and deliverables.
- Identify key stakeholders (designers, contractors, suppliers, and commissioning agents).

- **Documentation Review**:
  - Verify design intent, construction drawings, specifications, and equipment submittals.
  - Review operation and maintenance (O&M) manuals.
- **Schedule Coordination**:
  - Align commissioning activities with the construction timeline.
  - Identify critical milestones and dependencies (e.g., system start-up).
- **Assemble Commissioning Team**:
  - Assign responsibilities for commissioning tasks to relevant personnel.
  - Ensure all team members are trained and familiar with project requirements.

Pre-installation Verification

- **Inspection**:
  - Confirm that HVAC equipment and components are delivered as per specifications.
  - Verify storage conditions for sensitive equipment (e.g., coils, filters, controls).
- **Site Readiness**:
  - Ensure spaces (mechanical rooms, ceilings, and shafts) are prepared for installation.
  - Verify that utility connections (electricity, water, gas) are available.

Installation Phase

Installation Inspections

- **Equipment Installation**:
  - Verify correct placement, orientation, and anchoring of HVAC equipment.
  - Inspect ductwork, piping, and insulation for compliance with design specifications.
- **Controls and Sensors**:
  - Confirm installation of thermostats, sensors, and actuators in the correct locations.
  - Ensure wiring and connections comply with control diagrams.
- **System Protection**:
  - Check for protective measures during construction to prevent dirt and damage to HVAC components.

Pre-startup Checks

- **Mechanical Systems**:
  - Inspect ducts and pipes for leaks, alignment, and cleanliness.
  - Confirm that valves, dampers, and controls are installed and accessible.

- **Electrical Systems**:
  - Verify electrical connections, breaker sizing, and grounding.
  - Confirm that all equipment is powered and operational.

Start-up and Initial Testing Phase

Equipment Start-up

- **HVAC Equipment**:
  - Start major equipment (chillers, boilers, AHUs, fans, etc.) as per manufacturer guidelines.
  - Monitor start-up conditions (voltage, current, pressures, and temperatures).
- **Safety Systems**:
  - Test all safety devices (pressure relief valves, fire dampers, etc.) for functionality.

Preliminary Performance Checks

- **System Calibration**:
  - Set initial control parameters (set-points, schedules, and limits).
  - Verify calibration of sensors, meters, and actuators.
- **Air and Water Balancing**:
  - Perform preliminary air and water flow measurements.
  - Identify and correct major deviations from design values.

Functional Testing Phase

System Functionality Testing

- **Individual Components**:
  - Test individual system components (e.g., fans, pumps, VAV boxes) for performance and operation.
- **Integrated Systems**:
  - Verify that interconnected systems (e.g., AHUs, chillers, controls) function as a whole.
  - Simulate operating conditions (e.g., varying loads, occupancy levels).

Air and Water Balancing

- **Detailed Balancing**:
  - Perform final balancing for airflows and water flows to meet design requirements.
  - Document balancing results and identify corrections.

System Performance Verification Phase

Performance Testing

- **Load Testing**:
  - Test HVAC systems under full and partial load conditions.
  - Record system responses (energy use, temperatures, pressures, and flows).
- **Energy Efficiency**:
  - Verify equipment efficiency (COP, SEER, etc.) against design benchmarks.
  - Identify any energy performance deficiencies.

Indoor Environmental Quality

- **Comfort and Air Quality**:
  - Measure temperature, humidity, and CO_2 levels in conditioned spaces.
  - Test air distribution to ensure even airflow and occupant comfort.

Handover and Documentation Phase

Handover Preparation

- **Punch List Resolution**:
  - Address and rectify all deficiencies noted during testing and inspections.
- **System Walkthrough**:
  - Conduct walkthroughs with the client and project stakeholders.
  - Demonstrate system operation and key features.

Documentation and Training

- **As-Built Documentation**:
  - Provide updated as-built drawings, commissioning reports, and test certificates.
- **Training**:
  - Train client personnel on system operation, controls, and maintenance.

Post-commissioning Phase

Monitoring and Optimization

- **Initial Operation Period**:
  - Monitor system performance for a defined period after handover.
  - Address any issues reported by the client or identified through monitoring.
- **System Optimization**:
  - Adjust control parameters, set-points, and schedules based on operational data.

Maintenance Planning

- **Preventive Maintenance**:
  - Provide a schedule for regular maintenance of HVAC equipment.
  - Highlight critical maintenance tasks (e.g., filter replacement, coil cleaning)

Deliverables

- Commissioning Plan and Schedule
- Pre-Commissioning Checklists
- Test Reports (Leak Tests, Balancing, Functional Testing)
- Commissioning Summary Report
- As-Built Documentation
- Training Materials

Chapter 5
HVAC Load Components

5.1 Overview

Heating and cooling load calculations are performed to determine the necessary capacity of heating and cooling systems that can sustain the desired conditions within a conditioned space. To calculate these capacities, it is essential to gather details about the design indoor and outdoor conditions, building specifications, characteristics of the conditioned space (including occupancy, activity levels, appliances, and equipment in use), and any unique requirements of the application. For comfort-related purposes, indoor conditions are typically set based on thermal comfort criteria, whereas in industrial or commercial settings, they are defined by the specific processes carried out or the products being stored.
These calculations generally follow a structured, step-by-step approach that considers all energy flows within the building. In practice, methods range from simple rules of thumb to advanced techniques like transfer function methods to estimate the heating and cooling loads accurately.

5.2 Load Calculations

Accurate load calculations form the foundation of HVAC design. These calculations determine the heating and cooling requirements of a building, ensuring that systems are appropriately sized and energy-efficient. The key types of building loads are:

1. **Heating Load**: The amount of heat energy required to maintain indoor temperatures during cold weather.
2. **Cooling Load**: The energy needed to remove excess heat from a space during warm weather.
3. **Latent Load**: Associated with humidity control, such as dehumidification.

V. Vakiloroaya, *HVAC and Building Services Rules of Thumb*, Springer Tracts in Civil Engineering, https://doi.org/10.1007/978-3-032-10285-0_5

Factors Influencing Load Calculations

- Building orientation and envelope properties (e.g., insulation, glazing).
- Internal heat gains from occupants, lighting, and equipment.
- External factors like solar radiation, wind exposure, and climate conditions.

Methods of Calculation

- Manual methods like the Heat Balance Method (HBM) and the Cooling Load Temperature Difference (CLTD) method.
- Software tools such as HAP, Camel, EnergyPlus and TRACE 700 for detailed simulations.

Principles of Heat Transfer
HVAC systems rely on the principles of heat transfer to maintain thermal comfort. Understanding these mechanisms is essential for effective system design:

Conduction: Heat transfer through solid materials, such as walls, windows, and floors. Insulation minimizes conductive losses.
Convection: Heat transfer via fluid movement, including air and water. This principle governs ventilation and hydronic systems.
Radiation: Heat transfer through electromagnetic waves, such as solar radiation entering a building.
Evaporation: Cooling through the phase change of water to vapour, a key principle in evaporative cooling systems.

5.3 Cooling Load Components

Internal Heat Gain
Internal heat gains refer to the heat generated within a building from occupants, equipment, lighting, and other sources. Internal heat gains depend on the building's purpose and usage, which must be clearly identified before beginning any calculations. Calculating these gains is essential for properly sizing HVAC systems, as they directly impact cooling and heating load requirements. Followings are some rules of thumb for estimating internal heat gains in HVAC and building services.

Occupant Heat Gains

- As a general guideline:
  - **Residential Spaces**: Approximately **100 W/person**.
  - **Office Spaces**: Between **100 and 150 W/person**, depending on activity level and comfort conditions (more active workers generate more heat).
  - **High-Density Areas**: For densely populated areas like conference rooms or gyms, consider **150–200 W/person** or more.

Following table showing approximate internal heat gain values for people based on their gender, age, and activity level. These values are based on typical metabolic rates and can vary slightly depending on individual factors. This table is a general guideline. For HVAC design, actual measurements or specific guidelines (like ASHRAE standards) should be referred to.

Category	Age group (year)	Activity level	Heat gain (W/person)
Male adults	18–65	Resting	70–90
		Light Activity (office)	100–130
		Moderate Activity	140–200
Female adults	18–65	Resting	60–80
		Light Activity (office)	80–100
		Moderate Activity	120–150
Children	3–12	Resting	50–70
		Light Play	70–90
		Vigorous Play	100–150
Seniors	65+	Resting	50–70
		Light Activity (walking)	80–100
		Moderate Activity	100–130

Lighting Heat Gains

- Use the following estimates for internal lighting loads:
  - **General Office Space**: Approximately **10–15 W/m^2**.
  - **Retail Spaces**: Higher values around **20–30 W/m^2** due to increased lighting for product displays.
  - **Specialized areas (like laboratories or studios)**: Consider **30–50 W/m^2** or more depending on specific needs.

Following table showcasing the approximate internal heat gain for different types of lighting commonly used in residential, commercial, and industrial settings. These values represent heat emitted as a byproduct of light generation and can vary based on the specific product and its efficiency.

Lighting type	Power rating (W)	Heat gain (W)	Notes
Incandescent bulbs	40–100	40–100	Nearly all electrical energy converts to heat (90–95%)
Halogen bulbs	35–150	30–145	More efficient than incandescent, but still high heat output
Compact fluorescent (CFL)	5–23	4–20	Approximately 80% of energy converts to heat
Linear fluorescent tubes	14–85	12–75	Common in offices; less heat than CFLs or incandescent
LED bulbs	5–25	2–10	Highly efficient; minimal heat output.

Lighting type	Power rating (W)	Heat gain (W)	Notes
High-intensity discharge (HID)	100–1000	80–950	Used in industrial and outdoor applications; high heat gain
Metal halide	50–400	45–380	A type of HID lighting; significant heat production
High-pressure sodium (HPS)	50–1000	45–950	Common in street lighting; emits substantial heat

Key Points

Incandescent and Halogen lights generate significantly more heat compared to energy-efficient options like LEDs.

Fluorescent lights (CFLs and tubes) produce moderate heat, making them suitable for large indoor spaces.

LEDs are the most energy-efficient, converting a higher percentage of energy into light, resulting in minimal heat gain.

HID and **HPS** lighting are suited for outdoor and industrial uses but produce considerable heat.

Equipment Heat Gains

- Account for heat generated by electrical and mechanical equipment:
  - **Computers**: Approximately **200–300 W/workstation**.
  - **Office Equipment (printers, copiers)**: Around **500–800 W** for larger machines.
  - **Kitchen Appliances**: Vary widely; commercial kitchen equipment can exceed **5 kW** or more, depending on usage.

Appliances and Miscellaneous Loads

- Consider allowances for other miscellaneous heat sources:
  - **Refrigerators**: Typical heat gain is **100–200 W**.
  - **Televisions and Audio Equipment**: Estimate around **150–300 W**, depending on usage.
  - **Miscellaneous Loads**: Include allowances for various devices (e.g., chargers, projectors) at about **50–100 W** each.

Heat Gain from Office Equipment

Following table showing the approximate heat gain from various office equipment. These values are estimates and can vary based on the specific model, usage patterns, and energy efficiency.

Equipment	Typical power usage (W)	Heat gain (W)	Notes
Desktop Computer	100–300	100–300	Heat depends on workload; gaming or high-performance computers emit more
Laptop	30–90	30–90	Efficient designs result in lower heat output
Monitor (LCD/LED)	20–50	20–50	Modern monitors emit less heat than older CRT models
Printer (Laser)	300–800 (while printing)	10–20 (standby)	High heat during operation; low heat when idle
Printer (Inkjet)	20–50	20–50	Lower heat compared to laser printers
Photocopier	300–1500 (while copying)	20–50 (standby)	High heat during operation; standby heat depends on energy-saving modes
Server	200–800	200–800	High heat output; server rooms often require dedicated cooling
Network Switch/Router	20–100	20–100	Heat depends on the size and complexity of the device
Television/Display	50–200	50–200	Large displays, especially older plasma models, produce more heat
Phone Charger	5–10	5–10	Small but continuous heat output when plugged in
Task Lighting (LED)	5–20	5–20	Minimal heat gain with efficient lighting
Space Heater	500–1500	500–1500	Significant heat gain; should be factored carefully into HVAC calculations

Key Considerations

Operation vs. Standby: Devices like printers and photocopiers produce much more heat during operation than when idle.

Energy Efficiency: Modern equipment, particularly LED monitors and laptops, emit significantly less heat.

Server Rooms: Servers and network equipment require precise cooling solutions due to their continuous high heat output.

Heat Gain from Hospital Equipment

Following table that provides approximate heat gain values for common hospital equipment. These values depend on usage patterns, operational settings, and the specific models used.

Equipment	Typical power usage (W)	Heat gain (W)	Notes
MRI Machine	10,000–20,000	10,000–20,000	High heat gain; requires dedicated cooling systems
CT Scanner	5000–10,000	5000–10,000	Significant heat gain during operation
X-Ray Machine	2000–5000	2000–5000	Heat gain depends on usage frequency

Equipment	Typical power usage (W)	Heat gain (W)	Notes
Ultrasound Machine	100–500	100–500	Lower heat gain compared to other imaging equipment
Ventilator	50–200	50–200	Essential in ICU; relatively low heat output
Patient Monitor	50–100	50–100	Common in critical care; generates moderate heat
Dialysis Machine	500–2000	500–2000	Heat gain depends on operational settings
Infusion Pump	20–50	20–50	Minimal heat gain
Operating Room Lights	150–500	150–500	Heat gain varies with type (LEDs produce less heat)
Surgical Tools (Powered)	100–300	100–300	Includes drills, saws, and other electric tools
Autoclave (Sterilizer)	1500–5000	1500–5000	High heat output during sterilization cycles
Refrigerators/ Freezers	100–800	100–800	Medical-grade units may have higher heat rejection
Lab Centrifuge	200–1000	200–1000	Heat gain depends on size and speed
ECG/EKG Machine	50–100	50–100	Low heat gain; often used intermittently
Blood Analyser	500–1500	500–1500	Common in labs; significant heat output during continuous operation

Key Considerations

Large Imaging Equipment: MRI, CT scanners, and X-ray machines generate substantial heat and require dedicated cooling systems.

Surgical Equipment: Operating rooms often have heat-intensive equipment like lights and powered tools.

Lab Equipment: Continuous operation of centrifuges and analysers in labs contributes to overall heat gain.

Room-Specific Loads: Different hospital areas (e.g., ICUs, operating rooms, imaging centres) have unique heat gain profiles that must be accounted for in HVAC design.

Total Internal Heat Gains Calculation

- Your total internal heat gains can be approximated as follows:
  - **Total Internal Heat Gains (W)** = (Number of Occupants × Occupant Load) + (Area × Lighting Load) + (Number of Equipment × Equipment Load).
- Be sure to adjust for building usage patterns, such as weekdays versus weekends or peak occupancy hours.

Diversity Factor

- When calculating heat gains, consider using a **diversity factor** to account for the fact that not all equipment or occupants are active simultaneously:
  - A common practice is to apply a diversity factor of **0.7–0.9** depending on the application (higher for spaces with more unpredictable occupancy).

Impact of Internal Heat Gains

- Recognize that internal heat gains not only impact the cooling load but also can reduce the heating load in winter:
  - For example, during winter conditions, internal heat gains can help offset heating requirements, which should be reflected in your HVAC sizing calculations.

External Heat Load

External heat gains refer to the heat entering a building from outside sources, primarily through the building envelope (walls, roofs, windows, and doors). These gains are crucial for determining cooling loads in HVAC systems. Followings are some rules of thumb for estimating external heat gains in HVAC and building services:

Solar Heat Gains Through Windows

- For estimating solar heat gain through windows for Northern Hemisphere, consider the following:
  - **Window Area**: Calculate the total window area and apply a general rule of thumb of about **100–200 W/m²** depending on orientation and shading.
  - South-facing Windows: Receive the most direct sunlight, especially in winter, when the sun is lower in the southern sky. This explains the higher solar loads.
  - East and West-facing Windows: Experience significant solar gains in the morning (east) and afternoon (west), especially in summer when the sun is higher and more intense.
  - North-facing Windows: Typically receive only diffused light, resulting in much lower solar heat gains.
  - **Orientation**:
    - **South-facing Windows**: Higher loads due to direct sunlight—consider **200–300 W/m²** in peak conditions.
    - **East and West-facing Windows**: Also significant, around **150–250 W/m²**.
    - **North-facing Windows**: Generally lower solar gains, around **50–150 W/m²**.
  - **Shading**: Account for exterior shading devices (awnings, overhangs) which can reduce solar gains significantly (by around **30–50%**).
- For estimating solar heat gain through windows for Southern Hemisphere, consider the following:

- **Window Area**: Calculate the total window area and apply a general rule of thumb of about **100–200 W/m²** depending on orientation and shading.
- North-facing Windows: Would receive the most sunlight and have higher solar heat gains, similar to south-facing windows in the Northern Hemisphere.
- East and West-facing Windows: Similar behaviour as in the Northern Hemisphere, with peak gains in the morning and afternoon.
- South-facing Windows: Would have lower solar gains, akin to north-facing windows in the Northern Hemisphere, as they primarily receive diffused light.
- **Orientation**:
    - **South-facing Windows**: Higher loads due to direct sunlight—consider **50–150 W/m²** in peak conditions.
    - **East and West-facing Windows**: Also significant, around **150–250 W/m²**.
    - **North-facing Windows**: Generally lower solar gains, around **200–300 W/m²**.

Heat Gains Through Walls and Roofs

- **Heat Transfer through Exterior Walls**: Use the following estimates:
    - **Wall Area**: For calculating heat gains through walls, a rough estimate is to consider:
        - **Low Insulation Levels**: **15–30 W/m²** per degree temperature difference (ΔT) between inside and outside.
        - **Good Insulation Levels**: **5–15 W/m²** per degree ΔT.
- **Heat Transfer through Roofs**: Roofs generally experience higher temperature differences. Use:
    - **Flat Roofs**: Estimate around **30–50 W/m²** for moderate insulation levels.
    - **Sloped Roofs**: Typically similar to walls but can vary depending on orientation and materials.

Impact of External Environment

- Consider external environmental factors:
    - **Seasonal Variations**: Adjust estimates based on seasonal temperature differences.
    - **Urban Heat Island Effect**: For buildings in urban areas, consider higher external temperatures, especially during summer months.

Building Orientation and Shape

- Building orientation can impact heat gains significantly:
    - **Longer Facades Facing Sun**: More heat gain.
    - **Compact Building Forms**: Generally experience lower external heat gains compared to elongated forms.

Common U-Values for HVAC Load Calculations

U-Values (thermal transmittance) represent the rate of heat transfer through a building element per unit area and per degree of temperature difference. Lower U-values indicate better insulating properties. Following is a guide to commonly used U-values and rules of thumb for various building materials:

Walls

Wall type	Common U-value (W/ m^2 K)	Rules of thumb
Uninsulated brick wall (single layer)	2.0–2.5	High U-value; consider adding insulation
Cavity wall (uninsulated)	1.5–1.8	Moderate U-value; benefits from cavity insulation
Insulated cavity wall	0.3–0.6	Depends on insulation thickness and material
Solid concrete wall (uninsulated)	2.0–3.5	Poor thermal performance; needs external insulation
Insulated concrete wall	0.4–0.6	Good thermal performance with proper insulation

Roofs

Roof type	Common U-value (W/ m^2 K)	Rules of thumb
Flat roof (uninsulated)	2.0–3.0	High U-value; insulate to improve energy efficiency
Flat roof (insulated)	0.2–0.3	Typical target for energy-efficient buildings
Pitched roof (uninsulated)	2.0–3.0	Poor performance; benefits from insulation
Pitched roof (insulated)	0.2–0.4	Modern designs aim for U-values below 0.25

Floors

Floor type	Common U-value (W/ m^2 K)	Rules of thumb
Concrete floor (uninsulated)	0.7–1.0	Conductive; insulate in energy-efficient designs
Insulated concrete floor	0.2–0.4	Proper insulation significantly reduces U-value
Timber floor (uninsulated)	1.0–1.5	Wood offers moderate insulation but needs reinforcement
Insulated timber floor	0.3–0.5	Adequate for thermal comfort and load calculations

Doors

Door type	Common U-value (W/m^2 K)	Rules of thumb
Solid timber door	2.5–3.0	Moderate insulation; often uninsulated
Insulated door	0.6–1.0	Suitable for energy-efficient buildings
Metal door (uninsulated)	3.0–5.0	Very high U-value; requires thermal breaks or insulation
Metal door (insulated)	0.8–1.5	Much lower U-value with proper insulation

Windows

Window type	Common U-value (W/m^2 K)	Rules of thumb
Single-glazed window	4.5–6.0	High U-value; major source of heat loss/gain
Double-glazed window (air-filled)	2.0–3.0	Standard for most modern buildings
Double-glazed window (low-E coating)	1.0–1.8	Improved thermal performance
Triple-glazed window	0.6–1.2	Ideal for high-performance or passive buildings
Low-emissivity windows with argon gas	0.4–0.8	Premium performance for energy-efficient designs

Glazing Ratios

- **Rule of Thumb**: Glazing should not exceed **40% of the total wall area** to balance natural light and thermal performance.
- Use low-E or reflective coatings for large glazing areas in hot climates.

U-Value Rules of Thumb for HVAC Load Calculations

- **Climate-Specific Adjustments**:
  - Cold climates: Target U-values **<0.3 W/m^2 K** for walls, roofs, and floors.
  - Hot climates: Prioritize reflective materials and shading over ultra-low U-values.
  - Mixed climates: Balance insulation and heat retention.
- **Insulation Thickness**:
  - Add **50–100 mm** of insulation to walls, roofs, and floors to reduce U-values significantly.
  - Use high-performance materials like PIR or spray foam for thinner insulation layers.

- **Air Tightness**:
  - Reduce infiltration losses by designing for an air permeability of **3–5 $m^3/h\ m^2$** at 50 Pa for standard buildings.
- **Thermal Breaks**:
  - Introduce thermal breaks in metal components to minimize heat transfer.
- **Material Selection**:
  - Use lightweight materials (e.g., aerated concrete, insulated panels) for better U-values in walls.
  - Consider green roofs for additional thermal resistance.
- **Code Compliance**:
  - Align U-values with local building regulations (e.g., ASHRAE 90.1, IECC, EN standards).

Key Considerations in U-Value Selection

- **Thermal Comfort**: Lower U-values improve indoor temperature stability.
- **Energy Efficiency**: Buildings with better U-values have reduced HVAC energy demand.
- **Cost vs. Performance**: Balance upfront insulation costs with long-term savings.
- **Moisture Management**: Ensure materials do not trap moisture or cause condensation issues.

Ventilation Load
Fresh air is essential in buildings for several reasons related to health, comfort, and functionality:

- **Indoor Air Quality (IAQ):**
  - Fresh air dilutes indoor pollutants such as carbon dioxide (CO_2), volatile organic compounds, odours, and particulates.
  - Proper ventilation prevents the build-up of contaminants, reducing risks of respiratory issues and sick building syndrome.
- **Oxygen Supply:**
  - Occupants consume oxygen and release CO_2. Fresh air ensures a constant supply of oxygen to maintain safe and comfortable levels.
- **Humidity Control:**
  - Fresh air helps manage indoor humidity by introducing outdoor air with a potentially different moisture content, aiding in mold prevention and occupant comfort.
- **Temperature Regulation:**
  - Fresh air intake can aid in cooling (via economizers) when outdoor air conditions are favourable, reducing energy consumption in moderate climates.

Effect of Fresh Air on HVAC Load Calculation
Introducing fresh air has a significant impact on building HVAC load calculations:

- **Increased Cooling Load:**
  - **Sensible Load:** Outdoor air often enters the building at a higher temperature than the desired indoor conditions. The HVAC system must cool this air to maintain comfort.
  - **Latent Load:** Outdoor air may carry moisture. In humid climates, dehumidification adds to the cooling load.
- **Increased Heating Load:**
  - In cold climates, fresh air is typically colder than the indoor set-point. The HVAC system must heat this air to maintain desired temperatures.
- **Ventilation Equipment Sizing:**
  - The introduction of fresh air requires larger ducts, fans, and sometimes dedicated air handling units (AHUs) to accommodate the airflow and ensure proper mixing with recirculated air.
- **Energy Recovery Systems:**
  - **Heat Recovery Ventilators (HRVs) or Energy Recovery Ventilators (ERVs):** These systems reduce the impact of fresh air on HVAC loads by transferring heat and, in some cases, moisture between outgoing exhaust air and incoming fresh air.
- **Demand-Controlled Ventilation (DCV):**
  - Ventilation rates can be adjusted based on occupancy or CO_2 levels, optimizing energy use while maintaining IAQ.

Following table indicates required **Fresh Air per Person (L/s per person)** based on typical ventilation standards for different buildings:

Building type	Percentage of fresh air (%)	Fresh air (L/s/m^2)	Fresh air (L/s/Person)	Notes
Residential (apartments)	10–15	0.3–0.5	5–10	Depends on occupancy and natural ventilation availability
Office buildings	15–20	0.6–1.0	8–12	Higher rates for densely occupied spaces like meeting rooms
Educational facilities	20–30	1.0–2.0	10–15	Classrooms require higher ventilation due to occupancy density
Retail spaces	10–20	0.5–1.0	5–10	Includes shopping malls and department stores
Healthcare facilities	25–40	2.0–3.5	20–30	Operating rooms and ICUs have higher requirements for infection control

Building type	Percentage of fresh air (%)	Fresh air (L/s/m^2)	Fresh air (L/s/Person)	Notes
Hotels (guest rooms)	10–15	0.3–0.6	5–8	Based on intermittent occupancy and air quality needs
Restaurants	20–30	2.0–3.0	15–25	Higher rates due to kitchen exhaust requirements and odours
Theatres/cinemas	20–30	1.5–2.5	15–20	Ensures comfort in crowded spaces
Industrial buildings	15–25	1.0–2.5	10–15	Specific processes may demand higher ventilation rates
Laboratories	25–50	4.0–8.0	25–50	High rates needed for contaminant removal and safety
Gyms and fitness centres	25–35	2.0–4.0	20–40	Higher rates due to increased respiration from physical activity
Data centres	5–10	0.2–0.5	2–5	Low requirements due to minimal occupancy, but cooling is critical
Parking garages	50–100	4.0–8.0	N/A	Based on exhaust removal requirements for CO and pollutants

Key Notes

- **Fresh Air per Person** is based on average occupancy levels and activity types within each building type.
- **Parking Garages** are primarily ventilated to control pollutant concentrations, so air per person isn't directly applicable.
- Values may vary based on local regulations and specific design requirements.

Following table outlining the required fresh (outdoor) air ventilation rates for rooms designed to care for different animals:

Animal type	Room/facility type	Fresh air requirement (ACH)	Fresh air requirement (L/s m^2)	Special considerations
Dogs	Kennels, Dog Boarding Rooms	6–10	10–15	High odour control; consider using HEPA filters or additional air purifiers
Cats	Catteries, Cat Boarding Rooms	4–8	8–12	Minimize drafts; temperature stability is critical
Birds	Aviaries	10–20	15–25	Fine particulate filtration is essential due to feathers and dust
Reptiles	Terrariums, Reptile Rooms	2–6	4–8	Maintain humidity levels; ensure proper exhaust for heat and odours

Animal type	Room/facility type	Fresh air requirement (ACH)	Fresh air requirement (L/s m^2)	Special considerations
Fish	Aquariums, Fish Rooms	6–12	8–15	Manage humidity; consider water vapour control
Horses	Stables, Equine Care Facilities	4–8	8–12	High ventilation to control ammonia build-up; ensure fresh air reaches all areas
Rodents	Research Labs, Rodent Habitats	10–15	12–18	Odour and particulate control are critical; ensure tight exhaust systems
Farm Animals	Livestock Barns	3–8	6–10	Include provisions for high moisture and dust; adjust for large spaces
Zoo Animals	General Zoo Enclosures	6–12	10–18	Variable based on species and enclosure size; specialized filtration may be needed
Exotic Animals	Exotic Animal Rooms	8–15	12–20	Consider species-specific needs; humidity and air purity may be critical

Key Notes

Fresh Air Requirement (ACH):
Air changes per hour based on the volume of the room.
Fresh Air Requirement (L/s m^2):
Ventilation rate per square meter of room area, provided in litters per second.
Special Considerations:
Each animal type has unique needs, such as odour control, humidity regulation, or particulate filtration.

Infiltration Load
Infiltration refers to the uncontrolled entry of outdoor air into a building through leaks in the building envelope. This can significantly impact heating and cooling loads in HVAC systems. Proper management of infiltration is critical for energy efficiency and occupant comfort. This section provides some rules of thumb and considerations related to infiltration and its associated loads. It is generally assumed that a pressurized building prevents infiltration, although infiltration will often occur in the lower third of a building taller than 80 ft. [25 m] with an operating HVAC system and can occur throughout a building when the HVAC system is shut off.

Estimating Infiltration Rates

- A common rule of thumb for residential buildings is to estimate:
  - **0.3–0.5 air changes per hour (ACH)** for tight buildings.
  - **0.5–1 ACH** for moderately tight buildings.
  - **1–2 ACH** for leaky buildings.

- For commercial buildings, infiltration is often estimated at **2.5–5 L/s/100 m² of building area** (depending on building type and envelope quality).

Following table summarizing infiltration factors for different types of buildings, which are commonly used to calculate cooling load due to air leakage.

Building type	Infiltration rate (ACH)	Infiltration factor (L/s m²)	Description
Tightly sealed residential	0.1–0.3	0.1–0.3	Modern homes with energy-efficient construction
Typical residential	0.5–1.0	0.5–1.0	Average homes with standard construction
Office buildings	0.1–0.5	0.2–0.8	Well-sealed windows and doors; HVAC systems active
Retail spaces	0.5–1.5	1.0–2.5	Frequent door usage; moderate air sealing
Warehouses	1.0–3.0	1.5–4.5	Large door openings; low energy sealing
Laboratories	0.2–0.5	0.3–0.8	High ventilation requirements, but tightly sealed
Hospital buildings	0.2–0.6	0.3–1.0	Controlled infiltration to maintain air quality
Schools	0.3–1.0	0.5–1.5	Regular foot traffic; moderate air sealing
Restaurants	1.0–3.0	1.5–5.0	High air exchange due to kitchen operations
Industrial buildings	1.5–5.0	2.5–8.0	Large open spaces with frequent ventilation

Heat Load due to Infiltration

- The heat load due to infiltration can be estimated using the following formula: $Q = V \times \Delta T \times \rho \times c$
- Where:
  - Q = Heat load (kW),
  - V = Infiltration air volume (m^3/s),
  - ΔT = Temperature difference between indoors and outdoors (°C),
  - ρ = Density of air (approximately **1.2 kg/m³** at room temperature),
  - c = Specific heat of air (approximately **1.005 kJ/kg K**).

Fresh Air Requirements

- When calculating total fresh air requirements, consider infiltration as part of the overall ventilation strategy:
  - Ensure it is accounted for in compliance with AS1668 or ASHRAE Standard 62.1 or equivalent local codes.
  - The minimum fresh air rate can be adjusted based on infiltration rates to avoid over-ventilation.

Pressure Relationships

- Consider the pressure differences created between indoors and outdoors, particularly in high-rise buildings where stack effect can increase infiltration rates:
  - For every **25 mm of water column** pressure difference, expect substantial increases in infiltration.

Mitigation Strategies

- Use air sealing techniques during construction or renovations to minimize infiltration:
  - Aim for a target air leakage rate of around **1.5 ACH@50 Pa** for new residential buildings to enhance energy performance.
- Incorporate well-designed windows, doors, and weather stripping to further reduce leakage pathways.

Impact on HVAC Sizing

- Adjust HVAC load calculations to account for infiltration loads:
  - Increase heating capacity by approximately **10–15%** in cold climates where infiltration is significant.
  - Consider the seasonal changes in infiltration when sizing cooling and heating systems, as infiltration can vary with temperature and wind pressure.

Building Envelope Design

- Pay attention to the design and materials of the building envelope to reduce infiltration:
  - Use continuous insulation and well-sealed joints to minimize thermal bridging and air leakage.

5.4 Cooling Load Rules of Thumb

Rules of thumb for building cooling load calculation provide quick, approximate estimates of cooling requirements based on typical values for similar buildings, climates, and usage patterns. They save time in the early stages of a project by enabling designers to make informed decisions about system sizing and feasibility without the need for detailed simulations or precise inputs, which may not yet be available. However, as the project progresses into the developed stage, accurate simulations and detailed load calculations must be conducted by the designer to ensure the final system design is precise, efficient, and tailored to the specific building requirements. Following table provides load calculation rules of thumb for various types of buildings.

Application	ACH	Total load (kw/m^2)	Air flow rate ((l/s)/m^2)	SHR
Residential	6–10	0.14	7.5	0.75–0.95
Auditorium	8–15	0.28	7–15 (l/s)/Seat	0.65–0.75

Application	ACH	Total load (kw/m²)	Air flow rate ((l/s)/m²)	SHR
Banks/Court houses	6–10	0.18	9.7	0.75–0.95
Hairdresser	6–10	0.21	11.6	0.65–0.90
Beauty Shop/Jewellery/Barber	6–10	0.21	11.6	0.65–0.90
Cafeteria/Dining Hall/Lunchrooms	12–15	0.35	19	0.65–0.8
Restaurant	8–12	0.35	19	0.65–0.85
Supermarket	6–10	0.15	8	0.65–0.85
Fast Foods	8–10	0.16	8.6	0.65–0.8
Offices	6–10	0.15	8	0.75–0.9
Mall/Shopping Centre	6–10	0.25	13.5	0.65–0.85
Clinic/Medical Centre	6–10	0.19	10.2	0.75–0.9
Computer Room	10–20	0.5	27	0.8–0.95
Theatre	8–15	0.28	15	0.65–0.75
Hotels/Motels Guest rooms	6–10	0.14	7.5	0.8–0.95
Classrooms	8–12	0.16	8.6	0.65–0.8
Police Offices	6–10	0.16	8.6	0.75–0.9
Post Offices	6–10	0.18	9.7	0.75–0.85
Precision Manufacturing	10–40	0.5	27	0.8–0.95
Hospital Patient Room	6–10	0.15	8	0.75–0.85
Libraries/Museum	8–12	0.15	8	0.8–0.9
Retails	6–10	0.19	10.2	0.65–0.9
Bowling Alleys	10–15	7.5 kw/Alley		0.6–0.8
Churches/Masque	8–15	0.28	7–15 (l/s)/Seat	0.65–0.75
Conference Rooms		0.28	15	0.65–0.8

The **Sensible Heat Ratio (SHR) in above table** is the ratio of **sensible heat** (heat that changes air temperature) to the total heat load, which includes both **sensible** and **latent heat** (heat associated with moisture removal). It indicates the balance between temperature control and humidity control in an HVAC system. A higher SHR (e.g., 0.8) means most of the cooling is focused on lowering air temperature, while a lower SHR (e.g., 0.6) indicates significant dehumidification is required. SHR is crucial for selecting and designing HVAC equipment to meet the specific thermal comfort and moisture control needs of a space.

The following tables outline the rules of thumb for various types of buildings with more details.

Hospital Spaces

Space/room type	Cooling load (kW/m²)	Fresh air (% of Supply Air) (%)	Fresh air (L/s/m²)	Description
Patient Rooms	0.10–0.15	25–30	3–5	Standard occupancy with moderate heat gain from equipment and occupants

Space/room type	Cooling load (kW/m^2)	Fresh air (% of Supply Air) (%)	Fresh air (L/s/m^2)	Description
Operating Theaters	0.20–0.30	100	15–20	High ventilation requirement with strict temperature and humidity control
ICU (Intensive Care Units)	0.18–0.25	80–100	10–15	Equipment-heavy spaces with strict temperature and humidity control
Laboratories	0.15–0.20	50–70	5–10	Moderate equipment heat gain and controlled environment
Pharmacies	0.12–0.15	25–30	2–4	Lower cooling demand compared to medical spaces; typically less equipment
Diagnostic Imaging Rooms	0.20–0.25	30–50	3–5	Includes CT, MRI, and X-ray rooms with significant equipment-generated heat
Reception/Waiting Areas	0.10–0.12	15–25	2–3	Moderate occupancy with minimal equipment heat gain
Corridors and General Areas	0.08–0.10	10–20	1–2	Minimal heat load from occupants and lighting
Cafeterias	0.12–0.18	25–40	3–5	Moderate occupancy and cooking-related heat gain
Kitchens	0.25–0.40	50–75	10–15	High heat load from cooking appliances and equipment
Storage Rooms (Medical Supplies)	0.08–0.10	10–20	1–2	Low heat gain; typically unoccupied with limited equipment
Emergency Rooms (ER)	0.15–0.20	50–70	5–10	Moderate cooling demand due to occupancy and lighting
Sterile Supply Rooms	0.15–0.20	50–70	5–10	Controlled environment with moderate heat gain from sterilization processes
Laundry Rooms	0.25–0.35	30–50	5–10	High heat load due to dryers and washing equipment
Mechanical/ Electrical Rooms	0.20–0.30	20–30	2–4	Heat load dominated by equipment such as UPS and electrical panels
Morgues	0.10–0.15	25–40	3–5	Minimal heat load but may require precise temperature and humidity control

Computer Rooms (Data Centres, Server Rooms, IT Facilities)

Space/room type	Cooling load (kW/m^2)	Fresh air (% of Supply Air) (%)	Fresh air (L/s/m^2)	Description
Small Server Room	0.50–1.0	5–10	0.5–1.0	Low-density servers with minimal equipment heat load

Space/room type	Cooling load (kW/m²)	Fresh air (% of Supply Air) (%)	Fresh air (L/s/m²)	Description
Mid-Sized Data Centre	1.0–2.0	5–15	1.0–2.0	Moderate server density with dedicated cooling and ventilation requirements
High-Density Data Centre	2.0–4.0	5–15	1.5–3.0	High server density with substantial equipment-generated heat load
Telecom Rooms (IDF/MDF)	0.30–0.50	5–10	0.5–1.0	Small communication hubs with moderate heat from equipment
Workstation IT Rooms	0.10–0.15	10–15	2.0–3.0	General office spaces with limited IT equipment
Backup UPS/ Power Rooms	0.20–0.30	10–20	2.0–4.0	Heat load from uninterruptible power supplies and electrical systems
Network Operations Centres (NOCs)	0.15–0.20	10–20	2.0–3.0	Moderate occupancy with additional heat from monitors and servers
Testing Labs	1.5–3.0	15–20	2.5–5.0	High cooling demand due to testing equipment and servers

Notes

Cooling Load (kW/m²): Based on typical IT equipment density and room-specific activities.

Fresh Air (%): Limited fresh air introduction is standard to minimize cooling load, focusing on recirculated air.

Fresh Air (L/s per m²): Determined by operational and safety ventilation requirements.

Humidity Control: Maintain relative humidity between 40% and 60% to prevent static discharge or condensation.

Clean Rooms (Different ISO Classes)

Clean room class (ISO)	Cooling load (kW/m²)	Fresh air (% of Supply Air) (%)	Fresh air (L/s/m²)	Air change rate (ACH)	Description
ISO Class 1	1–2	10–40	5.0–10.0	300–600	Ultra-clean environments for semiconductor manufacturing and nanotechnology
ISO Class 2	1–2	10–40	4.0–8.0	300–600	Similar to Class 1, with slightly relaxed particle control
ISO Class 3	0.8–1.5	10–40	3.0–6.0	200–400	Used in high-end electronics manufacturing and pharmaceutical production

Clean room class (ISO)	Cooling load (kW/m²)	Fresh air (% of Supply Air) (%)	Fresh air (L/s/m²)	Air change rate (ACH)	Description
ISO Class 4	0.5–1	10–40	2.5–5.0	200–300	Typical for advanced pharmaceutical and precision optical manufacturing
ISO Class 5	0.5–1	10–40	2.0–4.0	100–250	Used in aseptic filling and critical pharmaceutical operations
ISO Class 6	0.4–0.8	10–40	1.5–3.0	60–150	Cleanroom environments for medical device production and assembly
ISO Class 7	0.4–0.6	10–40	1.0–2.0	30–80	General cleanrooms for manufacturing and research environments
ISO Class 8	0.2–0.4	10–40	0.5–1.0	20–60	Controlled environments for less critical processes
ISO Class 9	0.1–0.2	10–20	0.2–0.5	<20	Basic controlled environments, often antechambers to cleaner areas

Notes

Cooling Load (kW/m²): Depends on equipment heat load, lighting, and process activities. Higher classes often demand more robust cooling systems.

Fresh Air (%): 100% fresh air is required to maintain cleanliness and air purity standards.

Fresh Air (L/s/m²): Based on stringent contamination control requirements.

Air Change Rate (ACH): Reflects the number of times air is replaced per hour; higher for stricter cleanliness levels.

Humidity Control: Maintain relative humidity at 40–60% to prevent particle adhesion or excessive dryness. Adjustments depend on application (e.g., electronics require stricter humidity control).

For Cinemas and Theatres

Application	Cooling load (kW/m²)	Fresh air (% of total) (%)	Fresh air (L/s/m²)	Typical height (m)	ACH
Cinema Hall	0.15–0.25	15–20	3.0–5.0	5.0	4–10
Theatre Hall	0.20–0.30	20–25	4.0–6.0	6.0	4–10
VIP Lounge	0.20–0.35	20–30	4.0–8.0	3.0	4–10
Projection Room	0.30–0.40	5–10	2.0–4.0	3.0	4–10
Lobby/Waiting Area	0.10–0.15	20–25	3.0–5.0	3.5	4–10

Key Assumptions

Room Height: Typical room heights vary depending on the space (e.g., taller ceilings in theatres and halls).

Fresh Air (% of Total): Based on ASHRAE and other ventilation standards, depending on the occupancy type.

Cooling Load (kW/m²): Indicative values for typical design, accounting for internal heat gains (lights, equipment, and occupants).

Restaurants, Fast Foods, and Cafés

Application	Cooling load (kW/m²)	Fresh air (% of total) (%)	Fresh air (L/s/m²)	Typical height (m)	ACH
Fine Dining Restaurant	0.20–0.30	25–30	6.0–8.0	3.5	6–12
Fast Food Restaurant	0.25–0.40	30–40	8.0–12.0	3.0	6–12
Café	0.15–0.25	20–25	4.0–6.0	3.0	6–12
Kitchen (Cooking Area)	0.40–0.60	50–75	15.0–20.0	3.5	6–12
Dining Hall	0.20–0.30	20–30	6.0–8.0	3.5	6–12
Bar Area	0.20–0.30	25–30	6.0–8.0	3.0	6–12

Key Assumptions

Room Heights: Typical heights adjusted for each space type.

Cooling Load: Reflects internal heat gains from people, lighting, appliances, and cooking equipment.

Fresh Air: Calculated as a percentage of total supply air or in liters per second per square meter.

Kitchen Areas: High fresh air and cooling loads are required due to heat from cooking and exhaust.

Sports Salons, Bowling Alleys, Gyms, and Related Spaces

Application	Cooling load (kW/m²)	Fresh air (% of total) (%)	Fresh air (L/s/m²)	Typical height (m)	ACH
Gym/Fitness Area	0.25–0.50	20–30	8.0–10.0	3.0	8–12
Aerobic/Dance Studio	0.30–0.50	25–35	10.0–12.0	3.0	8–12
Bowling Alley	0.20–0.35	20–25	6.0–8.0	4.0	8–12
Indoor Sports Hall	0.25–0.40	20–25	6.0–8.0	6.0	8–12
Swimming Pool Hall	0.35–0.55	50–75	10.0–15.0	5.0	8–12
Locker Room	0.15–0.25	50–60	8.0–12.0	2.5	10–20
Spectator Area	0.20–0.30	15–25	5.0–6.0	4.0	8–12
Climbing Wall Area	0.20–0.30	20–25	5.0–8.0	8.0	10–15

Key Assumptions

Room Heights: Adjusted based on typical usage (e.g., sports halls and climbing areas have higher ceilings).
Cooling Load: Reflects internal heat gains from people, equipment, and lighting.
Fresh Air Requirements: Based on the activity level and occupancy density.
Swimming Pool Consideration: Includes humidity control in addition to cooling.

Notes

- Spaces with high physical activity, like gyms and aerobic studios, require higher ventilation and cooling loads.
- Swimming pools have higher fresh air and cooling demands due to evaporation and humidity levels.
- Spectator areas have lower fresh air and cooling requirements compared to active zones.

Universities, Schools, and Educational Institutes

Application	Cooling load (kW/m^2)	Fresh air (% of total) (%)	Fresh air ($L/s/m^2$)	Typical height (m)	ACH
Classrooms	0.10–0.20	15–25	5.0–8.0	3.0	6–10
Lecture Halls	0.15–0.25	20–30	6.0–10.0	3.5	6–10
Libraries	0.10–0.15	10–20	3.0–5.0	3.0	6–10
Computer Labs	0.20–0.30	20–30	8.0–12.0	3.0	8–12
Science Labs	0.25–0.40	30–40	10.0–15.0	3.5	8–12
Auditoriums	0.15–0.25	20–30	6.0–10.0	5.0	6–10
Cafeterias	0.20–0.30	25–35	8.0–12.0	3.0	6–10
Administration Offices	0.10–0.15	15–20	5.0–6.0	3.0	6–10
Gymnasiums (Schools)	0.25–0.40	20–30	8.0–10.0	6.0	6–10
Locker Rooms	0.15–0.25	50–60	10.0–15.0	2.5	10–20

Key Assumptions

Cooling Load: Includes internal heat from occupants, equipment, and lighting specific to each space type.
Fresh Air Requirements: Based on occupancy density and activity levels per space.
Science Labs and Computer Labs: Require additional ventilation for heat and fume removal.
Auditoriums: Designed for intermittent high-occupancy usage with higher ventilation rates.

Notes

- Spaces like science labs and computer labs have higher fresh air requirements due to heat generation and safety considerations.
- Large spaces such as lecture halls and auditoriums require additional fresh air for occupant comfort.
- Cafeterias and gymnasiums have higher cooling loads due to dense occupancy and physical activity levels.

Universities, Schools, and Educational Institutes (Including Specialized Areas)

Application	Cooling load (kW/m²)	Fresh air (% of total) (%)	Fresh air (L/s/m²)	Typical height (m)	ACH
Classrooms	0.10–0.20	15–25	5.0–8.0	3.0	6–10
Lecture Halls	0.15–0.25	20–30	6.0–10.0	3.5	6–10
Libraries	0.10–0.15	10–20	3.0–5.0	3.0	6–10
Computer Labs	0.20–0.30	20–30	8.0–12.0	3.0	8–12
Science Labs	0.25–0.40	30–40	10.0–15.0	3.5	8–12
Auditoriums	0.15–0.25	20–30	6.0–10.0	5.0	6–10
Cafeterias	0.20–0.30	25–35	8.0–12.0	3.0	6–10
Administration Offices	0.10–0.15	15–20	5.0–6.0	3.0	6–10
Gymnasiums (Schools)	0.25–0.40	20–30	8.0–10.0	6.0	6–10
Locker Rooms	0.15–0.25	50–60	10.0–15.0	2.5	10–20
Art Studios	0.20–0.30	20–30	6.0–8.0	3.5	6–10
Music Rooms	0.15–0.20	15–25	5.0–6.0	3.5	6–10
Dance Studios	0.20–0.35	20–30	8.0–10.0	3.5	8–12
Language Labs	0.15–0.25	15–20	4.0–6.0	3.0	6–10

Key Notes for Specialized Areas

Art Studios: Require higher ventilation to mitigate odours and fumes from paint and materials.

Music Rooms: Require good ventilation and moderate cooling loads for occupant comfort during practice or performance.

Dance Studios: Have higher cooling loads due to physical activity, necessitating additional ventilation.

Language Labs: Require standard ventilation and cooling to maintain a conducive learning environment.

Retail Spaces, Shopping Centres, and Similar Applications

Application	Cooling load (kW/m²)	Fresh air (% of total)	Fresh air (L/s/m²)	Typical height (m)	ACH
Small Retail Stores	0.15–0.25	15–25	5.0–8.0	3.0	6–10
Large Department Stores	0.20–0.30	20–30	6.0–10.0	4.0	6–10
Supermarkets	0.25–0.35	20–30	8.0–12.0	4.0	6–10
Shopping Malls (Common Areas)	0.15–0.20	20–30	6.0–8.0	4.5	6–10
Food Courts	0.30–0.40	25–35	10.0–15.0	3.5	8–12
High-End Boutiques	0.15–0.20	15–25	5.0–8.0	3.0	6–10
Grocery Stores	0.20–0.30	20–30	6.0–10.0	4.0	6–10
Jewellery Stores	0.10–0.15	15–20	4.0–5.0	3.0	6–10
Electronic Stores	0.15–0.25	20–30	5.0–8.0	3.5	6–10

Application	Cooling load (kW/m²)	Fresh air (% of total)	Fresh air (L/s/m²)	Typical height (m)	ACH
Cinemas Inside Malls	0.20–0.30	30–40	8.0–12.0	5.0	6–10
Storage Areas (Backrooms)	0.10–0.15	10–15	3.0–4.0	3.5	6–10
Parking Areas (Enclosed)	0.05–0.10	100	2.0–3.0	4.0	10–15

Key Notes for Specific Applications

Food Courts: Higher cooling loads and ventilation due to cooking equipment and occupant density.

Shopping malls (Common Areas): Moderate cooling with high ventilation rates for large open spaces with high occupancy.

Supermarkets: Require good ventilation for perishable goods and increased cooling loads due to refrigeration systems.

High-End Boutiques and Jewellery Stores: Lower cooling loads as these spaces prioritize ambiance and often have controlled occupancy.

Office Spaces (Post Offices, Police Offices, Banks, etc.)

Application	Cooling load (kW/m²)	Fresh air (% of total) (%)	Fresh air (L/s/m²)	Typical height (m)	ACH
Post Offices	0.10–0.15	15–25	4.0–6.0	3.0	6–12
Police Offices	0.10–0.20	20–30	5.0–8.0	3.0	6–12
Banks (General Areas)	0.15–0.20	20–30	6.0–8.0	3.5	6–12
Banks (Vault Areas)	0.05–0.10	10–15	2.0–4.0	3.0	6–12
Insurance Offices	0.10–0.15	15–25	4.0–6.0	3.0	6–12
Government Offices	0.10–0.20	20–30	5.0–8.0	3.0	6–12
Libraries (Office Areas)	0.10–0.15	15–20	4.0–5.0	3.5	6–12
Conference Rooms	0.20–0.30	25–35	8.0–12.0	3.0	6–12
Meeting Rooms	0.20–0.25	25–35	8.0–10.0	3.0	6–12
Archives/Storage Areas	0.05–0.10	10–15	2.0–4.0	3.0	6–12
Open Plan Workspaces	0.10–0.20	20–30	5.0–8.0	3.0	6–12
Executive Offices	0.10–0.15	15–20	4.0–5.0	3.0	6–12
Courts (Courtrooms)	0.15–0.25	25–35	8.0–10.0	4.0	6–12
Embassies (General Areas)	0.10–0.20	20–30	5.0–8.0	3.5	6–12
Embassies (Sensitive Areas)	0.10–0.15	20–25	5.0–6.0	3.5	6–12
Media Offices	0.10–0.15	15–25	4.0–6.0	3.5	6–12
Call Centers	0.10–0.20	20–30	5.0–8.0	3.0	6–12
Judiciary Chambers	0.10–0.15	15–25	4.0–6.0	3.5	6–12

Key Notes for Specialty Offices

Post Offices and Police Offices: Moderate fresh air and cooling loads due to periodic high occupancy and equipment use.

Banks: Higher fresh air requirements in general areas for customer traffic; vault areas have minimal occupancy and lower cooling needs.

Conference and Meeting Rooms: Require significant cooling and ventilation due to occupant density and usage patterns.

Archives and Storage Areas: Lower cooling and ventilation needs but require controlled humidity and temperature for preservation.

Open Plan Workspaces: Demand higher ventilation rates to manage air quality in shared spaces.

Courts (Courtrooms): Higher ventilation is required for occupant comfort and maintaining air quality during long sessions.

Embassies: General areas need moderate ventilation, while sensitive areas prioritize controlled fresh air to maintain security and comfort.

Media Offices: Moderate fresh air due to extended working hours and equipment heat loads.

Call Centres: Higher occupant density necessitates increased fresh air and cooling loads to ensure comfort.

Judiciary Chambers: Controlled and quiet environments require balanced cooling and noise

Residential Spaces

Application	Cooling load (kW/m^2)	Fresh air (% of total) (%)	Fresh air (L/s/m^2)	Typical height (m)	ACH
Living Rooms	0.08–0.12	15–20	3.0–4.0	2.8	6–10
Bedrooms	0.05–0.08	15–20	2.0–3.0	2.8	6–10
Kitchens	0.12–0.20	15–30	4.0–6.0	2.8	6–10
Bathrooms	0.06–0.10	20–30	3.0–5.0	2.5	6–10
Dining Rooms	0.10–0.15	15–20	3.5–4.5	2.8	6–10
Home Offices	0.10–0.15	15–20	3.5–4.5	2.8	6–10
Utility Rooms	0.05–0.08	10–15	1.0–2.0	2.5	6–10
Corridors and Hallways	0.03–0.05	10–15	1.0–1.5	2.5	6–10
Garage (Attached)	0.08–0.10	15–25	2.5–4.0	3.0	6–10
Basement (Habitable)	0.06–0.08	15–20	2.0–3.0	2.8	6–10
Basement (Storage)	0.03–0.05	10–15	1.0–2.0	2.5	6–10

Key Notes for Residential Buildings

Living Rooms and Bedrooms: Moderate cooling loads and fresh air requirements to ensure comfort during both active and passive use.

Kitchens: Higher cooling loads due to heat from cooking appliances; ventilation must also address odours and humidity.

Bathrooms: Moderate ventilation to manage moisture and maintain air quality; exhaust fans are typically used.

Utility Rooms: Minimal fresh air requirements; focus is on maintaining air changes to remove appliance heat.

Garages: Fresh air requirements aim to address fume dilution from vehicles.

Basements: Habitable basements require more cooling and ventilation compared to storage basements, which prioritize basic air changes.

ACH for Multi-Family Residential, Penthouses, and Luxury Villas

Multi-Family Residential Buildings

Application	Cooling load (kW/m^2)	Fresh air (% of total) (%)	Fresh air (L/s/m^2)	Typical height (m)	ACH
Living Rooms	0.07–0.10	15–20	3.0–4.0	2.8	6–10
Bedrooms	0.05–0.08	15–20	2.0–3.0	2.8	6–10
Common Areas (Lobbies)	0.10–0.15	15–25	4.0–5.0	3.0	6–10
Corridors	0.03–0.05	10–15	1.0–1.5	2.8	6–10
Basement Parking	0.08–0.10	20–30	5.0–7.0	3.0	6–10

Penthouses

Application	Cooling load (kW/m^2)	Fresh air (% of total) (%)	Fresh air (L/s/m^2)	Typical height (m)	ACH
Living Rooms	0.10–0.15	20–25	5.0–6.0	3.0	6–10
Master Bedrooms	0.08–0.12	15–20	4.0–5.0	3.0	6–10
Kitchens	0.12–0.18	20–30	6.0–8.0	2.8	6–10
Bathrooms	0.08–0.10	20–30	3.5–5.0	2.5	6–10
Balconies/Enclosures	0.04–0.06	10–15	1.5–2.5	3.0	6–10

Luxury Villas

Application	Cooling load (kW/m^2)	Fresh air (% of total) (%)	Fresh air (L/s/m^2)	Typical height (m)	ACH
Living Rooms	0.10–0.15	20–25	5.0–6.0	3.0	6–10
Master Bedrooms	0.08–0.12	15–20	4.0–5.0	3.0	6–10
Guest Bedrooms	0.06–0.08	15–20	2.0–3.0	2.8	6–10
Dining Rooms	0.10–0.12	15–20	4.0–5.0	2.8	8–12
Home Gyms	0.15–0.20	20–30	6.0–8.0	3.0	8–12
Home Theaters	0.12–0.15	15–20	4.0–5.0	2.8	6–10
Swimming Pool Areas	0.25–0.30	30–40	8.0–12.0	3.5	10–20

Key Considerations

- **Multi-Family Buildings**: Corridors and lobbies require ventilation for air quality, while individual units prioritize energy efficiency and comfort.

- **Penthouses**: Higher cooling loads reflect premium features like expansive glass areas and higher ceilings.
- **Luxury Villas**: Custom spaces like pools, gyms, and home theatres require specialized HVAC design with higher fresh air rates.

For Indoor Playgrounds

Activity zone	Cooling load (kW/m^2)	Fresh air (% of total) (%)	Fresh air (L/s/m^2)	Typical height (m)	ACH
Arcade Gaming Area	0.20–0.25	20–30	6.0–8.0	3.0	8–12
Bowling Lanes	0.18–0.22	20–25	5.0–7.0	4.0	8–12
Virtual Reality (VR) Zone	0.25–0.30	25–30	8.0–10.0	3.5	8–12
Laser Tag Arena	0.30–0.35	25–30	9.0–12.0	4.0	8–12
Bumper Cars Area	0.22–0.28	20–30	7.0–9.0	4.0	8–12
Party/Event Rooms	0.12–0.15	15–20	4.0–5.0	3.0	8–12
Snack Bar or Food Area	0.10–0.15	20–25	4.0–6.0	3.0	8–12
Relaxation Lounges/Seating	0.08–0.12	15–20	3.0–4.0	3.0	6–10
Climbing/Adventure Zones	0.18–0.22	20–30	7.0–9.0	4.0	8–12
Prize Redemption Counters	0.10–0.12	15–20	3.0–4.0	3.0	8–12

Key Considerations for Activity Zones

- **High Occupancy and Activity Levels**:
  - Cooling loads are higher due to heat generated by active children and parents.
  - Fresh air requirements increase to maintain air quality and control CO_2 levels.
- **Diverse Zones**:
  - Specific areas like climbing walls or trampolines demand tailored ventilation to manage dynamic heat loads.
- **Height Variations**:
  - Play structures and climbing zones might have elevated ceiling heights, requiring more effective air distribution.
- **Hygiene**:
  - Adequate air changes ensure control of odours and airborne particles, especially in enclosed play areas.
- **Special Zones**:
  - Cafeteria or snack areas must balance cooling with exhaust systems to handle food odours and moisture.
- **High Heat Load Areas**:
  - VR, laser tag, and bumper car zones often have high occupant activity, leading to increased cooling load and ventilation needs.

- **Indoor Air Quality**:
  - High ACH values ensure odour control in densely occupied areas like bumper cars and laser tag.
- **Diverse Heights**:
  - Bowling alleys and climbing zones require higher ceilings, influencing air distribution and cooling load.
- **Event Rooms**:
  - Lower cooling loads, but fresh air requirements remain critical during parties or events.
- **Energy Recovery**:
  - Consider using energy recovery ventilators (ERVs) to optimize energy usage for zones with high fresh air demand.

Hotels and Motels

Area/room type	Cooling load (kW/m^2)	Fresh air (% of total) (%)	Fresh air (L/s/m^2)	Typical height (m)	ACH
Guest Rooms	0.08–0.12	15–20	2.5–4.0	2.8–3.0	6–8
Luxury Suites	0.10–0.15	20–25	4.0–5.0	3.0	8–10
Lobby/Reception Area	0.12–0.18	20–30	5.0–7.0	3.5	10–12
Corridors	0.06–0.08	10–15	2.0–3.0	2.8–3.0	6–8
Banquet/Conference Rooms	0.20–0.25	25–35	8.0–12.0	4.0	10–15
Restaurants/Dining Areas	0.15–0.20	20–30	6.0–8.0	3.5	10–12
Kitchens	0.20–0.30	50–70	12.0–18.0	3.5	15–20
Fitness Centres/Gyms	0.20–0.25	30–40	8.0–12.0	3.5	12–15
Spa/Wellness Centres	0.18–0.22	25–30	7.0–9.0	3.5	10–12
Swimming Pool Area	0.20–0.25	50–60	15.0–20.0	4.0	6–8 (Moisture Control)
Laundry Rooms	0.15–0.20	50–60	12.0–15.0	3.0	12–15
Mechanical/Electrical Rooms	0.12–0.15	15–20	5.0–6.0	3.0	10–12
Storage Rooms	0.08–0.10	10–15	3.0–4.0	3.0	6–8

Key Considerations for HVAC Design in Hotels and Motels

- **Guest Comfort**:
  - Cooling loads in guest rooms are lower but must accommodate variations in occupancy and heat gain from equipment like TVs or mini-fridges.
- **High Fresh Air Areas**:
  - Kitchens, laundry rooms, and swimming pool areas require significantly higher fresh air due to exhaust and moisture control needs.

- **Air Quality**:
  - Use HEPA or MERV-rated filters in areas like spas or fitness centres to ensure high air quality.
- **Energy Recovery**:
  - Consider energy recovery ventilators (ERVs) or heat recovery ventilators (HRVs) for banquet halls, restaurants, and other spaces with high fresh air demand.
- **Humidity Control**:
  - Swimming pools and wellness areas require dedicated dehumidification systems to maintain moisture levels and prevent condensation.

Percentage Contribution of Cooling Load Components for Different Buildings

Building type	Internal heat gain (%)	External loads (%)	Infiltration load (%)	Ventilation load (%)
Residential Buildings	30–40	35–50	10–15	5–15
Office Buildings	50–60	20–30	5–10	10–20
Hospitals	40–50	20–30	5–10	20–30
Schools/Universities	40–50	25–35	5–10	10–20
Retail/Shopping Centres	50–60	20–30	5–10	10–20
Cinemas/Theatres	60–70	10–20	5–10	5–15
Restaurants/Cafes	50–60	20–30	5–10	10–20
Hotels/Motels	40–50	30–40	5–10	10–20
Clean Rooms (Class 1–10,000)	30–40	10–20	10–20	30–50
Laboratories	30–40	20–30	5–10	30–40
Factories/Workshops	50–60	20–30	5–15	10–20
Gyms/Sports Centres	50–60	20–30	5–10	10–20

Notes

Internal Heat Gain:
Includes occupants, lighting, equipment, and other internal sources of heat. It is predominant in high-occupancy or equipment-intensive buildings like offices, retail spaces, and gyms.

External Loads:
Solar heat gains through walls, roofs, and windows dominate in residential buildings or buildings with large glass façades.

Infiltration Load:
Higher in buildings with frequent door openings or poor envelope sealing, such as retail stores and factories.

Ventilation Load:
Significant in spaces requiring fresh air or high filtration standards, such as hospitals, laboratories, and clean rooms.

Cooling Load Components by Building Type and Climate Zone

Building type	Climate zone	Internal heat gain (%)	External loads (%)	Infiltration load (%)	Ventilation load (%)	Envelope recommendations	HVAC strategies
Residential	Hot/Humid (e.g., Miami)	30–40	40–50	10–15	5–15	High R-value walls, reflective roofs	SEER ≥16 split systems, dehumidification, proper window shading
	Temperate (e.g., Sydney)	30–40	35–45	10–15	5–15	Moderate insulation, efficient glazing	Zoning systems, ceiling fans, reverse-cycle air conditioning
	Cold (e.g., Stockholm)	20–30	50–60	10–20	5–10	High R-value windows, wall insulation	Heat pumps, hydronic systems, tight envelopes
Office Buildings	Hot/Dry (e.g., Phoenix)	50–60	30–40	5–10	10–20	High-performance glazing, light colours	DOAS with energy recovery, high-efficiency lighting
	Tropical (e.g., Singapore)	50–60	20–30	5–10	15–25	Airtight envelopes, minimal glazing	Dehumidification, economizers, energy-efficient lighting
	Cold (e.g., Toronto)	40–50	30–40	10–15	10–20	Triple-glazed windows, airtight walls	Demand-controlled ventilation (CO_2-based), efficient boilers
Hospitals	All (global standards)	40–50	20–30	5–10	20–30	Sealed construction, insulated ducts	High-capacity AHUs, HEPA filters, heat recovery ventilators
Clean Rooms	Hot (e.g., Dubai)	20–30	20–30	10–20	40–50	Ultra-sealed envelope, high insulation	Dedicated HVAC systems, precise humidity control, specialized HEPA/ULPA filters
	Cold (e.g., Oslo)	30–40	10–20	15–25	20–30	Airtight walls, insulated pipes	Energy recovery ventilators, efficient heating systems
Retail/ Shopping	Tropical (e.g., Jakarta)	50–60	20–30	5–10	10–20	High-performance glass, airtight doors	DOAS, vestibules, efficient VRF systems
	Temperate (e.g., London)	50–60	25–35	5–10	10–20	Double-glazing, moderate insulation	AHUs with heat recovery, zoning for large retail areas
Cinemas/ Theatres	Hot/Humid (e.g., Mumbai)	60–70	10–20	5–10	5–15	Insulated walls, sealed entries	Zoning for show times, VAV systems, advanced economizers
	Cold (e.g., Moscow)	50–60	15–25	10–20	5–10	Airtight insulation, sealed glazing	High-efficiency boilers, demand-controlled HVAC

Climate-Specific Envelope & HVAC Adjustments

- **Hot/Humid Zones**:
  - Focus on **dehumidification** to offset latent cooling loads.
  - Use **low-emissivity (low-E) coatings** and reflective roofing materials.
- **Temperate Zones**:
  - Optimize **natural ventilation** and mixed-mode cooling systems.
  - Employ **seasonal economizers** and adaptive comfort HVAC strategies.
- **Cold Zones**:
  - Focus on **thermal insulation** to reduce heat loss.
  - Employ **heat recovery systems** for ventilation air.

Internal Heat Gain Breakdown by Building Type

Building type	People (%)	Lighting (%)	Equipment (%)	Other (%) (e.g., Cooking, Misc.)	Notes
Residential	30–40	20–30	20–30	5–10	Equipment includes appliances like TVs, computers, and kitchen devices
Office Buildings	20–30	30–40	30–40	0–10	High equipment load due to computers, printers, and other devices
Retail/ Shopping Malls	10–20	40–50	20–30	5–10	High lighting intensity for product display
Hospitals	10–20	30–40	20–30	10–20	Includes specialized medical equipment (e.g., imaging devices)
Clean Rooms	10–20	30–40	30–40	5–10	Equipment includes specialized machinery
Cinemas/ Theatres	50–60	20–30	10–20	0–10	People density and lighting dominate
Restaurants	10–20	20–30	10–20	30–50	Cooking loads are significant contributors
Hotels	20–30	30–40	20–30	10–20	Includes guest room appliances and public areas
Schools/ Universities	30–40	30–40	10–20	0–10	Higher people density in classrooms
Sport Centres/ Gyms	40–50	20–30	10–20	0–10	People activity contributes significantly to heat gains
Data Centres	0–5	5–10	80–90	0–5	Equipment load (servers) is the primary source of heat

Building type	People (%)	Lighting (%)	Equipment (%)	Other (%) (e.g., Cooking, Misc.)	Notes
Healthcare Facilities	10–20	30–40	30–40	10–20	Includes heat from specialized medical equipment
Industrial Buildings	10–20	20–30	30–40	10–30	Equipment load varies based on the type of manufacturing processes

Notes

- **People Heat Gains**:
  - Based on **metabolic rates**, density, and activity level.
  - Significant in buildings like gyms, cinemas, and schools.
- **Lighting Heat Gains**:
  - Proportional to the lighting design (lux levels) and type (LED, fluorescent, or incandescent).
  - Higher in retail and display-intensive areas.
- **Equipment Heat Gains**:
  - Varies by equipment density and operating hours.
  - Dominant in **data centres**, hospitals, and offices.
- **Other Heat Gains**:
  - Include cooking, process heat, or incidental loads like miscellaneous plug loads.

External Heat Load Breakdown by Building Type

Building type	Solar heat gain (%)	Conduction (%) (Walls, Roofs, Windows)	Infiltration (%)	Other (%) (e.g., Heat from Adjacent Spaces)	Notes
Residential	40–50	30–40	10–20	0–10	Solar gains through windows dominate during the day
Office Buildings	40–50	30–40	10–20	0–10	Significant solar gain from large glazing areas
Retail/ Shopping Malls	30–40	30–40	20–30	0–10	Large entrances increase infiltration loads
Hospitals	20–30	30–40	20–30	10–20	Heat from infiltration and adjacent zones is significant

Building type	Solar heat gain (%)	Conduction (%) (Walls, Roofs, Windows)	Infiltration (%)	Other (%) (e.g., Heat from Adjacent Spaces)	Notes
Clean Rooms	10–20	30–40	40–50	0–10	High infiltration due to pressure differentials
Cinemas/ Theatres	20–30	40–50	20–30	0–10	Reduced solar gain due to limited window areas
Restaurants	20–30	30–40	30–40	0–10	High infiltration load due to frequent door opening
Hotels	30–40	30–40	20–30	10–20	Includes heat from adjacent spaces like kitchens and laundry
Schools/ Universities	30–40	30–40	20–30	0–10	Moderate solar gain and infiltration from high usage
Sport Centres/ Gyms	30–40	30–40	20–30	0–10	Infiltration load varies based on occupancy and ventilation requirements
Data Centres	5–10	30–40	10–20	40–50	Heat from adjacent spaces (e.g., UPS rooms) can be significant
Healthcare Facilities	20–30	30–40	20–30	10–20	Conduction load includes specialized zones like operating rooms
Industrial Buildings	20–30	30–40	30–40	0–10	Heat load varies based on the building envelope and process requirements

Notes

- **Solar Heat Gain**:
  - Solar gain through windows depends on orientation, glazing type, shading, and time of day.
  - Significant in buildings with large glass façades (offices, malls).
- **Conduction Heat Gain**:
  - Affected by the insulation level and thermal properties of the building envelope.
  - Higher in poorly insulated structures or areas with extreme temperature differences.

- **Infiltration**:
  - Depends on airtightness, door/window usage, and pressure differentials.
  - Dominant in clean rooms, malls, and restaurants.
- **Other External Loads**:
  - Include heat transfer from adjacent spaces or equipment-intensive zones.

General Time of Peak Load for Buildings

Hemisphere	Building type	Season	Typical peak load time	Reason
Northern Hemisphere	Residential	Summer	Late Afternoon (2 PM–6 PM)	Heat gain from solar radiation and occupant activity at home increases cooling loads
	Commercial (Offices)	Summer	Midday to Early Afternoon (11 AM–3 PM)	Solar radiation and heat gains from lighting, equipment, and occupants
	Retail	Summer	Late Afternoon to Evening (4 PM–8 PM)	Increased occupancy and activity levels contribute to higher cooling demand
	Industrial	Summer	Daytime Hours (Varies)	Dependent on operations, machinery, and heat production schedules
	Educational (Schools)	Summer	Midday to Early Afternoon (11 AM–3 PM)	Occupant density, lighting, and solar radiation coincide with school hours
Southern Hemisphere	Residential	Summer	Late Afternoon (2 PM–6 PM)	Similar reasons to the northern hemisphere, with cooling loads from solar heat and occupancy
	Commercial (Offices)	Summer	Midday to Early Afternoon (11 AM–3 PM)	Solar radiation and internal heat gains during business hours
	Retail	Summer	Late Afternoon to Evening (4 PM–8 PM)	Occupant-driven cooling demand peaks in the late day due to shopping activity
	Industrial	Summer	Daytime Hours (Varies)	Same as northern hemisphere, depends on operations and processes
	Educational (Schools)	Summer	Midday to Early Afternoon (11 AM–3 PM)	Similar to the northern hemisphere, with load dictated by school hours and environmental factors

Hemisphere	Building type	Season	Typical peak load time	Reason
Both Hemispheres	Healthcare (Hospitals)	Year-Round	Varies Based on Occupancy	Peaks depend on occupancy and environmental controls needed for patient care rather than external temperatures alone
	Hospitality (Hotels)	Year-Round	Evening and Night (6 PM–12 AM)	Guest occupancy and activity during evening hours drive demand.
	Mixed-Use Developments	Year-Round	Varies	Peak load timing is influenced by the mix of residential, retail, and commercial activities within the building complex

Chapter 6
HVAC System Selection Criteria

6.1 Overview

An HVAC system is a combination of components that work together to achieve heating, cooling, ventilation, and air quality objectives. Understanding these components and their configurations is crucial for designing, operating, and maintaining effective systems. In this chapter, we will explore the core elements of HVAC systems and the different ways they are configured to meet diverse building requirements.

6.2 Core Components of an HVAC System

- **Heating Equipment**:
  - **Furnaces**: Convert fuel (natural gas, oil, or electricity) into heat, distributing it via ducts.
  - **Boilers**: Heat water or generate steam for hydronic heating systems.
  - **Heat Pumps**: Provide both heating and cooling by transferring heat between indoor and outdoor environments.
  - **Radiant Heating Systems**: Use heated surfaces, such as floors or panels, to radiate warmth directly into the space.
- **Cooling Equipment**:
  - **Air Conditioners**: Extract heat from indoor air and release it outdoors.
  - **Chillers**: Remove heat from water, which is then circulated to air handling units or fan coil units.
  - **Evaporative Coolers**: Use water evaporation to cool air, suitable for dry climates.

V. Vakiloroaya, *HVAC and Building Services Rules of Thumb*, Springer Tracts in Civil Engineering, https://doi.org/10.1007/978-3-032-10285-0_6

- VRF/VRV Systems: Variable refrigerant flow/volume systems adjust refrigerant flow to provide precise temperature control.

- **Ventilation Systems**:
  - **Natural Ventilation**: Relies on architectural design to allow airflow through windows, vents, and openings.
  - **Mechanical Ventilation**: Uses fans and ductwork to ensure adequate air exchange.
  - **Exhaust Fans**: Remove stale air, odours, and pollutants from specific areas such as kitchens and bathrooms.
- **Air Distribution Systems**:
  - **Ductwork**: Channels that distribute conditioned air throughout a building.
  - **Diffusers and Grilles**: Devices that regulate and direct airflow into occupied spaces.
- **Control Systems**:
  - **Thermostats**: Basic controls for setting desired temperatures.
  - **Sensors**: Monitor temperature, humidity, and air quality.
  - **Building Management Systems (BMS)**: Advanced systems for centralized control of HVAC and other building services.
- **Filters and Air Purification**:
  - **Mechanical Filters**: Remove particulates from the air.
  - **Electronic Air Cleaners**: Use electrostatic charges to capture particles.
  - **UV-C Lamps**: Neutralize biological contaminants like bacteria and viruses.

6.3 HVAC System Configurations

HVAC systems can be configured in various ways depending on the building type, size, and operational needs. Here are the most common configurations:

- **Centralized Systems**:
  - Serve large buildings or campuses.
  - Use centralized equipment like boilers and chillers to provide heating and cooling.
  - Distribute conditioned air or water to individual zones.
  - Example: A district cooling system serving multiple buildings.
- **Decentralized Systems**:
  - Use localized units for heating and cooling individual zones or rooms.
  - Include window air conditioners, split systems, and packaged rooftop units.
  - Suitable for smaller buildings or spaces with varying usage patterns.

- **Hybrid Systems**:
  - Combine centralized and decentralized approaches.
  - Allow for flexibility and energy efficiency in multi-zone applications.
- **Zoned Systems**:
  - Divide a building into zones with independent temperature controls.
  - Ideal for buildings with diverse occupancy and usage patterns.
  - Example: A VRF system with multiple indoor units.

6.4 Special-Purpose HVAC Systems

- **Cleanroom Systems**:
  - Designed for environments requiring high levels of air cleanliness, such as laboratories and semiconductor manufacturing.
  - Use HEPA filters, laminar airflow, and pressurization.
- **Data Centre Cooling**:
  - Focused on maintaining optimal temperatures for IT equipment.
  - Use precision cooling units, chilled water systems, and hot/cold aisle containment.
- **Industrial HVAC Systems**:
  - Address unique challenges like managing process heat and ensuring worker safety.
  - Include specialized equipment like air washers, scrubbers, and industrial ventilation systems.
- **Hospital and Healthcare Systems**:
  - Prioritize infection control and patient comfort.
  - Include isolation room ventilation, humidification, and high-efficiency filtration.

6.5 Integration with Building Services

Effective HVAC design requires integration with other building services to achieve seamless operation. Key areas of integration include:

- **Electrical Systems**:
  - Ensuring adequate power supply for HVAC equipment.
  - Coordinating with backup power systems like generators.

- **Plumbing Systems**:
  - Supporting hydronic heating and cooling systems.
  - Managing condensate drainage and humidification.
- **Fire Safety Systems**:
  - Incorporating smoke control and fire dampers.
  - Ensuring compliance with fire codes and standards.
- **Building Automation Systems (BAS)**:
  - Centralizing control of HVAC, lighting, and security systems.
  - Enabling energy management and fault detection.

6.6 Vapour Comparison Systems

The **vapour compression cycle** is the most widely used refrigeration cycle in HVAC systems, forming the foundation of air conditioning, refrigeration, and heat pump technologies. Below is a comprehensive description of its components, the cycle's operation, and HVAC systems that utilise it.

Components of the Vapour Compression Cycle

Compressor

- **Function**: Compresses the refrigerant, increasing its pressure and temperature.
- **Process**: The refrigerant enters the compressor as a low-pressure vapour and exits as a high-pressure superheated vapour.
- **Types**:
  - Reciprocating
  - Scroll
  - Rotary
  - Screw
  - Centrifugal

Condenser

- **Function**: Rejects heat from the refrigerant to the surroundings, causing it to condense into a high-pressure liquid.
- **Process**:
  - The refrigerant enters as a high-pressure vapour.
  - Heat is transferred to the external medium (air, water, or both), converting the vapour into a liquid.
- **Types**:
  - Air-cooled
  - Water-cooled
  - Evaporative

Expansion Valve

- **Function**: Reduces the pressure and temperature of the refrigerant.
- **Process**:
  - The refrigerant enters as a high-pressure liquid.
  - The valve throttles the refrigerant to produce a low-pressure, low-temperature mixture of liquid and vapour.
- **Types**:
  - Thermostatic expansion valves (TEV)
  - Electronic expansion valves (EEV)
  - Capillary tubes

Evaporator

- **Function**: Absorbs heat from the space to be cooled, causing the refrigerant to evaporate.
- **Process**:
  - The refrigerant enters as a low-pressure liquid/vapour mixture.
  - Heat is absorbed from the air or water, and the refrigerant becomes a low-pressure vapour.
- **Types**:
  - Direct expansion (DX) coils
  - Flooded evaporators

Description of the Vapour Compression Cycle

Step-by-Step Cycle

- **Compression (Isentropic Process)**:
  - The compressor raises the refrigerant's pressure and temperature.
- **Heat Rejection (Isobaric Condensation)**:
  - The high-pressure refrigerant releases heat to the surroundings in the condenser.
- **Expansion (Isenthalpic Process)**:
  - The expansion valve reduces the refrigerant's pressure and temperature without changing its enthalpy.
- **Heat Absorption (Isobaric Evaporation)**:
  - The low-pressure refrigerant absorbs heat from the conditioned space in the evaporator.

Types of HVAC Systems Using the Vapour Compression Cycle

Air Conditioners

- **Window AC Units**:
  - Compact systems where all components are housed in a single unit.

- **Split AC Systems**:
  - The evaporator and fan are indoors, while the compressor and condenser are outdoors.
- **Packaged Units**:
  - Self-contained systems used for small commercial applications.

Chillers
- **Air-Cooled Chillers**:
  - Use ambient air to cool the condenser.
  - Common in smaller or decentralized systems.
- **Water-Cooled Chillers**:
  - Use cooling towers to reject heat.
  - Ideal for larger buildings requiring central cooling.

Heat Pumps
- Operate on the vapour compression cycle in reverse for heating.
- Types:
  - Air-source heat pumps (ASHP)
  - Water-source heat pumps (WSHP)
  - Ground-source heat pumps (GSHP)

Refrigeration Systems
- Used in cold storage, food preservation, and process cooling.
- Types:
  - Domestic refrigerators
  - Commercial freezers
  - Industrial refrigeration plants

Variable Refrigerant Flow (VRF) Systems
- Advanced systems where refrigerant is directly circulated to multiple indoor units.
- Allow simultaneous cooling and heating in different zones.

Rooftop Units (RTUs)
- Packaged systems installed on building roofs, combining the vapour compression cycle with ventilation.

Key Advantages of the Vapour Compression Cycle

- **Energy Efficiency**:
  - Modern systems incorporate high-efficiency compressors and variable speed drives.
- **Flexibility**:
  - Suitable for a wide range of applications, from small residential units to large industrial plants.

- **Refrigerant Variety**:
  - Compatibility with various refrigerants, including environmentally friendly options like R-32 and R-290.

Key Considerations for System Design

- **Refrigerant Selection**:
  - Choose refrigerants based on efficiency, environmental impact (GWP, ODP), and safety.
- **Compressor Type**:
  - Match the compressor to the application's size and load variability.
- **Heat Exchangers**:
  - Optimize evaporator and condenser design for efficient heat transfer.
- **Control Systems**:
  - Incorporate advanced control systems for efficient operation (e.g., thermostatic or electronic expansion valves).

Applications and Suitability of Vapour Compression Systems

- **Residential**: Split ACs, window units, VRF systems and small heat pumps.
- **Commercial**: VRF systems, packaged units, and chillers.
- **Industrial**: Large chillers and specialized refrigeration systems.
- **Transportation**: Vehicle air conditioning and cold storage transport.

Types of Compressors in HVAC Systems

Below is a detailed table that outlines the different types of compressors used in HVAC systems, their applications, cooling ranges, and other key details:

Type of compressor	Description	Applications	Cooling range	Advantages	Disadvantages
Reciprocating	Uses a piston and cylinder to compress refrigerant	Residential & commercial AC, refrigeration	0.5–200 kW	High efficiency, widely available	High vibration, not suitable for large loads
Scroll	Compresses refrigerant with two interleaving scrolls	Residential AC, heat pumps, VRF systems	5–100 kW	Quiet, reliable, fewer moving parts	Limited to medium capacity
Rotary	Uses a rotating mechanism (vane or blade) to compress refrigerant	Small residential units, window AC	1–10 kW	Compact, efficient for small systems	Limited to small applications

(Continued)

Type of compressor	Description	Applications	Cooling range	Advantages	Disadvantages
Screw	Uses two interlocking helical rotors to compress refrigerant	Industrial chillers, large HVAC systems	100 kW to 5 MW	High efficiency for large loads	Expensive, requires precise manufacturing
Centrifugal	Uses centrifugal force to compress refrigerant (dynamic compression)	Large commercial chillers	300 kW to 10+ MW	High capacity, compact for its size	High initial cost, sensitive to load changes
Scroll-variable speed	Advanced scroll compressor with variable-speed drive (VSD)	VRF systems, inverter AC systems	5–200 kW	High efficiency, adapts to variable loads	Expensive, complex control
Rotary screw-variable speed	Screw compressor with VSD for energy efficiency	Large commercial systems	200 kW to 3 MW	Energy-efficient for varying loads	High cost, complex electronics
Rotary vane	Uses rotating vanes in a cylinder to compress refrigerant	Medium-sized HVAC applications	5–100 kW	Quiet operation, durable	Limited to medium capacities
Digital scroll	Modulates capacity by bypassing a portion of refrigerant flow	Commercial AC, VRF systems	10–100 kW	Precise load matching, energy-efficient	Higher cost, complexity
Hermetic	Fully sealed unit containing both motor and compressor	Domestic refrigeration, small AC units	0.5–10 kW	Compact, low maintenance	Difficult to repair, limited capacity
Semi-hermetic	Motor and compressor in a sealed housing but can be serviced	Refrigeration, industrial HVAC	10–500 kW	Repairable, suitable for moderate loads	Expensive, larger footprint
Open-type	Compressor and motor are separate. Shaft connects them	Industrial refrigeration, process cooling	100 kW to 5 MW	Easy to repair, can handle larger capacities	Requires shaft sealing, risk of leaks

(Continued)

Type of compressor	Description	Applications	Cooling range	Advantages	Disadvantages
Magnetic bearing	Centrifugal compressor using magnetic bearings for frictionless operation	High-efficiency commercial chillers	1–10+ MW	High efficiency, no oil required	Very high cost, complex technology

Notes

- **Applications**:
  - Residential AC: Typically uses rotary or scroll compressors.
  - Commercial chillers: Often use screw or centrifugal compressors.
  - VRF systems: Scroll or digital scroll compressors dominate.
  - Industrial HVAC: Screw and centrifugal compressors are standard.
- **Cooling Range**:
  - Indicates the capacity range for cooling loads the compressor can handle.
- **Selection Criteria**:
  - System size, load profile, efficiency requirements, initial cost, and maintenance complexity.

Types of Evaporators in HVAC Systems

Following is a detailed table outlining different types of evaporators used in HVAC systems, their applications, cooling ranges, and other details:

Type of evaporator	Description	Applications	Cooling range	Advantages	Disadvantages
DX (Direct Expansion)	Refrigerant flows directly inside the coil, absorbing heat from air	Split AC, VRF systems, packaged units	1–200 kW	Compact, efficient, direct heat transfer	Limited to small and medium loads
Flooded	Maintains a liquid refrigerant pool; refrigerant absorbs heat and evaporates	Large chillers, industrial refrigeration	100 kW to 5 MW	High efficiency for large loads	Requires complex control, higher cost
Shell-and-Tube	Refrigerant flows through tubes or around a shell, exchanging heat with water	Chillers, industrial cooling systems	100 kW to 10 MW	Durable, handles high capacities	Bulky, requires regular maintenance

(Continued)

Type of evaporator	Description	Applications	Cooling range	Advantages	Disadvantages
Plate	Refrigerant flows between thin metal plates; compact and efficient heat exchange	Compact chillers, heat recovery systems	10–500 kW	Compact, efficient, easy to clean	Limited to moderate capacities
Finned-Tube Air Coil	Air passes over finned tubes carrying refrigerant	Residential and commercial AC systems	1–200 kW	High heat transfer efficiency, compact	Prone to fouling, requires regular cleaning
Bare-Tube	Simple design with refrigerant flowing through plain tubes	Refrigeration, low-capacity applications	1–50 kW	Simple design, cost-effective	Lower efficiency compared to finned tubes
Falling Film	Refrigerant spreads over tubes in a thin film, maximizing heat transfer	High-capacity chillers	500 kW to 10+ MW	Very efficient, minimal refrigerant charge	High cost, complex control
Spray-Type	Refrigerant sprayed onto the heat exchange surface for efficient cooling	Large chillers, special industrial systems	200 kW to 5 MW	High efficiency, reduced refrigerant load	Expensive, requires precise operation
Microchannel	Uses microchannels for refrigerant flow, increasing surface area	Modern AC systems, heat pumps	1–100 kW	Compact, lightweight, efficient	Prone to clogging, challenging to repair
Horizontal Coil	Coil positioned horizontally for direct air cooling	Air-handling units, ventilation systems	5–100 kW	Space-saving, efficient for air systems	Requires proper drainage to avoid flooding
Vertical Coil	Coil positioned vertically, typically for water cooling	Industrial and commercial HVAC systems	10–500 kW	Effective for vertical airflow designs	Bulky, requires adequate clearance
Double-Pipe	Refrigerant and water flow in adjacent pipes, transferring heat	Small chillers, heat recovery	5–50 kW	Compact, easy to install	Limited capacity, moderate efficiency

Notes

- **Applications**:
  - **Residential Systems**: DX coils and finned-tube coils are common.
  - **Commercial Systems**: Flooded, shell-and-tube, and plate evaporators are used.

  - **Industrial Systems**: Falling film and spray-type evaporators dominate for high capacities.
- **Selection Criteria**:
  - Cooling load, system size, type of refrigerant, and required efficiency.
- **Advantages and Disadvantages**:
  - Each type is suited to specific applications and cooling ranges, depending on the trade-offs between efficiency, cost, and complexity.

Types of Condensers in HVAC Systems

Following table detailing the various types of condensers used in HVAC systems, their applications, cooling ranges, and key features:

Type of condenser	Description	Applications	Cooling range	Advantages	Disadvantages
Air-Cooled Condenser	Uses ambient air to remove heat from the refrigerant via finned coils	Split AC, rooftop units, small chillers	1–1000 kW	Simple design, no water usage, easy to install	Lower efficiency in hot climates
Water-Cooled Condenser	Uses water to absorb heat from the refrigerant, typically in a shell-and-tube design	Large chillers, industrial cooling	50 kW to 5 MW	High efficiency, smaller size for given capacity	Requires cooling tower, water consumption
Evaporative Condenser	Combines water and air to enhance heat rejection by evaporating water	Industrial refrigeration, large AC systems	100 kW to 5 MW	High efficiency, lower operating temperature	Requires water treatment, higher maintenance
Shell-and-Tube Condenser	Refrigerant flows inside tubes, and water flows in the shell or vice versa	Industrial chillers, process cooling	100 kW to 10 MW	Durable, handles high capacities	Requires regular cleaning and maintenance
Plate Condenser	Refrigerant flows between thin plates, exchanging heat with water or air	Compact chillers, heat pumps	10–500 kW	Compact, high heat transfer efficiency	Limited capacity, prone to fouling
Microchannel Condenser	Features microchannels for refrigerant flow, increasing heat transfer efficiency	Modern air conditioners, heat pumps	1–100 kW	Lightweight, compact, and efficient	Susceptible to clogging, difficult to repair

(Continued)

Type of condenser	Description	Applications	Cooling range	Advantages	Disadvantages
Double-Pipe Condenser	Heat exchange occurs in a double-pipe system with refrigerant and coolant flows	Small cooling units, industrial cooling	5–50 kW	Compact, cost-effective	Limited capacity, moderate efficiency
Air-Water Condenser	Combines air and water for heat rejection without requiring a cooling tower	Specialized HVAC systems, remote applications	50 kW to 1 MW	Moderate efficiency, adaptable to conditions	Limited to specific applications
Spray-Type Condenser	Refrigerant is cooled by water sprayed over coils, enhancing heat rejection	Large refrigeration systems	500 kW to 5 MW	Highly efficient, low refrigerant charge	Higher cost, requires regular water treatment
Flooded Condenser	Refrigerant is submerged in a liquid, transferring heat to the cooling medium	Large process cooling	500 kW to 10+ MW	Highly efficient for large-scale applications	Expensive, complex controls

Notes

- **Applications**:
  - **Air-Cooled**: Ideal for areas with limited water supply.
  - **Water-Cooled**: Preferred for high capacities and applications requiring high efficiency.
  - **Evaporative**: Used where water and power efficiency are critical.
- **Selection Criteria**:
  - Based on cooling load, ambient conditions, water availability, and installation space.
- **Advantages and Disadvantages**:
  - The trade-offs depend on installation location, operational efficiency, and maintenance requirements.

Types of Expansion Valves in HVAC Systems

Following is a detailed table summarizing different types of expansion valves used in HVAC systems, their applications, operating ranges, and characteristics:

Type of expansion valve	Description	Applications	Operating range	Advantages	Disadvantages
Thermostatic Expansion Valve (TXV)	Regulates refrigerant flow based on evaporator temperature and pressure	Split AC, chillers, refrigeration units	−40 °C to +10 °C	Precise control of superheat, improves efficiency	Requires proper adjustment, costlier than capillaries
Electronic Expansion Valve (EEV)	Uses electronic sensors and a stepper motor to regulate refrigerant flow	VRF systems, advanced HVAC systems	−50 °C to +20 °C	High precision, adaptable to varying conditions	Expensive, requires complex controls
Capillary Tube	Fixed orifice that meters refrigerant based on pressure differential.	Window AC, small refrigeration systems	−20 °C to +10 °C	Simple design, inexpensive	Inefficient under varying load conditions
Automatic Expansion Valve (AXV)	Maintains constant evaporator pressure by regulating refrigerant flow	Small refrigeration units, dehumidifiers	−30 °C to +10 °C	Simple operation, low cost	Poor performance under varying loads
Float Valve (High-Side)	Regulates refrigerant level in the condenser, ensuring proper liquid transfer	Industrial refrigeration, chillers	−50 °C to +10 °C	Reliable for large systems	Limited to specific applications
Float Valve (Low-Side)	Maintains a constant liquid level in the evaporator	Large chillers, flooded systems	−50 °C to +10 °C	Ensures efficient liquid transfer	Requires proper installation and maintenance
Manual Expansion Valve	Manually adjusted to regulate refrigerant flow.	Specialized applications	−50 °C to +10 °C	Low cost, simple to use	Requires manual adjustment, lacks precision
Pressure-Actuated Expansion Valve	Operates based on pressure in the evaporator or condenser	Commercial refrigeration systems	−40 °C to +10 °C	Reliable under steady conditions	Limited flexibility, not ideal for dynamic loads

(Continued)

Type of expansion valve	Description	Applications	Operating range	Advantages	Disadvantages
Step Motor Valve (SMV)	Advanced valve type controlled by a stepper motor for precise operation	VRF systems, industrial cooling	−60 °C to +15 °C	High efficiency, precise modulation	Expensive, requires advanced control systems

Notes

- **Applications**:
  - **Capillary Tubes**: Best for simple, low-cost systems.
  - **TXVs and EEVs**: Suitable for systems requiring precise control and efficiency.
- **Advantages and Disadvantages**:
  - Advanced valves like EEVs offer precision but require higher costs and complex controls.
  - Simpler valves like capillaries are cost-effective but less efficient under varying loads.
- **Selection Criteria**:
  - Based on system capacity, operational efficiency, cost, and adaptability to load variations.

Thermostatic Expansion Valve (TXV)

Key Features
- Regulates refrigerant flow based on evaporator superheat.
- Uses a temperature-sensing bulb to monitor evaporator outlet temperature.

Advantages
- Improves system efficiency by preventing liquid refrigerant from reaching the compressor.
- Simple to install and maintain.

Set Superheat The TXV will regulate flow to maintain a 5 °C superheat (ensures efficient operation).

Electronic Expansion Valve (EEV)

Key Features
- Controlled by a stepper motor and sensors.
- Adjusts refrigerant flow dynamically based on real-time data.

Advantages
- More precise than TXVs.
- Adaptable to varying load conditions.

Valve Comparison Table

Feature	TXV	EEV
Control method	Temperature and pressure (mechanical)	Electronic sensors and motor
Response time	Moderate	Fast
Precision	Moderate	High
Cost	Lower	Higher
Applications	Split AC, small chillers	VRF systems, advanced HVAC systems

Fixed Orifice Valves

- **Control Mechanism**: Fixed-size orifice regulates refrigerant flow based on pressure differences.
- **Applications**: Economical option for small systems with constant loads.
- **Advantages**: Simple and low cost.
- **Disadvantages**: No flexibility for varying loads.

Capillary Tubes

- **Control Mechanism**: Long, narrow tube creates pressure drop.
- **Applications**: Small systems like refrigerators and small air conditioners.
- **Advantages**: Inexpensive and reliable.
- **Disadvantages**: Limited to fixed load conditions.

Refrigerant-Specific Performance

Different refrigerants require specific considerations due to their thermodynamic properties. For example:

Refrigerant	Common applications	Boiling point (°C)	Efficiency	Environmental impact
R410A	Split ACs, VRFs, chillers	−51.6	High	Moderate (GWP > 2000)
R134a	Chillers, refrigerators	−26.3	Moderate	Moderate (GWP > 1400)
R744 (CO_2)	Low-temperature applications	−78.4	High	Low (GWP = 1)
R32	VRFs, split ACs	−51.7	High	Lower than R410A

Types of Expansion Valves and Their Suitability

Valve type	Load flexibility	Efficiency	Cost	Applications
TXV	Moderate	High	Moderate	Split AC, small chillers
EEV	High	Very High	High	VRFs, large HVAC systems
Fixed orifice	Low	Low	Low	Small refrigeration units
Capillary tube	Low	Low	Very Low	Refrigerators, window ACs

Expansion Valve Maintenance

- **TXV Troubleshooting**:
  - Symptom: Low cooling output.
  - Cause: Improper superheat setting or clogged sensing bulb.
  - Solution: Adjust or replace the TXV.
- **EEV Troubleshooting**:
  - Symptom: Erratic operation.
  - Cause: Faulty sensors or stepper motor.
  - Solution: Diagnose with the control panel and replace faulty parts.

Refrigerant-Specific Expansion Valve Performance
Different refrigerants have unique thermodynamic properties that influence the operation and design of expansion valves. The key factors are:

- **Pressure-Temperature Relationship**:
  - Refrigerants like R410A have higher operating pressures, requiring robust valve designs.
  - Lower-pressure refrigerants, such as R134a, allow for simpler valve construction.
- **Latent Heat of Vaporization**:
  - Determines how much refrigerant must flow to achieve the desired cooling load.
  - Higher latent heat (e.g., CO_2/R744) means less refrigerant flow, leading to smaller orifices.
- **Environmental and Safety Concerns**:
  - Low-GWP refrigerants (R32, R744) are becoming more prevalent, driving demand for valves compatible with these refrigerants.
  - Flammability of refrigerants like R32 impacts valve safety features.

Comparison of Refrigerant Properties and Valve Implications

Refrigerant	Operating Pressure (Bar)	Latent Heat (kJ/kg)	Valve requirements	Applications
R410A	8–25	~200	High-pressure, precision control	Split ACs, VRFs, packaged units
R134a	1.5–12	~210	Moderate-pressure, widely compatible	Chillers, refrigerators
R744 (CO_2)	20–50	~150	High-pressure, robust design	Supermarket refrigeration, low-temp units
R32	8–22	~250	High-pressure, safety-enhanced	Split ACs, VRFs
R290 (Propane)	1–12	~350	Explosion-proof, specialized materials	Small commercial units

Refrigerant-Expansion Valve Pairing

R410A and TXVs
- TXVs designed for R410A feature stronger materials to withstand high pressures.
- Suitable for small-to-medium HVAC systems.

R134a and EEVs
- Ideal for variable load applications like chillers.
- R134a's moderate pressure range allows for flexible valve sizing.

R744 (CO_2) and Specialized Valves
- CO_2 operates at extremely high pressures (~30–50 bar).
- Expansion valves for R744 are compact yet robust, often with precision control for supermarket refrigeration.

R32 and EEVs
- R32's efficiency and low GWP make it a leading refrigerant for modern systems.
- EEVs paired with R32 ensure precise flow adjustment, enhancing energy savings.

R290 (Propane) and Capillary Tubes
- Often used in small systems where simplicity and low cost are essential.
- Safety is paramount due to R290's flammability.

Refrigerant Table with Key Properties and Valve Implications

Refrigerant	GWP	Flammability	Critical temperature (°C)	Valve type	Applications
R410A	~2088	Non-flammable	72	TXV, EEV	Split ACs, VRFs, packaged units
R134a	~1430	Non-flammable	101	TXV, EEV	Chillers, refrigerators
R744 (CO_2)	1	Non-flammable	31	Specialized EEVs	Supermarkets, low-temp units

(Continued)

Refrigerant	GWP	Flammability	Critical temperature (°C)	Valve type	Applications
R32	~675	Mildly flammable	78	TXV, EEV	Split ACs, VRFs
R290 (Propane)	~3	Highly flammable	97	Capillary Tube, TXV	Small commercial units

Refrigerants

Following is a comparison of different refrigerants commonly used in HVAC systems, considering key properties, applications, environmental impact, and other relevant factors:

Key Refrigerants and Their Characteristics

Refrigerant	Chemical type	Boiling point (°C)	ODP	GWP (100 years)	Applications
R-22	HCFC	−40.8	0.05	1810	Older residential and commercial systems (phased out in many regions)
R-134a	HFC	−26.3	0	1430	Chillers, refrigeration, automotive air conditioning
R-410A	HFC blend (R-32/R-125)	−51.6	0	2088	Residential and light commercial air conditioning
R-32	HFC	−51.7	0	675	Split air conditioners, heat pumps
R-290	Propane (HC)	−42.1	0	~3.3	Small-scale refrigeration, residential A/C (limited use due to flammability)
R-600a	Isobutane (HC)	−11.7	0	~3	Domestic refrigerators, small-scale cooling
R-744	CO_2 (natural)	−78.5 (sublimation)	0	1	Supermarket refrigeration, industrial systems, heat pumps
R-1234yf	HFO	−29.4	0	4	Automotive air conditioning, low-GWP applications
R-1234ze(E)	HFO	−19	0	<1	Chillers, low-temperature refrigeration
R-717	Ammonia (natural)	−33.3	0	~0	Industrial refrigeration, cold storage

Environmental Impact

Ozone Depletion Potential (ODP)

- HCFCs (e.g., R-22) contribute to ozone depletion and are being phased out globally under the Montreal Protocol.
- HFCs, HFOs, natural refrigerants (R-290, R-600a, R-744, R-717) have **zero ODP**.

Global Warming Potential (GWP)

- HFCs like R-410A and R-134a have high GWPs, contributing significantly to climate change.
- HFOs (R-1234yf, R-1234ze) and natural refrigerants (R-744, R-717) have ultra-low GWP values, making them preferred options in environmentally conscious applications.

Efficiency

- **HFCs (R-410A, R-134a, R-32)**: Offer high energy efficiency but may require additional mitigation for environmental concerns.
- **HFOs (R-1234yf, R-1234ze)**: Provide similar or better efficiency compared to HFCs with lower environmental impact.
- **Natural Refrigerants**
  - **R-744 (CO_2)**: High efficiency in heat pumps and refrigeration but requires higher operating pressures.
  - **R-717 (Ammonia)**: Excellent thermodynamic properties but toxic and requires careful handling.
  - **R-290 (Propane), R-600a (Isobutane)**: High efficiency in small-scale systems but are flammable.

Safety

- **Flammability**:
  - Non-flammable: R-134a, R-410A, R-744, R-717 (ammonia is non-flammable but toxic).
  - Mildly flammable: R-32, HFOs like R-1234yf.
  - Highly flammable: R-290 (propane), R-600a (isobutane).
- **Toxicity**:
  - R-717 (ammonia) is toxic and requires stringent safety protocols.
  - Other natural and synthetic refrigerants are generally safe under normal operation.

Operating Conditions

- **Pressure Requirements**:
  - High-pressure refrigerants: R-410A, R-744 (CO_2).
  - Moderate-pressure refrigerants: R-22, R-134a, R-32.
  - Low-pressure refrigerants: R-123, R-1234ze(E).
- **Temperature Suitability**:
  - Low-temperature applications: R-744, R-1234ze(E).
  - High-temperature heat pumps: R-717, R-290, R-600a.

Cost and Availability

- **HFCs (R-410A, R-134a)**: Widely available but increasingly regulated due to high GWP.
- **HFOs (R-1234yf, R-1234ze)**: Higher upfront cost but incentives for low-GWP refrigerants can offset costs.
- **Natural Refrigerants**:
  - R-290, R-600a: Low cost but require safety measures due to flammability.
  - R-717 (Ammonia): Economical for industrial applications but high initial safety compliance cost.
  - R-744 (CO_2): Higher equipment cost due to high-pressure requirements.

Applications Overview

Application	Preferred refrigerants	Considerations
Residential A/C	R-32, R-410A, R-290	Transitioning to R-32 and R-290 for lower GWP
Commercial A/C	R-410A, R-134a, R-1234ze	HFOs gaining traction for environmental compliance
Industrial Refrigeration	R-717, R-744, R-290	R-717 and R-744 for large-scale systems
Domestic Refrigerators	R-600a, R-134a	R-600a preferred for low-GWP, high-efficiency systems
Automotive A/C	R-134a, R-1234yf	R-1234yf is now standard for new vehicles
Heat Pumps	R-744, R-410A, R-1234ze	R-744 for high-temperature applications

Emerging Trends

- **HFOs** are rapidly replacing high-GWP HFCs in commercial and automotive sectors due to their low GWP and non-ozone-depleting characteristics.
- **Natural Refrigerants** are increasingly used for their eco-friendliness, particularly in Europe and Asia.
- **Blends** of refrigerants (e.g., R-407C, R-448A) are being developed to balance efficiency, environmental impact, and safety.

6.7 Direct Expansion (DX) Systems

Direct Expansion (DX) systems use refrigerants to directly cool air in the evaporator coil. They are widely used in residential, commercial, and industrial applications. The common types of DX systems are:

Split Systems

- **Description**: Consist of an outdoor condenser and an indoor evaporator connected via refrigerant lines.

- **Application**: Residential homes, small offices, and retail spaces.
- **Advantages**: Simple installation, low upfront cost, easy maintenance.
- **Sizing Rule of Thumb**:
  - Cooling load: **0.1–0.15 kW/m^2** for residential applications.
  - Typical capacity: **2–20 kW**.
- **Selection Consideration**: Match capacity to cooling load; ensure refrigerant piping lengths are within allowable limits.

Packaged Units

- **Description**: All components (compressor, condenser, and evaporator) housed in a single unit, usually installed on rooftops.
- **Application**: Medium to large commercial buildings, gyms, and schools.
- **Advantages**: Compact design, ease of maintenance, no refrigerant piping between units.
- **Sizing Rule of Thumb**:
  - Cooling load: **0.15–0.2 kW/m^2** for office spaces.
  - Typical capacity: **10–100 kW**.
- **Selection Consideration**: Verify rooftop structural support; ensure adequate ventilation.

Variable Refrigerant Flow (VRF) Systems

- **Description**: Multi-split systems that allow individual temperature control in multiple zones.
- **Application**: High-end offices, hotels, and multi-story buildings.
- **Advantages**: High efficiency, precise temperature control, scalability.
- **Sizing Rule of Thumb**:
  - Cooling load: **0.08–0.12 kW/m^2** for well-insulated buildings.
  - Typical capacity: **10–200 kW**.
- **Selection Consideration**: Evaluate zoning requirements; consider control system complexity.

Ducted DX Systems

- **Description**: Centralized systems where air is cooled and distributed through ducts.
- **Application**: Medium to large homes, office buildings, and retail spaces.
- **Advantages**: Uniform cooling, hidden components.
- **Sizing Rule of Thumb**:
  - Cooling load: **0.1–0.15 kW/m^2** for offices.
  - Typical capacity: **5–50 kW**.
- **Selection Consideration**: Consider duct losses and insulation quality.

DX Rooftop Units

- **Description**: Packaged systems installed on rooftops for large-scale cooling.
- **Application**: Warehouses, shopping malls, and manufacturing plants.
- **Advantages**: High capacity, no indoor equipment.
- **Sizing Rule of Thumb**:
  - Cooling load: **0.2–0.25 kW/m^2** for warehouses.
  - Typical capacity: **20–500 kW**.
- **Selection Consideration**: Ensure accessibility for maintenance; confirm structural capacity.

DX Split Units with Heat Recovery

- **Description**: Capable of providing simultaneous heating and cooling for different zones.
- **Application**: Hotels, hospitals, and multi-use spaces.
- **Advantages**: Energy-efficient, dual functionality.
- **Sizing Rule of Thumb**:
  - Cooling load: **0.1–0.15 kW/m^2**.
  - Heating load: Match to space requirements (typically lower than cooling).
- **Selection Consideration**: Ensure balanced heat recovery between zones.

General Rules of Thumb for Sizing and Selection

- **Cooling Load Estimation**:
  - **Residential**: **0.1–0.15 kW/m^2**.
  - **Commercial**: **0.15–0.25 kW/m^2**.
  - **Warehouses**: **0.2–0.3 kW/m^2**.
- **Airflow Requirement**:
  - **Typical Airflow**: **0.047–0.061 m^3/s/kW of cooling**.
- **Refrigerant Piping**: Follow manufacturer guidelines for length and diameter.
- **Ventilation Requirements**: Include fresh air demand for compliance with ASHRAE standards.
- **System Efficiency**: Aim for a COP (Coefficient of Performance) above **3.0** for energy efficiency.
- **Noise Control**: Evaluate dB levels for urban installations.

Optimal Superheat and Subcooling for DX HVAC Systems

Superheat indicates the temperature of the refrigerant vapor above its saturation temperature at the evaporator outlet. Ensures no liquid refrigerant enters the compressor, avoiding mechanical damage. Sub-cooling Measures the temperature of the liquid refrigerant below its saturation temperature at the condenser outlet.

Ensures sufficient refrigerant enters the expansion device, enhancing system performance. The table below provides typical superheat and sub-cooling temperature ranges for various refrigerants commonly used in direct expansion systems. These ranges ensure efficient system operation and help prevent issues like compressor damage (from low superheat) or insufficient cooling (from improper sub-cooling).

Refrigerant	Application	Superheat (°C)	Sub-cooling (°C)	Comments
R410A	Residential and commercial	5–8	5–10	High pressure; widely used in split systems, VRF, and packaged units
R134a	Chillers, refrigeration	3–5	5–8	Moderate pressure; suitable for chillers and small commercial refrigeration systems
R32	Residential and commercial	5–8	5–10	Efficient, low GWP refrigerant; increasingly replacing R410A in DX systems
R22	Older systems (phased out)	4–6	5–8	Still used in legacy systems; being replaced due to ozone depletion concerns
R744 (CO_2)	Low-temperature systems	2–4	2–6	Requires precise control due to very high operating pressures; used in supermarkets, etc.
R290 (Propane)	Small commercial systems	5–7	4–8	High efficiency and environmentally friendly but requires safety precautions (flammable)
R1234yf	Automotive, refrigeration	4–6	5–9	Low GWP refrigerant used in automotive and commercial applications
R600a (Isobutane)	Small appliances	4–6	5–8	Used in domestic refrigerators; highly efficient for low-capacity systems

Example Calculation

- **System**: DX system using R410A.
- **Target Cooling Load**: 12 kW.
- **Superheat Target**: 6 °C.
- **Sub-cooling Target**: 8 °C.

Steps

- **Determine Refrigerant Pressure and Temperature**:
  - Saturation Pressure (evaporator): ~8 bar.
  - Saturation Temperature (evaporator): ~2 °C.
  - Actual Outlet Temp (evaporator): 2 + 6 = 8 °C

- **Check Sub-cooling**:
  - Saturation Temp (condenser): ~45 °C.
  - Liquid Line Temp: 45–8 = 37 °C

Adjusting and Troubleshooting Superheat and Sub-cooling in DX HVAC Systems

Proper superheat and sub-cooling values are critical for optimal operation and efficiency of a direct expansion (DX) system. Below is a step-by-step guide to adjust and troubleshoot:

Superheat Adjustment

Step 1: Measure Superheat

1. **Tools Required**: Digital thermometer, manifold gauge set.
2. **Locate**: The suction line near the evaporator outlet.
3. **Procedure**:
   - Measure the suction line temperature (actual temperature of refrigerant vapor).
   - Measure the saturation temperature (using the suction pressure and a refrigerant pressure-temperature chart).
   - Calculate superheat:
     Superheat (°C) = Suction Line Temperature − Saturation Temperature

Step 2: Adjust Superheat

1. **Too Low (Risk of liquid flooding)**:
   - Turn the thermostatic expansion valve (TXV) adjustment screw clockwise to reduce refrigerant flow.
   - Recheck and repeat until the desired superheat is achieved.
2. **Too High (Insufficient refrigerant flow)**:
   - Turn the TXV adjustment screw counter clockwise to increase refrigerant flow.
   - Recheck and repeat.

Sub-cooling Adjustment

Step 1: Measure Sub-cooling

1. **Locate**: The liquid line near the condenser outlet.
2. **Procedure**:
   - Measure the liquid line temperature (actual temperature of refrigerant liquid).
   - Measure the saturation temperature (using discharge pressure and a pressure-temperature chart).
   - Calculate sub-cooling:
     Sub-cooling (°C) = Saturation Temperature − Liquid Line Temperature.

Step 2: Adjust Sub-cooling

1. **Too Low (Insufficient refrigerant)**:
   - Add refrigerant to the system slowly.
   - Recheck sub-cooling after each addition until the target value is achieved.
2. **Too High (Overcharged system)**:
   - Recover excess refrigerant carefully.
   - Recheck after every adjustment.

Common Troubleshooting Scenarios

- **High Superheat, Low Sub-cooling**:
  - Likely cause: Low refrigerant charge or restriction in the liquid line.
  - Solution: Check for leaks, add refrigerant, or clear the restriction.
- **Low Superheat, High Sub-cooling**:
  - Likely cause: Overcharged system or faulty expansion valve.
  - Solution: Remove excess refrigerant or inspect the TXV for issues.
- **Both Superheat and Sub-cooling Too High**:
  - Likely cause: Oversized compressor or low evaporator load.
  - Solution: Verify system sizing or adjust airflow across the evaporator.
- **Both Superheat and Sub-cooling Too Low**:
  - Likely cause: Faulty TXV or insufficient airflow.
  - Solution: Check and replace the TXV or inspect airflow and filters.

Tips for Optimization

- **Stable Conditions**: Allow the system to stabilize before taking readings.
- **Airflow**: Ensure correct airflow across the evaporator and condenser (check filters, fans).
- **Refrigerant Charge**: Follow manufacturer's specifications for refrigerant quantity.
- **System Size**: Verify the system is correctly sized for the application load.

DX System Pipe Sizes for Different Refrigerants and Capacities

Below is a table showcasing the approximate sizes of suction, discharge, and liquid lines in a DX system for various refrigerant types and capacities. The sizes are based on industry-standard guidelines and typical refrigerant properties. For simplicity, values are provided in millimetres for copper piping.

Capacity (kW)	Refrigerant	Suction Line (mm)	Discharge Line (mm)	Liquid Line (mm)
3.5	R-410A	9.5	9.5	6.4
	R-134a	9.5	9.5	6.4
	R-22	12.7	9.5	6.4
7	R-410A	12.7	12.7	9.5
	R-134a	15.9	12.7	9.5
	R-22	15.9	12.7	9.5
10.5	R-410A	15.9	15.9	9.5
	R-134a	19.1	15.9	9.5
	R-22	19.1	15.9	12.7
14	R-410A	19.1	19.1	12.7
	R-134a	19.1	19.1	12.7
	R-22	22.2	19.1	12.7
28	R-410A	22.2	22.2	12.7
	R-134a	25.4	22.2	15.9
	R-22	25.4	22.2	15.9
50	R-410A	28.6	25.4	15.9
	R-134a	28.6	25.4	15.9
	R-22	31.8	28.6	15.9
100	R-410A	35.0	31.8	22.2
	R-134a	38.1	31.8	22.2
	R-22	41.3	35.0	22.2

Notes

- **Pipe Sizing Assumptions**:
  - The table assumes standard pressure drops for each refrigerant, with pipe lengths up to 30 m.
  - For longer pipe runs, larger sizes may be needed to account for pressure losses.
- **Temperature Difference**:
  - Suction line sizing ensures proper superheat.
  - Discharge and liquid lines are sized for efficient heat transfer.
- **Refrigerant Properties**:
  - R-410A operates at higher pressures than R-22 and R-134a, requiring smaller pipe sizes.
  - Other refrigerants (e.g., R-407C, R-32) follow similar trends and can be interpolated.

Required Insulation Thickness for Refrigerant Suction Pipes
Here's a table showing the required insulation thickness for refrigerant pipes based on different insulation materials, pipe diameters, and standard temperature

conditions for HVAC applications. The values assume compliance with typical standards for energy efficiency and condensation prevention.

Pipe size (mm)	Material	R-value (m^2 K/W)	Ambient temp (°C)	Thickness (mm)
≤9.5 (Small)	Polyethylene Foam	0.34	30	13
	Elastomeric Rubber	0.3	30	10
	Fiberglass	0.43	30	15
10–25 (Medium)	Polyethylene Foam	0.53	30	20
	Elastomeric Rubber	0.51	30	19
	Fiberglass	0.71	30	25
26–50 (Large)	Polyethylene Foam	0.66	30	25
	Elastomeric Rubber	0.59	30	22
	Fiberglass	0.86	30	30
>50 (XLarge)	Polyethylene Foam	0.79	30	30
	Elastomeric Rubber	0.68	30	25
	Fiberglass	1	30	38

Assumptions and Notes

- **Insulation Types**:
  - **Polyethylene Foam**: Lightweight and cost-effective for small pipes.
  - **Elastomeric Rubber**: Offers excellent thermal resistance and flexibility.
  - **Fiberglass**: Higher R-value for larger pipes but requires external vapor barriers to prevent condensation.
- **Conditions**:
  - **Ambient Temperature**: 30 °C with a relative humidity of 70%.
  - **Pipe Temperature**: −10 °C to 5 °C for refrigerant suction lines and 35–50 °C for discharge/liquid lines.
- **Thickness Increase**:
  - For high humidity environments, increase insulation thickness by 25–50%.
  - For cold climates, consider frost protection.
  - In Australia, National Construction Code (NCC) must be followed to calculate the right thickness of insulation. Follow your local Codes and Standards for other countries.
- **R-Value Compliance**:
  - Ensure materials meet local building codes for minimum R-values.

Insulation Thickness for DX Refrigerant Lines (Suction, Discharge, Liquid)

Pipe size (mm)	Line type	Material	Thickness (mm)	Refrigerant examples
6–15	Suction	Rubber Foam	19–25	R-410A, R-22, R-134a
	Discharge	Rubber Foam	9–13	
	Liquid	Rubber Foam	6–9	
16–25	Suction	Polyurethane	25–30	R-410A, R-407C
	Discharge	Polyurethane	13–19	
	Liquid	Polyurethane	9–13	
26–40	Suction	Fiberglass	30–38	R-410A, R-404A
	Discharge	Fiberglass	19–25	
	Liquid	Fiberglass	13–19	
41–50	Suction	Elastomeric	38–50	R-22, R-410A, R-32
	Discharge	Elastomeric	25–30	
	Liquid	Elastomeric	13–19	
51–75	Suction	Rubber Foam	50+	R-410A, R-407C
	Discharge	Rubber Foam	30–38	
	Liquid	Rubber Foam	19–25	

Notes

- **Suction Lines**:
  - Have the thickest insulation to prevent condensation and energy loss.
  - The exact thickness depends on the ambient temperature, humidity, and refrigerant properties.
- **Discharge Lines**:
  - Require moderate insulation to retain heat and ensure safety.
  - Insulation thickness is typically lower than suction lines.
- **Liquid Lines**:
  - Require the least insulation since their temperature is closer to ambient.
  - Often insulated only in high-temperature environments.

Here's how the insulation needs differ for the three pipe types:

- **Suction Line:**
  - **Temperature Range**: Typically −10 °C to 5 °C.
  - **Priority**: Preventing condensation and minimizing thermal losses.
  - **Insulation Thickness**: The highest among the three lines due to the low temperature and condensation risk.
- **Discharge Line:**
  - **Temperature Range**: Typically, 35–100 °C, depending on the system.

  - **Priority**: Heat containment and safety (to prevent burns from high temperatures).
  - **Insulation Thickness**: Usually less than suction lines but may still require some insulation for energy efficiency or high ambient temperatures.

- **Liquid Line:**
  - **Temperature Range**: Typically, 25–50 °C.
  - **Priority**: Mainly energy efficiency and protection from external heat gains.
  - **Insulation Thickness**: The least among the three, often not insulated unless in very hot environments.

6.8 VRF Systems

In recent years, the field of air conditioning has witnessed a paradigm shift with the advent of Variable Refrigerant Flow (VRF) systems. These advanced systems offer unparalleled flexibility, energy efficiency, and comfort, making them a preferred choice for modern buildings ranging from commercial spaces to high-end residential complexes. Variable Refrigerant Flow (VRF) technology is an advanced climate control solution that allows simultaneous heating and cooling in different zones of a building. By adjusting the flow of refrigerant to multiple indoor units, VRF systems can maintain optimal comfort while minimizing energy wastage. This capability is particularly beneficial in environments with varying occupancy and thermal loads, such as office buildings, hotels, and educational institutions.

Understanding the Basics
At its core, a VRF system consists of an outdoor unit connected to multiple indoor units through refrigerant piping. Unlike traditional systems that rely on a fixed amount of refrigerant, VRF systems modulate the flow based on real-time demand. This modulation is achieved through advanced compressors and sophisticated control systems, ensuring precise temperature control and reduced energy consumption.

Key Features of VRF Systems Include

- **Zoning Capability:** Each indoor unit operates independently, allowing customized settings for different zones.
- **Energy Efficiency:** By only using the necessary amount of refrigerant, VRF systems significantly reduce energy waste.
- **Compact Design:** The modular nature of VRF systems makes them ideal for buildings with limited space for mechanical rooms.
- **Quiet Operation:** Advanced engineering ensures that VRF systems operate with minimal noise, enhancing occupant comfort.

Applications of VRF Technology
VRF systems are highly versatile and can be tailored to a wide range of applications, including:

- **Commercial Buildings:** Offices, retail stores, and malls benefit from the zoning and energy efficiency of VRF systems.
- **Residential Complexes:** High-rise apartments and luxury homes leverage VRF systems for personalized comfort.
- **Hospitality Industry:** Hotels and resorts appreciate the ability to provide individualized climate control for guest rooms and common areas.
- **Educational Institutions:** Schools and universities utilize VRF technology to manage varying thermal loads across classrooms and auditoriums.

6.9 Chillers

Chillers are essential components of HVAC (Heating, Ventilation, and Air Conditioning) systems, designed to remove heat from a liquid via a refrigeration cycle. This cooled liquid is then used to lower the temperature of a building or industrial process. By efficiently managing heat, chillers help create comfortable environments while minimizing energy consumption.

Types of Chillers and Cooling Towers

Air-Cooled Chillers
- Air-cooled chillers are known for their simplicity and suitability for smaller spaces. They don't require a water source, making them ideal for areas with limited water availability. However, air-cooled chillers tend to be less efficient than water-cooled systems, especially in hot climates, due to their reliance on ambient air for heat dissipation.

Water-Cooled Chillers
- Water-cooled chillers are more energy-efficient and capable of handling larger cooling loads, making them suitable for extensive commercial and industrial applications. These systems demand regular maintenance, including water treatment, to prevent scaling and corrosion that can compromise performance.

Selecting the Right Type of Chiller
- Choosing between air-cooled and water-cooled chillers involves evaluating the facility's needs, budget, and environmental conditions. Air-cooled systems are easier to install, while water-cooled systems excel in energy efficiency for large-scale applications.

Determine Cooling Load

- **Rule of Thumb**: Calculate the total cooling load in kilowatts (kW) or refrigeration tons (1 RT = 3.517 kW) · $Q = m \cdot C_p \cdot \Delta T$
  - Where:
    - Q: Cooling load (kW)
    - m: Water mass flow rate (kg/s)
    - C_p: Specific heat capacity of water (4.186 kJ/kg°C)
    - ΔT: Temperature difference of chilled water supply and return (°C)

Estimate Chiller Capacity

- Match chiller capacity to the cooling load, considering a safety margin of 10–15%.
  - Example: If the cooling load is 500 kW, select a chiller with a capacity of 550–575 kW.

Chilled Water Temperature Difference

- Maintain a temperature difference (ΔT) of 5–7 °C between supply and return chilled water for efficient operation.

Cooling Tower Sizing

- **Rule of Thumb**: Cooling tower capacity should be 15–20% higher than the chiller capacity due to heat rejection requirements.
  - Heat rejection capacity: $Q_{rejected} = Q_{chiller} + Q_{compressor}$
  - Typical water flow rate through the cooling tower:
- Rule of Thumb for Cooling Tower Water Flow Rate:
  - SI Units: 0.05 L/s/kW of heat rejection.
  - British Units: 3 GPM/Ton of refrigeration.

Condenser Water Flow

- Design condenser water flow with a temperature rise (ΔT) of 5–10 °C.

Pump Sizing

- Determine the pump flow rate and head:
  - Chilled water pump: Flow rate matches chilled water flow with 10–20 m of head loss.
  - Condenser water pump: Flow rate matches condenser water flow with 20–40 m of head loss.

Pipe Sizing

- Design for a water velocity of 1.5–3 m/s to balance efficiency and pressure drop.

Efficiency and Part Load Performance

- Choose chillers with high Coefficient of Performance (COP) and good part-load efficiency:
  - Typical COP for water-cooled chillers: 5–7.

System Redundancy

- **Rule of Thumb**: For critical applications, design with N + 1 redundancy, where an additional chiller, pump, or tower can take over during failure or maintenance.

Cooling Tower Approach Temperature

- Maintain an approach temperature (difference between cooling tower water outlet temperature and wet-bulb temperature) of 3–5 °C.

System Controls

- **Rule of Thumb**: Implement Variable Frequency Drives (VFDs) on pumps, cooling towers, and compressors to optimize energy efficiency.

Safety Factor

- Include a safety margin of 10–15% in calculations to account for unforeseen changes in cooling demand or environmental conditions.

6.10 Cooling Towers

Cooling towers complement chillers by dissipating the heat extracted from the system into the atmosphere. Typically used in large-scale applications like factories, power plants, and commercial buildings, cooling towers ensure the effective transfer of heat, enabling the system to maintain optimal performance.

An illustration of a Cooling Tower

Types of Cooling Towers

Crossflow Cooling Towers

- Crossflow cooling towers feature a design where the airflow moves horizontally through the fill media, and water flows downward. This configuration simplifies maintenance and reduces operational noise, making it a popular choice for medium to large installations.

Counterflow Cooling Towers

- Counterflow cooling towers direct both water and air flow in opposite directions. This design maximizes heat transfer efficiency, making it suitable for applications where space is limited, though it often comes with a higher operational cost.

Hybrid Cooling Towers

- Hybrid cooling towers combine elements of wet and dry cooling. They minimize water usage while maintaining effective heat dissipation, offering a sustainable option for facilities striving to reduce environmental impact.

Cooling Towers Rules of Thumb

Determine Heat Load

- **Rule of Thumb**: Calculate the total heat load (Q) that the cooling tower needs to reject in British Thermal Units per hour (BTU/h), Kilowatts (kW), or Tons of Refrigeration. $Q = m \cdot C_p \cdot \Delta T$
  - Where:
    - m: Mass flow rate of water
    - C_p: Specific heat capacity of water (1 BTU/lb. °F or 4.186 kJ/kg °C)
    - ΔT: Temperature difference of water (inlet-outlet)

Define Water Flow Rate

- Estimate water flow rate using:
  - 1 ton of refrigeration ≈ 3 GPM (gallons per minute)
  - **kW ≈ 0.054 L/s** (litters per second)

Temperature Range

- Choose based on process requirements:
  - Typical range: 5–8 °C (10–15 °F) drop between inlet and outlet water.
  - Ensure the tower can handle your required range effectively.

Approach Temperature

- Approach temperature is the difference between the outlet water temperature and the ambient wet-bulb temperature.
  - Typical approach temperature: 3–6 °C (5–10 °F).
  - Lower approach temperatures require larger or more efficient cooling towers.

Wet-Bulb Temperature

- Design for the highest expected ambient wet-bulb temperature.
  - Wet-bulb temperature is critical as it determines the cooling tower's theoretical cooling limit.

Tower Efficiency

- Higher efficiency systems often have higher capital costs but lower operating costs.
  - Crossflow and counterflow designs are common; counterflow is generally more efficient but may have higher initial costs.

Fan Selection

- Choose between induced draft or forced draft fans:
  - Induced draft: Energy efficient and quieter.
  - Forced draft: Smaller footprint but less efficient.

Material Selection

- Match materials with the environment and water quality:
  - Corrosion-resistant materials like stainless steel or fiberglass-reinforced plastics (FRP) for harsh environments.
  - PVC fill material for good thermal performance.

Drift and Blowdown Losses

- Drift: Keep below 0.002% of water flow for compliance.
  - Blowdown: Ensure proper water treatment and cycling to prevent scale and fouling.

Sizing

- Use manufacturer sizing charts or software for precision.

- Cooling towers are sized based on:
  - Heat rejection rate (BTU/h, kW, or Tons)
  - Wet-bulb temperature
  - Approach and range
  - Required flow rate

Safety Factor

- Add a safety margin of 10–15% to ensure reliable performance during peak conditions.
- Avoid placing cooling towers near outdoor air intakes to prevent high entering air wet-bulb temperatures and reduce the risk of bacterial contamination, such as Legionella, in intake air.
- Avoid positioning cooling towers near the building facade to prevent staining and minimize noise intrusion into occupied spaces.

6.11 Evaporative Cooling System

An illustration of a Evaporative Cooling System

Evaporative cooling technology has been widely used for more than a century. Direct evaporative cooling (DEC) systems have low set-up and running costs and have been proven to significantly improve a building's cooling and ventilation capacity with minimal energy use. Using water as the working fluid, one can avoid the use of ozone-destroying chlorofluorocarbons and hydro chlorofluorocarbons. Other benefits from this system include easy maintenance, easy installation, and operation as well as obviating CO_2 and other emissions.

Direct evaporator cooling plants are simple, inexpensive and can easily be integrated into built-up systems; however, its cooling effect is insufficient for indoor thermal comfort when **ambient wet-bulb temperature is higher than 21 °C**. Evaporative cooling systems can provide thermal comfort via the conversion of sensible heat to latent heat; however, the lowest temperature DEC systems can reach is the wet-bulb temperature of the outside air. Therefore, the temperature of the supply air after cooling would be just on the edge of comfort and could rise a few degrees in passing through space, taking the temperature beyond the comfort zone. Therefore, the idea is to investigate both the possibility of increasing the utilization potential of the evaporative cooling system by combination of different components with this system and the capability of improving the performance of other HVAC systems when integrating with evaporative cooling system.

Evaporative cooling systems are gaining traction as an efficient and environmentally friendly alternative to traditional air conditioning systems. These systems harness the natural process of water evaporation to cool air, offering a sustainable solution that aligns with modern efforts to reduce carbon footprints and combat climate change. Unlike refrigerant-based cooling methods that rely on complex chemical processes, evaporative cooling utilizes water and air—two of the most abundant and renewable resources. This process not only reduces energy consumption but also minimizes environmental impact, making it a preferred choice for eco-conscious individuals and organizations.

Types of Evaporative Cooling Systems

Evaporative cooling systems come in various configurations, each designed to cater to specific needs and applications. By understanding the different types of systems available, users can make informed decisions that align with their cooling requirements, budget, and environmental considerations.

Direct Evaporative Cooling Systems

Direct evaporative cooling systems are the most straightforward and widely used type. In these systems, warm air passes through a water-saturated medium, such as a cooling pad or filter. As the air flows through, water evaporates, reducing the air temperature and increasing humidity levels. The cooled air is then distributed into the desired space.

Advantages of Direct Systems

- Simple design and operation
- Low energy consumption compared to traditional air conditioning
- Cost-effective installation and maintenance

Limitations

- Increases indoor humidity, which may not be suitable for all climates
- Less effective in regions with high ambient humidity

Direct systems are ideal for dry climates where increased humidity can enhance comfort levels while providing efficient cooling.

Indirect Evaporative Cooling Systems
Indirect evaporative cooling systems use a heat exchanger to cool the air without adding moisture. In this setup, warm air passes through one side of the heat exchanger, while water evaporates on the other side. The evaporation process cools the heat exchanger, which in turn cools the air passing through it. The cooled air remains dry, making it suitable for environments where humidity control is essential.

Advantages of Indirect Systems

- Does not increase indoor humidity
- Provides efficient cooling in a wider range of climates

Limitations

- Slightly higher initial cost and complexity compared to direct systems

Indirect systems are often used in commercial and industrial settings where maintaining specific humidity levels is critical.

Hybrid Evaporative Cooling Systems
Hybrid systems combine elements of both direct and indirect cooling to maximize efficiency and flexibility. In a hybrid system, air is first pre-cooled using an indirect method and then further cooled using a direct method. This dual-stage process enhances cooling performance while keeping energy consumption low.

Advantages of Hybrid Systems

- High cooling efficiency
- Adaptable to various climatic conditions
- Reduces energy costs compared to traditional systems

Estimate Airflow Requirements

- Use approximately 20–40 air changes/h for the target space.
  - Residential: ~20 air changes/h.
  - Commercial/industrial: ~30–40 air changes/h.
  - Higher air changes may be required in hotter, drier climates.

Match the Cooler Size to the Space

- **Rule of Thumb**: Match cooler airflow (m^3/s) to the space volume and desired air change rate.
  - Example: For a 100 m^2 area with 3 m ceilings in a hot, dry climate:

Calculate the Volume of the Space

$$\text{Volume}\left(m^3\right) = \text{Area}\left(m^2\right) \times \text{Ceiling Height}\left(m\right) = 100\,m^2 \times 3\,m = 300\,m^3$$

Determine the Air Changes Per Hour (ACH)
In hot, dry climates, evaporative cooling typically requires **20–40 ACH** for effective cooling. We'll use an average of **30 ACH** for this calculation.

Convert ACH to Airflow Rate

To calculate airflow in cubic meters per second (m^3/s), use the formula:

$$\text{Airflow}\left(\text{m}^3/\text{s}\right) = \text{Volume}\left(\text{m}^3\right) \times \text{ACH}/3600\,\text{s}$$

Substitute the values:

$$\text{Airflow}\left(\text{m}^3/\text{s}\right) = 300\,\text{m}^3 \times 30/3600\,\text{s} = 2.5\,\text{m}^3/\text{s}$$

The required airflow rate for the evaporative cooler is **2.5 m^3/s.**

Wet-Bulb Depression

- **Rule of Thumb**: Consider the wet-bulb depression (difference between dry-bulb and wet-bulb temperatures in °C).
  - Larger wet-bulb depression = better cooling potential.
  - Evaporative cooling is most effective when the wet-bulb temperature is significantly lower than the dry-bulb temperature.

Climate Suitability

- Use evaporative coolers in dry climates with a relative humidity of less than 50%.
  - For regions with higher humidity, performance drops significantly, and alternative cooling methods may be better.

Water Supply

- Ensure adequate water supply for continuous operation.
  - Typical water consumption: 10–60 L/h depending on size and humidity.
  - Use water-efficient models to reduce wastage.

Pad Efficiency

- Choose high-efficiency pads for better cooling performance:
  - Aspen pads: Affordable but require frequent replacement.
  - Rigid cellulose pads: Durable, high cooling efficiency, and less maintenance.

Fan Type

- Select the fan type based on airflow and noise requirements:
  - Centrifugal fans: Quieter, good for residential or office use.
  - Axial fans: Higher airflow, suitable for industrial applications.

Power Consumption

- Select energy-efficient models to minimize operating costs.
  - Evaporative coolers typically use **0.2–0.3 kW/h** of operation, which is much lower than air conditioners.

Maintenance and Durability

- Ensure easy access for pad cleaning/replacement and motor maintenance.
- Corrosion-resistant materials (e.g., galvanized steel or plastic) for long service life.

Noise Levels

- Choose quieter models for residential or office use.
  - Noise levels should be below 70 dB for comfort in indoor settings.

Add Safety Margins

- Add 10–20% to the calculated airflow (m^3/s) to account for inefficiencies or peak cooling needs.

Evaluating Site Conditions
Before designing an evaporative cooling system, it is essential to evaluate the site where it will be installed. Key factors include:

Climate

- Assess the regional climate to determine the system's suitability.
- Evaporative cooling is most effective in arid and semi-arid regions with low relative humidity.

Water Availability

- Consider the availability and quality of water at the site.
- Use filtration or water treatment systems if necessary to ensure clean water supply.

Building Orientation and Layout

- Evaluate the building's orientation to maximize airflow and natural ventilation.
- Ensure the layout supports effective distribution of cooled air.

Environmental Impact

- Account for potential environmental concerns, such as water consumption and energy use.

Evaporative-Cooled Air Conditioning System
Recent research reveals that air conditioning systems based on mechanical vapor compression consume significant amounts of electricity. Therefore, increasing the coefficient of performance (COP) of these air conditioning systems with air-cooled condensers is a challenging problem. By pre-cooling the air before it reaches the condenser coil, the condenser is able to reject more heat. As a result, cooling capacity increases while energy demand and usage falls. As condensing temperatures are lowered, head pressure is reduced. This allows the compressor to run less frequently,

resulting in an energy saving. The standard design for these systems requires a frame to be built and filled by evaporative media pads which are installed in front of the air-cooled condenser. A water circulation system, consisting of a small pump, a tank and pipes, is added. The water then is injected on the top of the media pad. Hot ambient air passes the wet pad and then the condenser to improve the system performance. As the hot, ambient air is drawn through the media, the water absorbs heat and evaporates, lowering the temperature of the ambient air and creating a cooler operating environment for the air-cooled condenser which allows the condenser to reject additional heat into the atmosphere. The compression ratio is then reduced, resulting in reduced energy usage when the compressor is run.

Below figure shows an evaporative cooler coupled to the existing air-cooled condenser of a split air-conditioner [Hajidavallo and Eghtedari, Performance Improvement of Air-Cooled Refrigeration System by Using Eveporatively Cooled Condenser, International Journal of Refrigeration, 2010, 33: 982-988]. Their experimental results showed that the power consumption of the air-conditioner can be reduced up to 20% and the system COP can be improved around 50%.

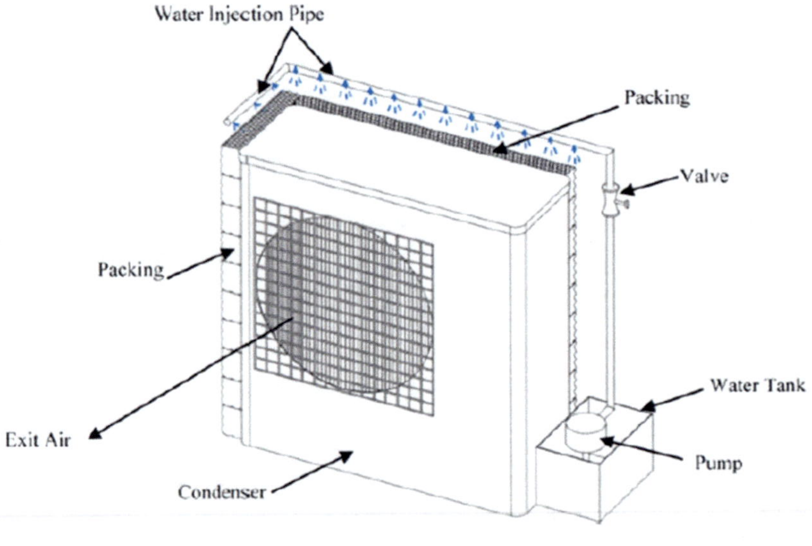

Schematic view of the evaporative-cooled air conditioner

In another design, mist ware is directly sprayed into the ambient air before passing through the air-cooled condenser. Therefore, the compressor head pressure is reduced from point (3) to point (b) at steady state conditions as shown in below figure.

As a result, the refrigeration effect of the evaporator is increased enabling the system to deal with higher load demand. As demonstrated in Fig. 2, this is occurred by replacing the point (1) to point (a) in the refrigeration cycle. As a result, the refrigeration effect of the evaporator is increased enabling the system to deal with higher load demand. As demonstrated in Fig. 3, this is occurred by replacing the point (1) to point (a) in the refrigeration cycle [Vakiloroaya, Zhu, Ha, Modelling and

Optimisation od Direct Expansion Air Conditioning Systems for Commercial Building Energy Savings, The 28th International Symposium on Automation and Robotics in Construction, Seoul, Korea, pp 232-237, 2011].

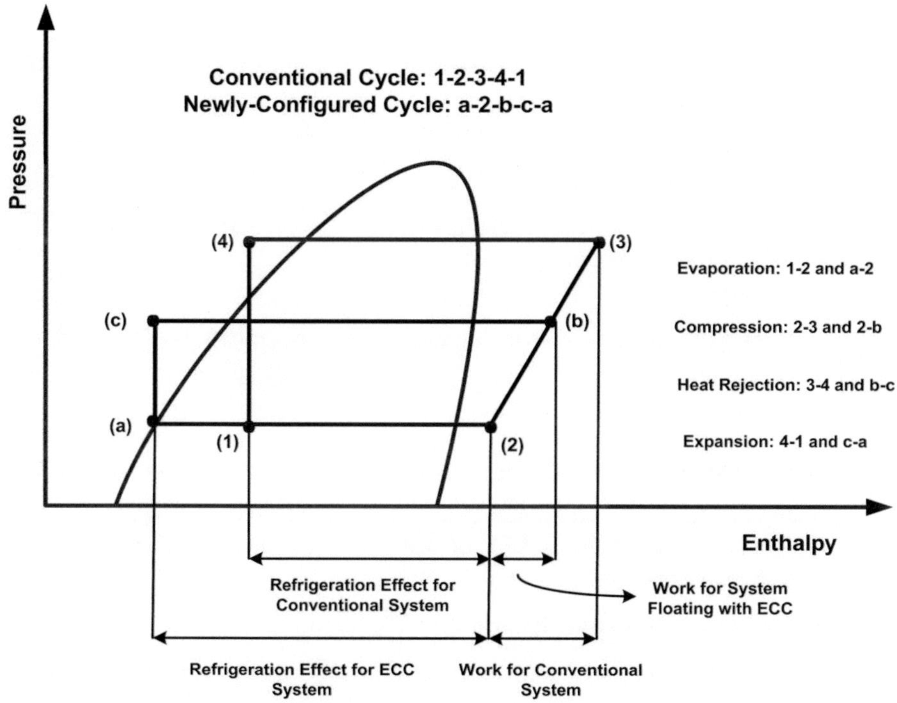

P-h diagram of the developed DX air conditioning system with ECC strategy

Benefits and Guidelines for Evaporative Condensing Units in HVAC

Evaporative condensing units (ECUs) are efficient components often used in refrigeration and air conditioning systems to reject heat. By leveraging the principle of evaporative cooling, these units offer significant energy savings compared to air-cooled condensers, especially in hot and dry climates.

Benefits of Evaporative Condensing Units

- **Improved Energy Efficiency**:
  - **Lower Condensing Temperatures**: ECUs reduce condensing temperatures by using water evaporation, which requires less compressor energy.
  - Efficiency gains of 10–30% compared to air-cooled condensers are common.
- **Compact Design**:
  - Smaller footprint than air-cooled systems due to higher heat transfer efficiency.
  - Reduces space requirements, making them ideal for urban or constrained installations.

- **Lower Operating Costs**:
  - Reduced compressor energy consumption leads to lower electricity bills.
  - Savings are more pronounced in areas with high ambient air temperatures.
- **Adaptability to High Ambient Temperatures**:
  - Performs better than air-cooled systems in hot climates, where air cooling becomes less effective.
  - Maintains stable operation even during peak heat.
- **Extended Equipment Life**:
  - Lower compressor workload results in less wear and tear, extending the lifespan of the HVAC system.
- **Sustainability**:
  - Consumes less electricity, reducing greenhouse gas emissions associated with energy production.

Sizing and Selection

- **Cooling Capacity**:
  - Ensure the unit's capacity matches or exceeds the load demand of the system.
  - Typical sizing: Provide **25–35 L of water/h/kW** of heat rejected, depending on ambient conditions.
- **Ambient Temperature and Humidity**:
  - Best suited for regions with hot, dry climates where evaporation is more effective.
  - Use where wet bulb temperature (WBT) is significantly lower than dry bulb temperature (DBT).
- **Water Flow Rate**:
  - Ensure proper water supply for consistent evaporation.
  - **Rule of Thumb**: 0.04–0.06 L/s/kW of condensing capacity.
- **Airflow Requirements**:
  - Ensure sufficient airflow across the evaporative media for optimal heat rejection.
  - **Rule of Thumb**: 200–300 cubic meters per hour (m^3/h) per kilowatt of capacity.
- **Water Quality**:
  - Use treated water or implement a water treatment system to prevent scale build-up and corrosion.
  - Regular maintenance is crucial in areas with hard water.

- **Maintenance Considerations**:
  - Clean and inspect evaporative media and nozzles regularly.
  - Replace or treat media every 1–3 years depending on water quality and usage.
- **Energy and Water Trade-Off**:
  - While ECUs save energy, they consume water for evaporation. Evaluate the trade-off between water and energy costs, especially in areas with water scarcity.
- **Material Selection**:
  - Opt for corrosion-resistant materials like stainless steel, galvanized steel, or coated surfaces to withstand wet conditions.
- **Environmental Conditions**:
  - Protect against freezing in colder climates by incorporating antifreeze systems or shutting down the water supply in winter.

Applications of Evaporative Condensing Units

- **Commercial Refrigeration**:
  - Supermarkets and food processing facilities often use ECUs to maintain low operating costs.
- **Industrial Processes**:
  - Chemical plants, pharmaceutical facilities, and data centres benefit from reliable cooling in large-scale applications.
- **HVAC Systems**:
  - Efficiently rejects heat in central cooling plants or large-scale air conditioning systems.
- **Combined Heat and Power (CHP) Plants**:
  - Used to improve efficiency in cogeneration applications by efficiently rejecting excess heat.

Comparison with Other Condenser Types

Aspect	Evaporative condenser	Air-cooled condenser	Water-cooled condenser
Efficiency	High, especially in hot climates	Moderate, decreases in high heat	Very high, but depends on water availability
Initial cost	Moderate	Low	High
Operating cost	Lower energy, moderate water	Higher energy, no water	Lowest energy, high water
Maintenance	Regular cleaning and water treatment	Minimal	High due to water treatment
Water consumption	Moderate	None	High
Best suited for	Hot and dry climates	Moderate climates	Areas with reliable water supply

6.12 Ground-Couple HVAC Systems

Ground-coupled HVAC systems, also known as ground-source heat pump (GSHP) systems, use the stable temperatures of the ground to provide heating, cooling, and sometimes hot water. Proper sizing and selection are critical to ensure efficiency, longevity, and cost-effectiveness. Ground-coupled technology relies on the fact that, at depth, the Earth has a relatively constant temperature that is colder than the air temperature in summer and warmer than the air temperature in winter. In this system, under cooling mode, operation heat is discharged to a ground loop that provides a lower temperature heat sink than ambient outdoor air temperature. During winter heating operations, heat is extracted from a source that is at a higher temperature than ambient outdoor air. This system has been used on a residential and commercial scale since the 1920s. The COP of a ground source heat pump (GSHP) is usually higher than that of air source heat pump (ASHP), due to lower condensing temperatures in the GSHP system. Also, while the initial cost for GSHP is more than ASHP, the operating cost of the GSHP is less with a payback time of about 2 years. However, GSHPs capture only a small percentage of the heating and cooling market due to the high cost of installing the ground heat exchanger as well as high initial capital cost when compared to air source units. GSHP was widely used in Europe and USA between 2004 and 2007, for public HVAC installations in Korea in recent decades. Compared with standard technologies such as vapour compression systems, ground-coupled heat pumps create less noise, reasonable environmental safety and can reduce greenhouse gas emissions compared with conventional heating and cooling systems.

Understand the Application

- **Building Load Requirements**: Calculate heating and cooling loads using software or standards like ASHRAE.
- **Climate Zone**: Consider the local climate's heating and cooling balance to optimize system design.
- **Ground Conditions**: Assess the ground's thermal properties (conductivity and diffusivity) and water table depth.

System Capacity

- **Cooling Capacity**: Approximately **35–50 W/m^2** of building floor area.
- **Heating Capacity**: Approximately **30–40 W/m^2** of building floor area.
- Adjust based on insulation, glazing, and occupancy.

Ground Loop Sizing

- **Closed-Loop Systems**:
  - Horizontal loops: **15–25 m^2** of ground area per kW of heating/cooling.
  - Vertical loops: **15–30 m** of borehole depth per kW.
  - Ensure pipe material (e.g., high-density polyethylene) is durable and thermally efficient.

- **Open-Loop Systems**:
  - Water flow: **0.03–0.05 L/s/kW** of system capacity.
  - Ensure compliance with local regulations for water usage and discharge.

Heat Pump Selection

- Select heat pumps with a high **Coefficient of Performance (COP)**:
  - Heating: COP **3.5–4.5**.
  - Cooling: Energy Efficiency Ratio (EER) **15–25**.
- **Sizing Rule**: Match the heat pump capacity to 100% of the design heating or cooling load. Avoid oversizing, as it can lead to short cycling and reduced efficiency.

Ground Temperature

- Use the stable ground temperature as the basis for system design:
  - Typical ground temperatures: **10–20 °C**, depending on location.
  - Ensure the heat exchanger is designed to work efficiently with local ground temperatures.

Antifreeze Solution

- Use antifreeze solutions (e.g., propylene glycol) for climates where ground loop fluid temperatures may fall below freezing.

Pumping Requirements

- Ensure the loop circulation pump is energy-efficient:
  - Flow rate: **2.5–3.0 L/min/kW** of system capacity.
  - Head pressure: Keep within manufacturer recommendations for loop design.

Energy Efficiency

- **Seasonal Performance Factor (SPF)**: Target SPF values **>4.0** for overall system efficiency.
- Integrate with **variable-speed drives (VSDs)** to improve part-load performance.

Installation Considerations

- **Horizontal Loops**:
  - Require significant land area.
  - Typically buried **1.5–2.5 m** deep to ensure stable ground temperature.
- **Vertical Loops**:
  - Require less land but involve higher installation costs.
  - Drilled to depths of **50–200 m**, depending on design requirements.

Cost and Feasibility

- **Initial Costs**:
  - Horizontal loops: Lower installation cost than vertical loops.
  - Vertical loops: Higher installation cost but suitable for limited land.
- **Operating Costs**:
  - Typically **30–50% lower** than conventional HVAC systems.
- Evaluate **payback period**: Typically **5–10 years** depending on energy savings.

6.13 Thermal Storage Systems

Thermal storage HVAC systems are designed to shift cooling or heating loads to off-peak hours by storing energy in the form of chilled water, ice, or hot water. Proper sizing and selection are essential to maximize energy efficiency, reduce operational costs, and meet load demands effectively. Thermal storage systems (TSS) shift the energy usage of the HVAC systems from on-peak to off-peak periods to avoid peak demand charges. Compared to conventional HVAC systems, TSS offers various advantages for heating and cooling systems, such as energy and capital cost savings, system operation improvements, system capacity extending and equipment size reduction, resulting in a technology that is widely used. Ice and chilled-water storage systems are two most commons TSS. In these systems, ice or chilled water is stored in tanks to cool buildings during peak electricity usage periods. In an ice storage system, ice is usually generated using glycol or brine solutions. There are various types of ice storage systems. An ice harvester system uses an open insulated storage tank and a vertical plate surface which is located above the tank. During the charging period, water flows on the outside surface of the evaporator and forms ice sheets. Ice slurry is another type of the ice storage system in which a glycol-water solution passes through pipes submerged in an evaporating refrigerant to form the ice. The generated ice particles are then dropped into the storage tank. In an ice-on-coil system, the refrigerant is evaporated by a coil submerged in water. The remaining cold water in the storage tank is then used for cooling in air-handling units. Often, this system is employed for industrial applications as it requires significant amounts of refrigerant.

Understand the Application

- **Purpose**: Determine the primary goal:
  - Reducing peak electrical demand.
  - Taking advantage of off-peak electricity rates.
  - Enhancing system resilience.

- **Cooling or Heating Load**:
  - Applications include commercial buildings, data centers, district cooling systems, and industrial processes.

System Type

- Select the appropriate thermal storage medium:
  - **Chilled Water Systems**:
    - Best for large-volume applications with moderate space constraints.
    - Typical temperature range: **4–12 °C**.
  - **Ice Storage Systems**:
    - High energy density, suitable for space-limited applications.
    - Typical ice melt temperature: **0 °C**.
  - **Hot Water Storage Systems**:
    - Used for heating applications in district energy systems.

Storage Capacity

- **Cooling Storage**:
  - Estimate cooling load using the building's peak demand:
    - Rule of thumb: **30–50 W/m²** for comfort cooling in office buildings.
    - Data centres or process cooling: Higher specific loads based on equipment.
  - Storage tank sizing: Storage Capacity (kWh) = $Q = m \cdot C_p \cdot \Delta T$
  - Where:
    - Q: Cooling energy stored.
    - m: Mass of water or ice.
    - C_p: Specific heat capacity of the medium.
    - ΔT: Temperature difference for water or phase change for ice (~334 kJ/kg).
- **Heating Storage**:
  - Calculate based on building heating requirements:
    - Offices: **50–100 W/m²** for space heating.
    - Industrial processes: Variable based on specific applications.

Charge and Discharge Rates

- Ensure charging equipment (chillers, boilers) can generate the required energy during off-peak periods:
  - Chilled water systems: Charging over **6–12 h**.
  - Ice storage systems: Charging over **6–10 h**.

- Discharge rate must meet peak demand loads:
  - Chilled water flow: **2–3 L/s/kW** of cooling load.
  - Ice storage: Ensure sufficient flow to achieve design capacity.

Storage Tank Sizing

- **Chilled Water Storage**:
  - Tank volume: **3–5 m^3/ton-h** of cooling.
  - Large tanks may require stratification baffles to maintain thermal layers.
- **Ice Storage**:
  - Volume: **0.03–0.04 m^3/ton-h** of cooling.
  - Modular ice banks are common for flexibility in installation.
- **Hot Water Storage**:
  - Volume: Depends on temperature range, typically **40–60 L/kW**.

Thermal Storage Efficiency

- Evaluate storage system efficiency:
  - Chilled water systems: **90–95% efficiency**.
  - Ice storage systems: **85–90% efficiency** (accounts for phase change and thermal losses).
  - Hot water systems: Depends on insulation and standby losses.

Insulation

- Proper insulation is critical to minimize heat gain or loss:
  - Use high-performance materials (e.g., polyurethane or polystyrene).
  - Insulation thickness: **100–150 mm** for chilled water tanks and **150–300 mm** for ice storage tanks.

Integration with HVAC Systems

- Ensure the storage system integrates seamlessly with HVAC components:
  - Chillers: Sized for charging the thermal storage.
  - Pumps and valves: Sized to handle storage charging/discharge flow rates.
  - Control systems: Enable automated switching between thermal storage and direct cooling/heating.

Peak Load Reduction

- Design to offset **50–100%** of the peak load, depending on objectives:
  - Full storage: Meets the entire peak demand using stored energy.

– Partial storage: Reduces peak demand while operating chillers during peak periods.

Energy Rates and Cost Savings

- Evaluate electricity rates to determine economic viability:
  – Off-peak rates: Typically **30–50% lower** than peak rates.
  – Calculate potential savings based on demand charges and energy tariffs.

Pumping and Distribution

- Ensure pumps are sized for both charging and discharging:
  – Flow rate: **2.5–3.0 L/min/kW** for chilled water systems.
  – Ice systems may require glycol mixtures for low-temperature operation.

6.14 Heat Recovery Systems

ASHRAE standards recommend the amount of required fresh air for different buildings. Unconditioned air greatly increases the building's cooling needs, which ultimately leads to an increase in the overall energy consumption of the building's HVAC systems. Heat recovery systems in HVAC improve energy efficiency by capturing waste heat and using it to precondition incoming air. Proper sizing and selection are critical for optimal performance.

Heat recovery techniques can be used to recover energy that might otherwise be wasted. The objective of heat recovery is to reduce the cost of operating an HVAC system by transferring heat between two fluids such as exhaust air and fresh air. According to ASHRAE handbook of HVAC Systems and Equipment (2008) there are three types of heat recovery systems: comfort-to-comfort, process-to-comfort, and process-to-process. Comfort-to-comfort systems use exhaust air that is captured and reused as the waste heat energy to precondition the fresh air coming into the HVAC system. This type of heat recovery system can be used as the sensible heat recovery mode and in total heat transfer mode. Usually, rotary wheel heat exchangers are used for comfort-to-comfort heat recovery systems. Process-to-comfort and process-to process systems perform sensible heat recovery. Different types of heat recovery systems, which are used to recovery energy between supply and exhaust airflows, consist of fixed plate, rotary wheel, heat pipe and run-around coil.

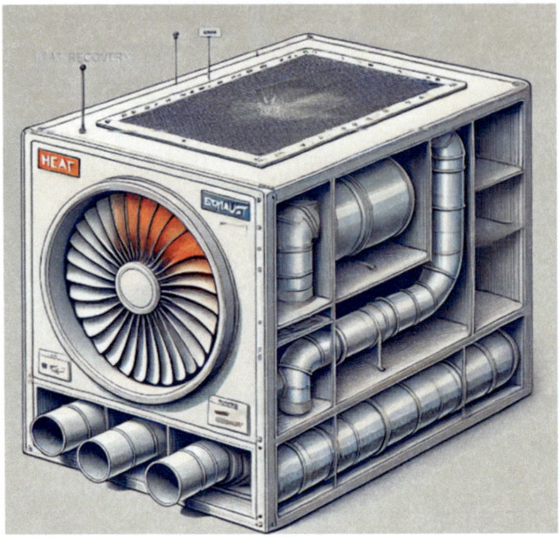

An illustration of a Heat Recovery System

Understand the Application

- **Purpose**: Heat recovery systems are used to:
  - Preheat or precool outdoor air using exhaust air.
  - Recover waste heat from processes for space or water heating.
  - Reduce energy costs in systems with high ventilation requirements.
- Common applications include commercial buildings, hospitals, laboratories, schools, and industrial facilities.

Airflow Requirements

- Match the heat recovery system's airflow capacity to the ventilation system:
  - **Typical Airflow Rates**:
    - Offices: 6–10 L/s/person.
    - Hospitals: 10–20 L/s/person.
    - Industrial facilities: Varies based on exhaust requirements.
- The heat recovery unit should handle at least **90–100% of the system's design airflow** to maximize efficiency.

Effectiveness

- Effectiveness (η) measures the system's ability to transfer heat:
  $\eta = \Delta T_{air}/\Delta T_{max}$
- Where
  - ΔT_{air}: Temperature difference achieved by the heat exchanger.
  - ΔT_{max}: Maximum possible temperature difference (exhaust vs. outdoor air).

- Typical effectiveness values:
  - Plate heat exchangers: **50–70%**.
  - Rotary wheels: **60–80%**.
  - Heat pipes: **50–65%**.
  - Run-around coils: **40–60%**.

Temperature and Humidity

- Select a system suitable for the climate:
  - **Hot Climates**: Focus on systems with good cooling recovery efficiency.
  - **Cold Climates**: Prioritize systems with high heating recovery efficiency and frost protection.
- Ensure humidity control, especially in humid climates. Consider systems with **enthalpy recovery** to manage latent heat (e.g., rotary wheels or enthalpy plates).

Pressure Drop

- Keep the pressure drop across the heat exchanger low to reduce fan energy consumption:
  - Typical pressure drop: **50–150 Pa**.
  - Ensure fans are sized to overcome the added resistance.

Energy Recovery System Types

- Select the appropriate type based on the application:
  - **Plate Heat Exchangers**:
    - No moving parts; low maintenance.
    - Suitable for clean environments.
  - **Rotary Heat Exchangers (Wheels)**:
    - High effectiveness for sensible and latent heat.
    - May require periodic cleaning.
  - **Heat Pipes**:
    - Passive system; ideal for systems with aligned airstreams.
    - Compact and reliable.
  - **Run-Around Coil Systems**:
    - Suitable for systems with non-adjacent supply and exhaust ducts.
    - Flexible but less effective.

Frost Protection

- For cold climates, ensure the system has frost protection to prevent freezing:
  - Bypass dampers or preheaters.
  - Adjust control sequences to limit the exhaust air temperature drop.

Construction Material

- Choose corrosion-resistant materials, especially for humid or industrial exhaust applications:
  - **Aluminium**: Lightweight and corrosion-resistant for dry conditions.
  - **Stainless Steel**: Suitable for corrosive or moist environments.
  - **Epoxy Coatings**: For additional protection in harsh conditions.

Energy Efficiency

- Evaluate the system's **energy recovery ratio (ERR)**:
  - Higher ERR indicates a more efficient system.
- Target ERR >5 for high-efficiency systems.

Noise Considerations

- Ensure the heat recovery unit's noise levels comply with design criteria:
  - Office environments: **35–45 dB(A)**.
  - Industrial settings: Tolerances are higher but should not exceed limits.

6.15 Absorption HVAC Systems

Absorption cooling systems are the best suitable for solar thermal applications. The idea of the absorption cooling system was introduced in the 1700s and first patent about ammonia-water absorption system was registered by Ferdinand Carre in 1859. However, the commercial development of absorption chillers powered by solar energy began in the 1940s. Among solar air conditioning systems, solar absorption systems are most widely used as they include about 60% of installed systems in Europe while this amount is respectively 10% and 22% for adsorption and desiccant cooling systems. In China nearly all large-scale solar cooling systems during the last 20 years have been based on absorption system. Absorption chillers are classified as direct- or indirect-fired and as single, double and triple effect. The COP of the single effect absorption chillers are within the range of 0.6–0.8 while the double and triple effect absorption chillers have respectively COP of 1 and 1.4–1.6.

To select of the refrigerant and absorbent for the absorption chillers, several properties should be considered. The solubility of the refrigerant in the absorber must be as low as possible at the generator temperature and pressure and as high as possible at the absorber temperature and pressure. Additionally, the absorbent should be non-volatile to avoid moving the absorbent into the condenser and evaporator. The working fluid should be non-flammable, non-toxic and chemically stable. In terms of working fluids used mostly in commercially available units, this system is divided to LiBr-water and NH_3-water systems. In the LiBr-water system, LiBr is the absorbent and water is the refrigerant while in the NH_3-water system, water is the absorbent and NH_3 is the refrigerant. However, the COP of the LiBr-water system is higher than that of NH_3-water system. Moreover, due to low freezing point of NH_3 around −77 °C, the NH_3-water system can be used for low temperature applications. However, most solar absorption HVAC systems use LiBr-water. Among different types of absorption chillers, single-effect LiBr-water absorption chillers can be powered by ordinary flat plate or evacuated solar collectors due to low required operating temperature ranges. For this reason, the single-effect LiBr-water absorption chillers are most popular machine in solar cooling. Double-effect LiBr-water absorption chillers need the heat source temperature about 150 °C and thus the flat plate and vacuum solar collectors cannot meet their request; a parabolic trough solar collector is required. Absorption chiller can also be classified based on the method of providing heat as:

- Direct fired chiller in which the absorbent is reconcentrated from the heat input through the combustion of gas.
- Indirect fired chiller in which the absorbent is reconcentrated from heat input from steam or hot water.

The ANSI/ARI Standard 560 has specified the following rang of operating limits for the absorption chiller:

- Cooled-water temperature entering the absorber and condenser should be in the range of **26.7–32.2 °C.**
- Cooled-water mass flow rate entering the absorber and condenser limits are **0.05–0.11 L/s** per system refrigeration kW.
- Chilled-water flow rate in the evaporator should be in the range of **0.03–0.05 L/s per system refrigeration kW**.

In ASHRAE Refrigeration Handbook, the heat input rate has been quantified for the indirect fired single-effect LiBr-water absorption chiller as **1.43–1.54 kW/kW of refrigeration**. The heat input for indirect fired double effect LiBr-water absorption chiller is specified as **0.83 kW/kW of refrigeration**.

For lithium bromide absorption chillers, the chilled water temperature leaving the evaporator is normally between 4 °C and 15 °C. The lower limit is because water freezes at 0 °C. In addition, there are limitations for cooling water entering and leaving the LiBr absorption chiller. In the very low cooling water temperature the condenser pressure drops too low and unnecessary vapour velocities convey over solution to the refrigerant in the condenser.

The LiBr-water working fluid appears more promising for solar application, however LiBr crystallise out of solution when the concentration is higher and temperature is lower than at saturation and thus blocking pipes and passages. In addition, LiBr is corrosive to metals. Nowadays, absorption chillers are designed for operation away from crystallisation. LiBr-water absorption chillers operate under vacuum. In contrast, ammonia-water absorption chillers operate at pressure higher than atmospheric pressure. Although, ammonia-water absorption chillers can always operate as air-cooled chillers, LiBr-water absorption chillers are essentially water-cooled due to the crystallisation problem. In solar applications, water-cooled condenser are preferred compared with air-cooled condenser.

There are two ways to connect the cooling tower to the water-cooled condenser in the absorption chillers. Normally, absorber and condenser are connected in series to the cooling tower where the absorber is fed first as low temperature is more critical for absorber. In another way, the absorber and condenser are connected in parallel so that both receive the cooled-water at its lowest temperature. A larger cooling tower will be required when connecting the absorber and condenser in parallel.

Understand the Application

- **Heat Source**: Determine the availability of heat for the generator:
  - Waste heat (e.g., from turbines, boilers, or industrial processes).
  - Renewable energy (e.g., solar thermal systems).
  - Direct-fired systems (if no waste heat is available).
- **Cooling Purpose**: Comfort cooling, process cooling, or combined heat and power (CHP) applications.
- **System Integration**: Compatibility with cooling towers, pumps, and other system components.

Coefficient of Performance (COP)

- COP measures the chiller's efficiency
- Typical COP values:
  - Single-effect absorption chillers: **0.6–0.8**.
  - Double-effect absorption chillers: **1.0–1.2**.
- **Rule of Thumb**: Use single-effect chillers for lower-grade heat sources (~80–120 °C) and double-effect chillers for higher-grade heat sources (~140–200 °C).

Heat Source

- Ensure the heat source provides sufficient and consistent energy:
  - Waste heat recovery: Ensure availability aligns with cooling demand.
  - Direct-fired systems: Ensure fuel availability and cost-effectiveness.
- **Heat Input**: Approximate required heat input for single-effect chillers is **1.25–1.7 kW of heat/kW of cooling**.

Chilled Water Temperature

- Typical supply temperatures:
  - Comfort cooling: **7 °C supply, 12 °C return**.
  - Process cooling: Specific to application but often lower (e.g., **4–7 °C**).
- Ensure compatibility with the HVAC distribution system.

Heat Rejection System

- Absorption chillers require more heat rejection compared to electric chillers:
  - Cooling towers or other heat rejection systems must handle **1.5–2.5 times** the cooling capacity.
  - Ensure adequate water flow and cooling tower capacity.

Operating Conditions

- **Generator Temperature**:
  - Single-effect: **80–120 °C**.
  - Double-effect: **140–200 °C**.
- **Absorber and Condenser Temperatures**:
  - Typically require **25–35 °C** cooling water.
- Check the cooling water temperature to prevent crystallization in the absorber.

Refrigerant and Absorbent

- Ensure the chiller uses appropriate refrigerant and absorbent:
  - **Water-Lithium Bromide (H_2O-LiBr)**: Common for air conditioning.
  - **Ammonia-Water (NH_3-H_2O)**: Used for lower temperature applications (e.g., refrigeration).

Capacity Sizing

- Select a chiller with a capacity slightly above the calculated cooling load to accommodate peak demands (typically **10–15% oversizing**).

Working Principle
The absorption system comprises major components such as generator, condenser, evaporator, absorber, solution heat exchanger, solution pump as well as expansion valves and connection tubes. In the absorption cooling cycle, when the lithium bromide as the solid salt is dissolved in the water, the solution is produced. Since the lithium bromide is virtually non-volatile, then when the aqueous solution of the LiBr-water is boiled the pure water vapour is produced. In solar absorption refrigeration system, hot water provided from the solar collector is the heat source in the generator. There are two pressure levels in the cycle. The pressure in the generator and condenser is high while the pressure in the absorber and evaporator is low. However, both pressure are below the atmospheric pressure and thus the specific volume of the water vapour is large. Therefore, to avoid the vapour pressure drop, the diameter of the piping tubes should be big. To minimise these losses the generator and condenser are combined in one shell and the evaporator and absorber are located in one shell. In actual system, the generator pressure is greater than condenser pressure and the evaporator pressure is greater than absorber pressure. As the

vapour pressure of the solution in the absorber and at the absorber temperature is lower than that of the refrigerant at the evaporative temperature, the refrigerant is evaporated in the evaporator. It causes the water temperature inside the evaporator tubes to decrease and produces the chilled water.

In this system the chiller is pumped into deep vacuum by a vacuum pump which creates the necessary condition for boiling of water at a low temperature. The resulting refrigerant vapour is attracted to the absorber by a pressure difference between the solution and the refrigerant water and then absorbed by concentration of the solution, and therefore performing continuous boiling of the refrigerant water. The diluted solution in the absorber is then pumped into the generator via a solution pump and then concentrated within the concentrator when heating by hot water or direct fire. The refrigerant vapour generated at the same time is condensed into water in the condenser. The refrigerant water then enters the evaporator and then pumped to the spray through a spraying device with the use of a refrigerant pump. The transfer of heat from the air-handling unit water to the refrigerant causes the refrigerant water to vaporize again, producing the chilled water. Finally, the chilled water is pumped to the air-handling unit to produce cooled air for the building. On the other side, the concentrated solution from the generator directly enters the absorber and the cycle is repeated. The schematic of a single-effect vapour-absorption cooling system is shown in the following figure, however the number of generator in a double-effect absorption chiller is two.

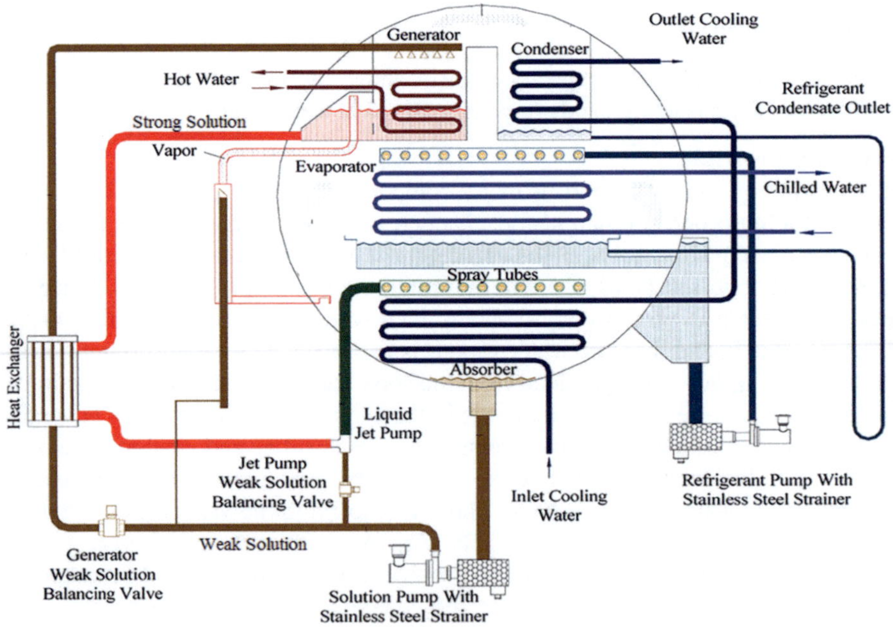

Operation cycle of the single-effect absorption chiller

6.16 Adsorption HVAC Systems

Adsorption chillers are a thermally driven cooling technology that uses solid adsorbents (e.g., silica gel, zeolites) to provide cooling. They are particularly suitable for applications with waste heat, renewable energy (solar or biomass), or where low electricity consumption is a priority. Proper sizing and selection are critical for optimal performance and efficiency. Below are the rules of thumb for adsorption chillers sizing and selection:

Understand the Application

- **Heat Source**: Identify the source and quality of thermal energy:
  - Waste heat from industrial processes, engines, or turbines.
  - Renewable heat sources like solar thermal or biomass.
  - Hot water, steam, or direct-fired systems.
- **Cooling Requirement**:
  - Comfort cooling, process cooling, or combined cooling and heating (CCHP) applications.

Coefficient of Performance (COP)

- Adsorption chillers typically have lower COPs than traditional chillers:
  - COP ranges from **0.5 to 0.7**, depending on the design and operating conditions.
- Select a chiller with a COP that matches the heat source quality and cooling demand.

Heat Source Temperature

- Adsorption chillers require moderate-temperature heat sources:
  - Typical range: **60–90 °C** (low-grade heat).
  - Ensure the heat source provides a stable supply within this range.

Chilled Water Temperature

- Standard chilled water temperature ranges:
  - Supply temperature: **7–10 °C**.
  - Return temperature: **12–15 °C**.
- Ensure compatibility with the HVAC distribution system.

Cooling Water Requirements

- Cooling water is used to reject heat from the adsorption process:
  - Typical cooling water supply temperature: **25–35 °C**.
  - Flow rate: **2.5–3.0 L/min/kW** of cooling capacity.
  - Cooling towers or other heat rejection systems must be sized to handle this load.

Heat Input

- Approximate required heat input is **1.5–2.0 kW of heat/kW of cooling**.
- Ensure sufficient thermal energy is available during peak cooling demand.

Adsorbent Material

- Select the appropriate adsorbent material based on application needs:
  - **Silica Gel**: Ideal for low-temperature applications (60–90 °C heat sources).
  - **Zeolites**: Suitable for higher temperatures and specific industrial needs.
  - Consider the material's adsorption/desorption efficiency and longevity.

Pumping and Distribution

- Ensure pumps are sized for both chilled and cooling water circuits:
  - Chilled water flow: **2.5–3.0 L/min/kW** of cooling capacity.
  - Cooling water flow: Sized to reject heat effectively.
- Use variable-speed pumps for improved energy efficiency.

Cost and Feasibility

- Initial cost: Higher than conventional chillers due to system complexity.
- Operating cost: Typically lower when waste heat or renewable energy is used.
- Payback period: Typically **5–10 years**, depending on energy savings and operating conditions.

Typical Applications

- Industrial facilities with waste heat recovery potential.
- Buildings with access to solar thermal or biomass energy.
- Locations with high electricity costs or limited grid capacity.

6.17 Underfloor Hydronic Heating Systems

Underfloor hydronic heating systems provide radiant heating by circulating hot water through pipes embedded in the floor. These systems are energy-efficient, comfortable, and suitable for a variety of applications.

An illustration of a Underfloor Hydronic Heating System

Types of Underfloor Hydronic Heating Systems

- **Embedded in Concrete (Wet System)**:
  - **Description**: Pipes are embedded in a concrete slab that acts as a thermal mass.
  - **Applications**: Ideal for new constructions or spaces with sufficient floor height to accommodate the slab.
  - **Advantages**: High thermal mass ensures consistent and long-lasting heat distribution.
  - **Disadvantages**: Slower heat-up time; not suitable for retrofit projects.
  - **Example**: Residential homes, warehouses, and commercial buildings.
- **Suspended Between Joists (Dry System)**:
  - **Description**: Pipes are installed between floor joists with heat transfer plates.
  - **Applications**: Suitable for retrofit projects or wooden floor constructions.
  - **Advantages**: Faster response time compared to wet systems.
  - **Disadvantages**: Less thermal mass; requires insulation for efficiency.
  - **Example**: Renovations, small offices, or residential buildings with existing floors.
- **Floating System**:
  - **Description**: Pipes are embedded in lightweight panels that sit above the subfloor.
  - **Applications**: Best for retrofit applications with minimal structural modifications.
  - **Advantages**: Lightweight and quick to install; minimal height addition.

  - **Disadvantages**: Lower thermal mass and potentially higher operational cost.
  - **Example**: Apartments, retrofits in older homes.

- **Staple-Up System**:
  - **Description**: Pipes are attached beneath an existing floor, with heat transfer plates.
  - **Applications**: Retrofit projects where adding floor height is not feasible.
  - **Advantages**: Avoids floor height changes.
  - **Disadvantages**: Requires effective insulation to prevent heat loss.
  - **Example**: Older homes and small office spaces.

Applications of Underfloor Hydronic Heating Systems

- **Residential**:
  - Provides even heat distribution and eliminates the need for visible radiators.
  - Common in bathrooms, living rooms, and bedrooms.
- **Commercial**:
  - Enhances comfort in spaces with high ceilings by ensuring uniform heat at occupant level.
  - Suitable for retail spaces, offices, and conference rooms.
- **Industrial**:
  - Effective in warehouses or factories where other heating systems may interfere with operations.
- **Healthcare**:
  - Ideal for hospitals and clinics due to reduced airborne dust and allergens.
- **Sports Facilities**:
  - Used in indoor sports halls to provide comfort without obstructing the play area.

Rules of Thumb for Sizing and Selection

Heat Output Requirements

- Determine the required heat output based on building heat loss: $Q = U \cdot A \cdot \Delta T$
  - Q: Heat load (W)
  - U: Overall heat transfer coefficient (W/m2 K)
  - A: Floor area (m2)
  - ΔT: Temperature difference between inside and outside.

Pipe Spacing

- Typical pipe spacing: 150–300 mm.
- Closer spacing (150 mm): For high heat load areas or low-temperature supply systems.
- Wider spacing (300 mm): For well-insulated spaces with lower heat loads.

Floor Material
- Use materials with good thermal conductivity (e.g., tile, concrete).
- Carpets and wood may reduce efficiency unless designed accordingly.

Water Temperature
- Supply water temperature: 35–55 °C (lower temperatures for well-insulated buildings).
- Maintain a 10 °C difference between supply and return temperatures.

Floor Thickness
- Concrete slab thickness (wet systems): 75–100 mm.
- Panel height (floating systems): 20–30 mm.

Insulation
- Provide adequate insulation beneath the pipes to prevent downward heat loss.
- Insulation thickness: 20–50 mm, depending on building and system type.

Zoning
- Divide the system into zones to optimize control and comfort.
- Typical zone size: 20–40 m^2.

Manifolds
- Use a manifold to distribute water evenly across zones.
- Include flow meters and control valves for individual zone adjustment.

Comparison of Systems

System	Thermal mass	Response time	Ease of installation	Applications
Wet system	High	Slow	Complex for retrofits	New constructions
Dry system	Low	Moderate	Easier than wet systems	Retrofits, wood constructions
Floating system	Very Low	Fast	Simple	Retrofits, apartments
Staple-up system	None	Fast	Moderate	Retrofit without floor lift

Examples

- **Residential Bathroom in a New Home**:
  - **System**: Wet system with pipes embedded in concrete.
  - **Reason**: High thermal mass provides consistent heat for tile floors.
- **Office Renovation in a Temperate Climate**:
  - **System**: Dry system with panels between joists.
  - **Reason**: Easy installation without disrupting existing structures.

- **Apartment Retrofit**:
  - **System**: Floating system with lightweight panels.
  - **Reason**: Minimal floor height change and quick installation.

6.18 Chilled Beam Systems

Chilled beams are energy-efficient HVAC systems that use ceiling-mounted units to provide cooling (and sometimes heating) via natural convection and radiant heat exchange. They are ideal for achieving precise temperature control with minimal air movement.

Types of Chilled Beam Systems

Passive Chilled Beams

- **Description**: Uses natural convection to cool or heat the space. Chilled water flows through pipes in the beam, cooling the surrounding air.
- **Applications**: Suitable for spaces with low cooling loads and good natural ventilation.
- **Advantages**:
  - Energy-efficient (no fans).
  - Quiet operation.
  - Simple design and maintenance.
- **Disadvantages**:
  - Requires separate ventilation systems for fresh air.
  - Limited cooling capacity compared to active systems.
- **Example**: Office buildings, libraries, or areas with minimal occupant density.

Active Chilled Beams

- **Description**: Combines natural convection with mechanically supplied primary air to enhance cooling/heating capacity.
- **Applications**: Used in spaces with moderate to high cooling loads.
- **Advantages**:
  - Higher cooling capacity than passive beams.
  - Provides both ventilation and thermal comfort.
- **Disadvantages**:
  - Slightly higher energy consumption due to primary air supply.
- **Example**: Hospitals, laboratories, and densely occupied offices.

Multi-Service Chilled Beams (MSCBs)

- **Description**: Integrates chilled beam functions with lighting, fire sprinklers, and sensors into a single unit.

- **Applications**: High-end commercial spaces or areas requiring reduced ceiling clutter.
- **Advantages**:
  - Aesthetic design.
  - Space-saving and multipurpose.
- **Disadvantages**:
  - Higher initial cost.
  - Complex maintenance.
- **Example**: Corporate boardrooms, luxury hotels, or conference centres.

Applications of Chilled Beam Systems

- **Office Spaces**:
  - Provide quiet and efficient cooling for open-plan or private offices.
  - Reduces energy consumption compared to traditional systems.
- **Healthcare Facilities**:
  - Used in patient rooms and operating theatres where noise and air quality are critical.
  - Minimizes air movement to reduce contamination risk.
- **Educational Institutions**:
  - Ideal for classrooms and lecture halls where quiet operation is essential.
- **Hotels**:
  - Offers a discreet and efficient cooling solution for guest rooms and suites.
- **Laboratories**:
  - Ensures precise temperature control and improved air quality in controlled environments.

Sizing and Selection

Cooling Capacity
- Passive chilled beams: **100–300 W/m^2**.
- Active chilled beams: **300–600 W/m^2**.
- Adjust based on space load requirements, insulation levels, and occupant density.

Water Temperature
- Supply chilled water temperature: **14–18 °C**.
- Avoid temperatures below dew point to prevent condensation.
- Heating water temperature (if applicable): **35–45 °C**.

Primary Air Supply (Active Systems Only)
- Air supply rate: **10–25 L/s/beam**, depending on space ventilation requirements.
- Use dehumidified air to prevent condensation issues.

Beam Placement and Coverage

- Mount beams at **2.5–3.5 m** above the floor.
- Space beams approximately **1.5–3.0 m apart** for even temperature distribution.
- Avoid obstructions like large pendant lights or decorative fixtures below beams.

Room Design and Layout

- Ensure sufficient ceiling space for installation.
- Avoid installation in spaces with high ceilings (above **4 m**) unless specifically designed for such conditions.

Humidity Control

- Maintain indoor relative humidity below **60%** to avoid condensation.
- Use a dedicated dehumidification system if the outdoor air is humid.

Maintenance and Accessibility

- Allow access for cleaning and maintenance.
- Ensure control valves and manifolds are easy to service.

Load Matching

- Match the chilled beam system's capacity with the room's peak load.
- Account for internal heat gains (lighting, equipment, occupants) and external loads (solar radiation, envelope losses).

Comparison of Chilled Beam Types

Type	Cooling capacity (W/m^2)	Ventilation	Noise level	Energy efficiency	Applications
Passive Chilled Beams	100–300	Requires separate system	Silent	Very High	Libraries, offices, residential
Active Chilled Beams	300–600	Integrated	Low	High	Hospitals, labs, dense offices
Multi-Service Chilled Beams	300–600	Integrated	Low	High	High-end commercial spaces

Examples

- **Office in a Temperate Climate**:
  - **System**: Active chilled beams.
  - **Reason**: Handles ventilation and moderate cooling loads efficiently.
- **Hospital Patient Room**:
  - **System**: Passive chilled beams.
  - **Reason**: Quiet operation with minimal air movement, ensuring comfort and hygiene.

- **Luxury Hotel Suite**:
  - **System**: Multi-service chilled beams.
  - **Reason**: Combines aesthetics with efficient cooling and integrated lighting.

Advantages of Chilled Beams

- Energy efficiency (low water pumping energy compared to air-based systems).
- Quiet operation, enhancing occupant comfort.
- Improved indoor air quality due to reduced air movement.

Disadvantages

- Initial cost can be high.
- Requires careful humidity control to prevent condensation.
- Limited cooling capacity in humid climates.

Chapter 7
HVAC Systems Applications and Comparison

7.1 Overview

This chapter provides an overview of HVAC system applications across various building types, including residential, commercial, and industrial settings. It explores the design considerations for each application, emphasizing energy efficiency, occupant comfort, and system reliability. The chapter also compares common HVAC systems such as split systems, variable refrigerant flow (VRF) systems, and centralized chilled water plants, highlighting their advantages, limitations, and suitability for different scenarios. It offers insights into selecting the right system based on factors like building size, climate, and operational requirements.

7.2 Comparison of HVAC Systems by Type, Applications, Benefits, Limitations, and Key Factors

Centralized HVAC Systems

System type	Description	Applications	Advantages	Disadvantages
Chilled water system	Uses a central chiller to cool water, which is circulated to air handling units (AHUs) or fan coil units (FCUs)	Large commercial buildings, hospitals, universities	High efficiency, scalable, suitable for large loads	High upfront cost, complex maintenance
Boiler system	Provides heating by circulating hot water or steam to radiators or air-handling units	Schools, industrial facilities, cold climates	Reliable heating, suitable for cold climates	Requires a fuel source, higher emissions

V. Vakiloroaya, *HVAC and Building Services Rules of Thumb*, Springer Tracts in Civil Engineering, https://doi.org/10.1007/978-3-032-10285-0_7

System type	Description	Applications	Advantages	Disadvantages
Variable air volume (VAV)	Central system varies airflow to different zones based on demand	Office buildings, commercial spaces	Energy efficient, good zoning control	Requires careful design and controls
Constant air volume (CAV)	Supplies a constant airflow while varying temperature to meet load requirements	Small commercial buildings	Simple design and operation	Less energy efficient than VAV systems

Decentralized HVAC Systems

System type	Description	Applications	Advantages	Disadvantages
Split system	Consists of an outdoor compressor and an indoor evaporator coil	Residences, small offices	Cost-effective, easy installation and maintenance	Limited to small areas, no centralized control
Packaged units	Self-contained units that provide heating and cooling	Small commercial buildings, rooftops	Compact, easy to install	Less efficient for large buildings
Variable refrigerant flow (VRF)	Uses refrigerant as the cooling/heating medium, distributed among multiple indoor units	Hotels, offices, retail spaces	High energy efficiency, excellent zoning	High initial cost, complex controls
Window units	Single-unit systems installed in windows or walls	Residences, small spaces	Affordable, easy to install	Limited capacity, less efficient

Combined HVAC Systems

System type	Description	Applications	Advantages	Disadvantages
Heat pumps	Transfers heat for both heating and cooling	Residences, moderate climates	Energy efficient, dual-purpose system	Less effective in extreme climates
Geothermal heat pumps	Uses the earth's stable temperature for heating and cooling	Residences, sustainable buildings	Highly efficient, low operating cost	High installation cost, site-dependent
Hybrid systems	Combines heat pumps with traditional systems (e.g., boilers)	Residential, cold climates	Flexibility, improved efficiency	Higher installation cost

Specialty HVAC Systems

System type	Description	Applications	Advantages	Disadvantages
Dedicated outdoor air systems (DOAS)	Provides fresh air ventilation separately from heating and cooling	Commercial buildings, hospitals	Excellent ventilation and air quality	Higher initial cost, requires integration
Radiant heating/ cooling	Uses heated or cooled surfaces (floors, ceilings) to condition spaces	Residences, hospitals, luxury buildings	Comfortable, energy efficient	Slower response time, higher initial cost
Underfloor air distribution (UFAD)	Distributes conditioned air through a raised floor system	Offices, data centres	Energy efficient, improved air quality	Requires specific floor design, higher cost
Displacement ventilation	Supplies air at low velocity and near floor level	Auditoriums, theatres	Energy savings, improved air quality	Requires careful design for effectiveness

Hybrid and Renewable Systems

System type	Description	Applications	Advantages	Disadvantages
Solar HVAC systems	Use solar energy for heating or cooling	Sustainable buildings, off-grid applications	Renewable, reduces operating costs	High installation cost, weather dependent
Thermal storage systems	Store energy (e.g., chilled water, ice) for use during peak demand	Large commercial buildings, district cooling	Reduces peak energy demand, operational savings	High initial cost, space requirements
Heat recovery systems	Recover waste heat for reuse	Industrial facilities, hospitals	Improves energy efficiency, reduces costs	Complexity, additional equipment required

Comparison by Key Factors

Factor	Centralized systems	Decentralized systems	Combined systems	Specialty systems
Initial cost	High	Moderate	Moderate to High	High
Operating cost	Moderate to Low	Moderate to High	Low to Moderate	Variable
Energy efficiency	High	Moderate	High	High (varies by system)
Flexibility	Low	High	Moderate	High
Scalability	High	Low	Moderate	Variable

Factor	Centralized systems	Decentralized systems	Combined systems	Specialty systems
Maintenance	Centralized, requires expertise	Easier, localized maintenance	Depends on components	Variable
Space requirements	Requires plant room	Minimal	Moderate	Varies (e.g., radiant systems need space for piping)

Selecting the Right System

Key considerations when choosing an HVAC system:

- **Building Size and Usage**: Centralized systems for large-scale buildings, decentralized for smaller spaces.
- **Energy Efficiency Goals**: Geothermal and hybrid systems for sustainability.
- **Budget**: Decentralized systems for cost-sensitive projects.
- **Climate**: Heat pumps for moderate climates; boilers for cold regions.
- **Zoning Requirements**: VRF or VAV systems for precise control.

7.3 HVAC System Analysis by Application and Building Size

Residential Buildings

Building type	HVAC systems	Reasons for suitability
Single-family homes	Split Systems, Packaged Units, Heat Pumps	Affordable, easy to install, provides heating and cooling, compact for limited space
Multi-family buildings	Split Systems, VRF Systems, Heat Pumps, Boilers (for centralized heating in cold climates)	Zoning options for individual units, central boilers for efficient heating in shared spaces
High-end residences	Geothermal Heat Pumps, Radiant Heating, VRF Systems	Energy efficiency, luxury comfort, integration with green building standards

Office Buildings

Building type	HVAC systems	Reasons for suitability
Small offices	Split Systems, Packaged Units	Cost-effective, suitable for smaller spaces with moderate loads
Medium to large offices	Variable Air Volume (VAV) Systems, VRF Systems, Chilled Water Systems	Energy efficiency, zoning flexibility, scalability for varied building sizes
High-rise office towers	Chilled Water Systems, VAV Systems, Thermal Storage Systems	Scalable, handles large cooling and heating loads, integration with energy-saving systems

Commercial Buildings

Building type	HVAC systems	Reasons for suitability
Retail stores	Packaged Units, Split Systems, DOAS (Dedicated Outdoor Air Systems)	Easy installation, ensures air quality and comfort for shoppers
Shopping malls	Chilled Water Systems, VRF Systems, DOAS	Large cooling capacity, zoning for diverse tenant needs, improved ventilation
Restaurants	Packaged Units, Split Systems, Heat Recovery Systems	Compact systems to handle high heat loads from kitchens, energy recovery from exhaust air

Industrial Facilities

Building type	HVAC systems	Reasons for suitability
Warehouses	Packaged Units, Rooftop Units, Radiant Heating	Cost-effective solutions for open spaces; radiant heating for localized comfort
Manufacturing plants	Chilled Water Systems, Heat Recovery Systems, Industrial Ventilation Systems	Handles high cooling loads, recovery of waste heat, and effective air circulation
Data centres	Precision Cooling Systems, Chilled Water Systems, CRAC (Computer Room Air Conditioning) Units	High precision cooling for sensitive equipment, redundancy for reliability

Healthcare Facilities

Building type	HVAC systems	Reasons for suitability
Hospitals	Chilled Water Systems, DOAS, Heat Recovery Systems	High cooling and heating demands, critical ventilation needs, energy recovery for efficiency
Clinics and labs	VRF Systems, Packaged Units, Chilled Water Systems	Precise temperature and humidity control, zoning for different spaces
Pharmaceutical facilities	Chilled Water Systems, Clean Room HVAC Systems	Critical temperature and humidity control, compliance with regulatory standards

Educational Institutions

Building type	HVAC systems	Reasons for suitability
Schools	Packaged Units, Split Systems, DOAS	Reliable and cost-effective systems with good ventilation for large occupancy
Colleges and universities	Chilled Water Systems, VAV Systems, Geothermal Heat Pumps	Scalability for large campuses, zoning for diverse spaces like labs, dorms, and classrooms

Hospitality and Recreation

Building type	HVAC systems	Reasons for suitability
Hotels	VRF Systems, Chilled Water Systems, DOAS	Zoning for individual rooms, high cooling and heating demands, superior indoor air quality
Theatres and auditoriums	Displacement Ventilation, Packaged Units	Quiet operation, effective air distribution in large spaces
Gyms and fitness centres	DOAS, Packaged Units	Enhanced ventilation and moisture control for high humidity spaces

High-Tech and Specialized Buildings

Building type	HVAC systems	Reasons for suitability
Laboratories	Chilled Water Systems, Dedicated Lab HVAC Systems	Precision temperature and humidity control, stringent air quality standards
Green buildings	Geothermal Heat Pumps, VRF Systems, Solar HVAC Systems	Compliance with sustainability goals, reduced carbon footprint
Airports	Chilled Water Systems, Heat Recovery Systems, DOAS	Large-scale cooling and heating capacity, high ventilation standards

Multi-Use and Mixed-Use Buildings

Building type	HVAC systems	Reasons for suitability
Mixed-use developments	Chilled Water Systems, VRF Systems, Thermal Storage Systems	Flexible zoning for diverse spaces, energy-efficient cooling and heating
District heating/ cooling	Centralized Chilled Water and Boiler Systems	Economies of scale, energy efficiency for urban developments

Comparison by Climate

Climate zone	Suitable HVAC systems	Reasons
Hot and humid	Chilled Water Systems, DOAS, Heat Recovery Systems	High cooling loads, dehumidification needs
Cold climates	Boilers, Radiant Heating, Geothermal Heat Pumps	Efficient heating solutions, leveraging renewable energy
Temperate climates	Heat Pumps, VRF Systems	Balanced heating and cooling needs, energy-efficient solutions
Arid climates	Evaporative Cooling, Chilled Water Systems	Cost-effective cooling, effective for dry air conditions

Selection Considerations

- **Building Size**: Large buildings require centralized systems; smaller buildings benefit from split or packaged units.
- **Occupancy Type**: High-occupancy spaces need robust ventilation (DOAS, VAV).
- **Energy Efficiency Goals**: Opt for geothermal systems, VRF, or thermal storage where efficiency is a priority.
- **Budget**: Packaged and split systems are affordable for smaller applications; centralized systems require higher initial investments but offer long-term savings.
- **Zoning Needs**: VRF and VAV systems excel at controlling conditions in different zones.

7.4 HVAC System Comparison for Power Consumption

Power consumption in HVAC systems depends on the type, efficiency, capacity, and operating conditions. Following is a detailed comparison of power consumption for common HVAC systems, along with examples.

Key Factors Affecting Power Consumption

System Efficiency: Metrics such as Seasonal Energy Efficiency Ratio (SEER), Energy Efficiency Ratio (EER), Coefficient of Performance (COP), and Integrated Part Load Value (IPLV).

Capacity: Larger systems consume more power but may operate more efficiently under full load.

Operating Hours: Continuous operation in commercial or industrial settings increases consumption.

Climate: Systems in extreme climates require more energy for heating or cooling.

Split Systems

Power Consumption

- **Residential**:
  - Small (1–5 Tons): 1–1.5 kW/ton.
- **Commercial**:
  - Larger units: 1–1.5 kW/ton.
- Efficiency: 13–20 SEER for residential; higher SEER indicates lower power usage.

Example

- A 3-ton residential split system operating 8 h/day consumes: 3 tons × 1.5 kW/ton × 8 h = 36 kWh/day

Packaged Rooftop Units (RTUs)

Power Consumption
- Small (5–20 Tons): 0.8–1.2 kW/ton.
- Medium (20–50 Tons): 0.7–1.1 kW/ton.
- Efficiency: 9–15 EER, depending on model and climate.

Example
- A 20-ton RTU operating 10 h/day at 1.2 kW/ton consumes: 20 tons × 1.2 kW/ton × 10 h = 240 kWh/day

Variable Refrigerant Flow (VRF) Systems

Power Consumption
- Highly efficient: 0.8–1.2 kW/ton.
- Multi-zone systems allow for part-load operation, reducing overall power consumption.
- Efficiency: COP of 3.5–4.5 (heating) and 3–4 (cooling).

Example
- A 10-ton VRF system operating 12 h/day consumes: 10 tons × 1 kW/ton × 12 h = 120 kWh/day

Chiller-Based Systems

Air-Cooled Chillers
- Power Consumption: 1.1–1.3 kW/ton.
- Efficiency: COP of 3–4.

Water-Cooled Chillers
- Power Consumption: 0.5–0.8 kW/ton.
- Efficiency: COP of 5–6.
- Includes cooling tower and pump energy (additional 0.3–0.5 kW/ton).

Example
- A 300-ton water-cooled chiller operating 24 h/day consumes: 300 tons × 0.7 kW/ton × 24 h = 5040 kWh/day

Heat Pumps

Air-Source Heat Pumps
- Power Consumption: 1.0–1.2 kW/ton for cooling; 0.8–1.0 kW/ton for heating.
- Efficiency: COP of 3–4.

Geothermal Heat Pumps
- Power Consumption: 0.6–0.8 kW/ton.
- Efficiency: COP of 4–5.

Example
- A 5-ton geothermal heat pump operating 10 h/day consumes: 5 tons × 0.7 kW/ton × 10 h = 35 kWh/day

Fan Coil Systems

Power Consumption

- Dependent on chilled/hot water source efficiency.
- Fan motor power: 0.2–1.0 kW per unit, depending on size and airflow rate.

Example

- A building with 20 fan coils operating 8 h/day at 0.5 kW/unit consumes: 20 units × 0.5 kW × 8 h = 80 kWh/day

Dedicated Outdoor Air Systems (DOAS)

Power Consumption

- Ventilation-only: 0.5–2.0 kW/ton.
- Higher for systems with dehumidification or energy recovery.

Example

- A DOAS serving 50 occupants and operating 12 h/day consumes: 2.0 kW × 12 h = 24 kWh/day

Hydronic Heating Systems

Power Consumption

- Boiler pumps: 0.3–1.0 kW/pump.
- Circulation pumps: 0.5–2.0 kW/system.

Example

- A hydronic heating system with 3 pumps operating 8 h/day consumes: 3 pumps × 0.8 kW × 8 h = 19.2 kWh/day

Comparison Table

System type	Power consumption	Efficiency (COP/SEER/EER)	Example application
Split systems	08–1.2 kW/ton	13–20 SEER	Homes, small offices
Packaged rooftop units	0.7–1.5 kW/ton	9–15 EER	Commercial spaces, malls
VRF systems	0.8–1.2 kW/ton	COP 3–4.5	Hotels, offices, luxury apartments
Air-cooled chillers	1.1–1.3 kW/ton	COP 3–4	Medium commercial, industrial
Water-cooled chillers	0.5–0.8 kW/ton	COP 5–6	Large commercial, hospitals
Air-source heat pumps	1.0–1.2 kW/ton	COP 3–4	Homes, moderate climates
Geothermal heat pumps	0.6–0.8 kW/ton	COP 4–5	Eco-conscious buildings, schools
Fan coil systems	0.2–1.0 kW/unit	Dependent on source	Hotels, high-rise buildings
DOAS	0.5–2.0 kW/ton	High for energy recovery	Hospitals, schools

Key Insights

Energy Efficiency: Water-cooled chillers and geothermal heat pumps are the most energy-efficient systems.
Operational Costs: Systems with higher upfront costs (e.g., VRF, geothermal) often have lower operational costs due to better efficiency.
Part-Load Operation: Systems like VRF and DOAS save energy by operating efficiently at partial loads.
Application-Specific Choices: Match system type to building size, climate, and operational demands.

7.5 HVAC System Comparison Based on Coefficient of Performance (COP)

The **Coefficient of Performance (COP)** is a key metric for evaluating the efficiency of HVAC systems. It is defined as the ratio of useful heating or cooling provided to the work input (energy consumption). The higher the COP, the more efficient the system. Below is a detailed comparison of various HVAC systems based on their COP, applications, advantages, and limitations.

Comparison Table

System type	Typical COP range	Factors affecting COP	Applications
Air-cooled chiller	2.5–4	Ambient temperature, part-load performance, compressor type, refrigerant used	Small to medium-sized commercial buildings and where water availability is limited
Water-cooled chiller	4.0–7.0	Cooling tower performance, condenser water temperature, compressor technology	Large commercial and industrial buildings
Air-source heat pump (ASHP)	2.5–4.0 (cooling)	Outdoor temperature, refrigerant type, defrost cycles	Residential and small commercial spaces
	3.0–4.5 (heating)		
Ground-source heat pump (GSHP)	4.0–6.0	Ground loop design, soil thermal properties, pump efficiency	High-efficiency residential and commercial applications, especially in cold climates
VRF (variable refrigerant flow)	3.0–5.0	Indoor and outdoor unit efficiency, zoning, refrigerant distribution efficiency	Commercial buildings with diverse load profiles
Packaged rooftop units (RTU)	2.5–3.5	Ambient conditions, maintenance, economizer use	Retail spaces, small offices, and industrial facilities

System type	Typical COP range	Factors affecting COP	Applications
Absorption chillers	0.6–1.2 (single-effect)	Heat source temperature, working fluid, system configuration	Where waste heat or renewable heat is available (e.g., cogeneration plants)
	1.0–1.5 (double-effect)		
Evaporative cooling systems	10.0–30.0 (effective COP)	Ambient air humidity, air temperature, and water quality	Arid or semi-arid regions
Thermal storage systems	N/A (Efficiency-Based)	Charging efficiency, storage medium properties, and system integration	Peak shaving and demand management for large facilities

Detailed Analysis

Air-Cooled vs. Water-Cooled Chillers

- **COP**:
  - Air-cooled chillers have a lower COP due to the limited heat rejection capability of air.
  - Water-cooled chillers benefit from more efficient heat rejection through cooling towers.
- **Applications**:
  - Air-cooled systems are preferred where water is scarce or maintenance resources are limited.
  - Water-cooled systems are suitable for large-scale applications with high cooling loads.

Heat Pumps (Air-Source vs. Ground-Source)

- **COP**:
  - GSHPs generally have a higher COP than ASHPs because ground temperatures are more stable than air temperatures.
- **Applications**:
  - ASHPs are cost-effective for moderate climates.
  - GSHPs are ideal for cold climates where ASHP efficiency drops significantly in winter.

VRF Systems

- **COP**:
  - High COP, especially when operating at partial loads, due to inverter-driven compressors and precise refrigerant control.
- **Applications**:
  - Suitable for buildings with varying thermal zones, such as offices, hotels, and multi-family residences.

Absorption Chillers

- **COP**:
  - Single-effect systems have a low COP, making them less efficient compared to mechanical systems.
  - Double-effect systems improve efficiency but require higher temperature heat sources.
- **Applications**:
  - Ideal for facilities with excess heat (e.g., from industrial processes) or renewable heat sources.

Evaporative Cooling Systems

- **COP**:
  - Exceptionally high effective COP but limited to dry climates due to reliance on low ambient humidity.
- **Applications**:
  - Used in warehouses, factories, and buildings in arid regions.

Packaged Rooftop Units (RTUs)

- **COP**:
  - Moderate COP compared to other systems due to simpler construction and operation.
- **Applications**:
  - Cost-effective solution for small and medium-sized buildings.

Thermal Storage Systems

- **COP**:
  - Indirectly improve system performance by shifting loads and reducing peak demand.
- **Applications**:
  - Used in large facilities requiring load management and operational cost savings.

Factors Influencing COP

- **Ambient Conditions**: Air-cooled systems lose efficiency in high temperatures, while water-cooled systems maintain consistent performance.
- **Load Conditions**: Systems like VRF and chillers perform better at part-load due to modulating components.
- **Maintenance**: Poorly maintained systems suffer from degraded performance and lower COP.
- **Design and Integration**: Proper system design and integration into building controls can maximize COP.

Recommendations Based on COP

Building type	Recommended system	Reasoning
Residential homes	ASHP, GSHP	High COP and cost-effectiveness for heating and cooling
Large office buildings	Water-Cooled Chillers, VRF	High efficiency for large loads, zoning flexibility
Industrial facilities	Absorption Chillers, Water-Cooled Systems	Utilize waste heat and high-efficiency cooling
Retail spaces	Packaged RTU, VRF	Moderate COP, lower initial cost, flexibility
Data centers	Water-Cooled Chillers, Thermal Storage Systems	High COP systems for consistent cooling loads and peak load management
Buildings in dry climates	Evaporative Cooling Systems	Extremely high COP in low-humidity environments

7.6 HVAC Systems Analysis for Various Climates

Selecting the right HVAC system depends heavily on climatic conditions, building type, usage, and energy efficiency goals. Below is a detailed analysis and comparison of HVAC systems suited for various climates, with examples to guide system selection.

Climatic Zones and HVAC System Suitability

Hot and Dry Climate

- **Characteristics**: High temperatures, low humidity, large diurnal temperature variation.
- **Key HVAC Needs**: Cooling, energy efficiency, and some humidification.

Best Systems

- **Evaporative Coolers**:
  - Advantages: Low energy consumption, natural humidification.
  - Disadvantages: Limited cooling capacity, ineffective in high-humidity conditions.
  - Example: Ideal for warehouses or small residential buildings in deserts.
- **Split or Packaged DX Systems**:
  - Advantages: Effective cooling, precise control.
  - Disadvantages: Higher operational costs compared to evaporative cooling.
  - Example: Offices or residential buildings in hot, dry regions.
- **Variable Refrigerant Flow (VRF) Systems**:
  - Advantages: High efficiency, precise control, scalable.
  - Example: Multi-story office buildings in desert climates.

Hot and Humid Climate

- **Characteristics**: High temperatures, high humidity levels.
- **Key HVAC Needs**: Dehumidification, efficient cooling, mold prevention.

Best Systems

- **Chilled Water Systems**:
  - Advantages: Superior cooling, effective dehumidification, large capacity.
  - Disadvantages: High initial costs, requires maintenance.
  - Example: Hotels or shopping malls in tropical areas (e.g., Miami, Florida).
- **Split and Multi-Split Systems**:
  - Advantages: Easy installation, good dehumidification.
  - Disadvantages: May not be efficient for large spaces.
  - Example: Residential apartments in coastal cities.
- **Heat Pumps with Dehumidification Features**:
  - Advantages: Cooling and dehumidification in a single unit.
  - Example: Resorts or villas in humid coastal zones.

Cold Climate

- **Characteristics**: Low temperatures, occasional high winds, snow/ice.
- **Key HVAC Needs**: Heating efficiency, freeze protection, minimal cooling.

Best Systems

- **Hydronic Heating Systems**:
  - Advantages: Uniform heat distribution, efficient for long-term use.
  - Example: Multi-family residential buildings in northern regions (e.g., Chicago, Illinois).
- **Air-to-Air Heat Pumps (Cold Climate Models)**:
  - Advantages: High heating efficiency, can operate in subzero temperatures.
  - Example: Modern residential homes in cold climates.
- **Geothermal (Ground-Coupled) Heat Pumps**:
  - Advantages: Extremely efficient, provides both heating and cooling.
  - Disadvantages: High installation cost.
  - Example: Net-zero homes in cold regions.

Temperate Climate

- **Characteristics**: Moderate temperatures, seasonal variations.
- **Key HVAC Needs**: Versatility to handle both heating and cooling efficiently.

Best Systems

- **VRF Systems**:
  - Advantages: Efficient, heating and cooling in one system.
  - Example: Office buildings or retail spaces in temperate climates.

- **Packaged Rooftop Units**:
  - Advantages: Compact, cost-effective for medium-sized spaces.
  - Example: Commercial buildings in urban areas.
- **Air-to-Air Heat Pumps**:
  - Advantages: Good efficiency for both heating and cooling.
  - Example: Suburban homes in regions with moderate seasonal variation.

Cold and Humid Climate

- **Characteristics**: Low temperatures, high humidity, risk of condensation and mold.
- **Key HVAC Needs**: Effective heating, humidity control, good insulation.

Best Systems

- **Dehumidification Heat Pumps**:
  - Advantages: Simultaneous heating and humidity control.
  - Example: Public buildings in coastal northern climates.
- **Hydronic Radiant Heating**:
  - Advantages: Prevents cold drafts, avoids air movement that could lead to condensation.
  - Example: Museums or libraries in cold and humid climates.
- **Geothermal Heat Pumps**:
  - Advantages: Reliable in cold and damp conditions.
  - Example: Environmentally conscious developments.

Hot and Arid Climate

- **Characteristics**: Extremely high daytime temperatures, cool nights, minimal humidity.
- **Key HVAC Needs**: High cooling capacity, energy efficiency.

Best Systems

- **Evaporative Cooling Systems**:
  - Advantages: Utilizes dry air for effective cooling.
  - Example: Industrial buildings in desert regions (e.g., Middle east).
- **VRF with Heat Recovery**:
  - Advantages: Efficient in variable load conditions.
  - Example: Large office complexes in arid climates.

Comparison of HVAC Systems

HVAC system	Best climate	Advantages	Disadvantages	Example
Evaporative cooling	Hot and Dry	Low energy, natural humidification	Ineffective in humid climates	Desert regions, warehouses
Split/ multi-split	Hot and Humid, Temperate	Easy to install, cost-effective	Limited capacity for large buildings	Residential homes, offices
VRF systems	All climates	High efficiency, precise control	High initial cost	Multi-story buildings
Chilled water systems	Hot and Humid	Superior cooling and dehumidification	Expensive to install and maintain	Malls, hospitals
Geothermal heat pumps	Cold, Temperate, Mixed	High efficiency, renewable energy	High installation cost	Net-zero homes, schools
Hydronic heating	Cold and Humid, Cold	Uniform heat distribution, efficient for heating-only scenarios	Installation complexity	Apartments, libraries
Packaged rooftop units	Temperate	Compact, versatile	Less efficient for large spaces	Retail, medium offices
Jet nozzles	Large Spaces	Long-distance air throw, effective for large spaces	Requires precise placement and higher initial cost	Auditoriums, stadiums

Examples of System Selection

- **Tropical Coastal Hotel (Hot and Humid)**:
  - **System**: Chilled water system with dehumidification coils.
  - **Reason**: Effective cooling and humidity control for guest comfort.
- **Suburban Home in a Temperate Region**:
  - **System**: Air-to-air heat pump.
  - **Reason**: Handles both seasonal heating and cooling efficiently.
- **Net-Zero Office in a Cold Climate**:
  - **System**: Geothermal heat pump.
  - **Reason**: Sustainable energy source with long-term efficiency.
- **Desert Industrial Facility (Hot and Dry)**:
  - **System**: Evaporative cooler.
  - **Reason**: Low operational cost and high cooling effectiveness.

7.7 Comparison of Air-Cooled Vs. Water-Cooled HVAC Systems

Comparison Table

Factor	Air-cooled systems	Water-cooled systems
Operating cost	– Higher energy consumption in large systems due to lower efficiency – No water costs	– Lower energy consumption because of better heat transfer efficiency – Additional water cost for cooling tower operation
Required space	– Smaller footprint for equipment – Does not require cooling towers – Suitable for rooftop installations	– Larger overall footprint due to cooling towers and water treatment equipment
Maintenance cost	– Lower maintenance costs – Simpler systems with fewer components	– Higher maintenance costs due to water treatment (anti-scaling, anti-corrosion) and cooling tower upkeep
Efficiency	– Typically lower efficiency in hot climates – COP ranges **3.0–4.0**	– Higher efficiency, especially in large systems – COP ranges **4.5–7.0**
Noise	– Generates more noise from fans and compressors – Noise can be mitigated with proper design	– Generally quieter as compressors and heat rejection occur away from occupied areas
Lifecycle cost	– Lower initial investment – Shorter lifespan (**15–20 years**)	– Higher initial investment – Longer lifespan (**20–30 years**)
Water use	– No water required for operation	– Requires significant water for cooling and blowdown cycles
Environmental impact	– Less environmentally friendly due to higher energy use	– Potential water wastage and chemical use but more energy-efficient
Temperature sensitivity	– Less effective in high ambient temperatures	– Performance is consistent regardless of ambient air temperature
Flexibility	– Easier to install in small-scale applications or retrofits	– Best suited for large-scale applications where operational efficiency outweighs installation complexity

Advantages and Disadvantages

Air-Cooled Systems

- **Advantages:**
- Lower upfront cost and simpler installation.
- Requires no water treatment or cooling towers.
- Ideal for small- to medium-sized applications or where water is scarce.
- Easy to maintain with minimal specialized expertise.
- Suitable for rooftop or outdoor installations.

Disadvantages:

- Lower efficiency, especially in hotter climates.
- Higher energy consumption leading to higher operating costs.
- Noisier than water-cooled systems.
- Limited capacity for large-scale applications.

Water-Cooled Systems

- **Advantages:**
- Superior efficiency due to the high heat transfer capacity of water.
- Lower energy consumption for large-scale systems.
- Quieter operation as cooling towers and compressors are away from the occupied spaces.
- Longer equipment lifespan when properly maintained.

Disadvantages:

- Higher upfront cost due to cooling towers and water treatment systems.
- Requires a reliable water source and periodic water treatment.
- Larger installation space and complexity.
- Risk of Legionella bacteria in poorly maintained cooling towers.

Final Analysis

Operating Cost:
Water-cooled systems win for large-scale applications due to lower energy use. Air-cooled systems might be more economical for small installations.
Required Space:
Air-cooled systems are more compact and better for tight spaces or rooftops, whereas water-cooled systems need dedicated mechanical rooms and space for cooling towers.
Efficiency:
Water-cooled systems are more efficient, especially in climates with high ambient temperatures.
Maintenance Cost:
Air-cooled systems are simpler and cheaper to maintain, whereas water-cooled systems incur costs for water treatment and cooling tower maintenance.
Lifecycle:
Water-cooled systems have a longer lifespan, which might offset the higher initial investment over time.
Noise:
Water-cooled systems are generally quieter, especially important for noise-sensitive environments.
Environmental Impact:
Water-cooled systems are more energy-efficient, but their water usage can pose sustainability concerns in water-scarce regions.

Which Is the Winner?

- **Air-Cooled Systems** are better for:
  - Small- to medium-scale applications.
  - Areas with limited water resources or strict water regulations.
  - Retrofits and rooftop installations.
- **Water-Cooled Systems** are better for:
  - Large-scale facilities (hospitals, malls, data centres) where efficiency outweighs initial costs.
  - Locations with moderate water availability and lower water costs.
  - Long-term investments where lifecycle cost and performance matter.

Example Scenarios

- **Office Building in a Dense Urban Area**:
  - Air-cooled systems are suitable due to space constraints and easier maintenance.
- **Hospital or Data Centre**:
  - Water-cooled systems are ideal because of high cooling loads and the need for efficiency and reliability.
- **Small Retail Store in Hot Climate**:
  - Air-cooled systems offer practicality without the complexity of water-cooled systems.

In summary, **water-cooled systems** are generally more advantageous for large-scale and high-efficiency applications, while **air-cooled systems** are preferred for simpler, smaller, or space-constrained projects. The final choice depends on the specific requirements, budget, and constraints of the project.

7.8 Comparison Between All-Air, All-Water, and Air-Water HVAC Systems

HVAC systems are categorized into **all-air systems**, **all-water systems**, and **air-water systems**, depending on the medium used to transfer heating or cooling to the conditioned space. Below is a comprehensive comparison, along with examples for each system.

All-Air Systems

Definition

- Entirely air-based systems where conditioned air is distributed to the spaces through ductwork.

Key Features

- **Heating/Cooling Medium**: Air
- **Ductwork**: Extensive network for air distribution.
- **Control**: Centralized or decentralized control of temperature and air quality.

Examples

- **Variable Air Volume (VAV) Systems**:
  - Adjust air volume to match zone loads.
  - Common in commercial offices and hospitals.
- **Constant Air Volume (CAV) Systems**:
  - Maintain constant airflow; adjust air temperature.
  - Used in spaces requiring steady conditions, such as theaters.
- **Single-Duct or Dual-Duct Systems**:
  - Single-duct: Uniform temperature air distributed.
  - Dual-duct: Separate hot and cold air ducts for better flexibility.

Advantages

- Complete control over temperature, humidity, and air quality.
- Centralized filtration and ventilation.
- Suitable for spaces requiring precise air quality (e.g., hospitals).

Disadvantages

- High initial and operational costs due to ductwork.
- Requires significant space for ducts and mechanical rooms.
- Energy-intensive for large-scale applications.

All-Water Systems

Definition

- Use water as the heat transfer medium to deliver heating or cooling to the spaces.

Key Features

- **Heating/Cooling Medium**: Water
- **Distribution**: Piping system instead of ductwork.
- **Terminal Units**: Fan coil units, radiators, or chilled beams in the conditioned space.

Examples

- **Fan Coil Unit (FCU) Systems**:
  - Water is supplied to fan coil units, and air is circulated locally.
  - Common in hotels and multi-zone residential buildings.

- **Hydronic Radiators**:
  - Deliver heating through convection or radiation.
  - Common in colder climates for residential heating.

Advantages
- Requires less space than all-air systems (no large ducts).
- More energy-efficient for large buildings.
- Reduced noise since water distribution is silent.

Disadvantages
- Limited humidity control (requires a separate dehumidification system).
- Requires multiple units in large spaces.
- Maintenance of distributed terminal units can be challenging.

Air-Water Systems

Definition
- Combine air and water as the mediums for heating, cooling, and ventilation.

Key Features
- **Heating/Cooling Mediums**: Both air and water.
- **Distribution**: Smaller ducts (for air) and piping (for water).
- **Control**: Balances the advantages of both systems.

Examples
- **Chilled Beam Systems**:
  - Active or passive chilled beams use water for sensible cooling and ventilation air for latent load management.
  - Common in modern offices and educational institutions.
- **Induction Units**:
  - Use high-pressure air to induce circulation and mix with local air.
  - Found in older commercial buildings.
- **Hybrid Systems**:
  - Combine a central air handler with localized water-based units (e.g., radiant floors or panels).

Advantages
- Efficient for large buildings by combining air and water benefits.
- Smaller ducts and pipes reduce space requirements.
- Improved humidity and temperature control compared to all-water systems.

Disadvantages
- Higher initial cost due to dual systems.
- Complex installation and maintenance.
- Requires precise balancing for optimal performance.

Comparison Table

Feature	All-Air systems	All-Water systems	Air-Water systems
Medium	Air	Water	Air + Water
Distribution	Extensive ductwork	Piping system	Small ducts + pipes
Energy efficiency	Moderate (high fan energy consumption)	High (water has better heat transfer efficiency)	Moderate to high
Space requirements	Large (ducts require significant space)	Small (pipes require minimal space)	Medium
Control	Centralized	Decentralized (terminal units)	Balanced
Humidity control	Excellent	Limited	Good
Noise levels	Moderate to high (fan noise)	Low	Low
Maintenance	Centralized equipment	Distributed terminal units	More complex
Examples	VAV, CAV, Dual-Duct Systems	Fan Coils, Radiators, Hydronic Systems	Chilled Beams, Induction Units

Applications

All-Air Systems

- Suitable for spaces requiring tight control over air quality, such as:
  - Hospitals
  - Theatres
  - Data centres

All-Water Systems

- Ideal for buildings with space constraints and moderate load diversity, such as:
  - Hotels
  - Apartments
  - Office buildings

Air-Water Systems

- Well-suited for modern, energy-efficient buildings requiring a balance of air quality and space efficiency, such as:
- Green buildings
- Universities
- Laboratories

7.9 Comparison Between 2-Pipe, 3-Pipe, and 4-Pipe Fan Coil Systems

Fan coil systems are widely used for heating and cooling in buildings. The selection of a **2-pipe**, **3-pipe**, or **4-pipe** system depends on the building's needs for simultaneous heating and cooling, operational flexibility, energy efficiency, and cost. Below is a comparison of these systems with examples.

System Descriptions

2-Pipe Fan Coil System
- **Design**:
- Uses two pipes: one for supply and one for return.
  - Can operate in either cooling mode or heating mode, but not simultaneously.
  - The mode is determined centrally and applies to all connected units.
- **Applications**:
  - Suitable for buildings with uniform thermal loads or consistent seasonal usage, such as hotels or small offices.

3-Pipe Fan Coil System
- **Design**:
  - Uses three pipes: a common return pipe and separate supply pipes for heating and cooling.
  - Allows simultaneous heating and cooling, but less efficient due to mixing of return water streams.
- **Applications**:
  - Used in buildings where simultaneous heating and cooling are occasionally needed, such as spaces with diverse exposure to sunlight.

4-Pipe Fan Coil System
- **Design**:
  - Uses four pipes: two supply pipes (hot and chilled water) and two return pipes.
  - Provides true simultaneous heating and cooling without mixing.
- **Applications**:
  - Ideal for high-performance or mixed-use buildings, such as hospitals, hotels, or large office spaces with diverse thermal zones.

Key Features Comparison

Feature	2-pipe system	3-pipe system	4-pipe system
Number of pipes	2 (1 supply, 1 return)	3 (1 hot supply, 1 cold supply, 1 common return)	4 (2 supplies, 2 returns)
Simultaneous heating & cooling	No	Yes (but less efficient)	Yes (efficient and independent)
Flexibility	Low (entire system in one mode)	Moderate	High (independent operation per zone)
Energy efficiency	High when in single-mode operation	Lower due to mixed water streams	High (no mixing of water streams)
Cost	Lowest (fewer pipes and simpler controls)	Moderate (extra pipe and mixing losses)	Highest (more pipes and complex controls)
Control complexity	Simple	Moderate	Complex
Thermal comfort	Moderate (limited by single mode)	Better (limited by mixing losses)	Excellent (true zone control)
Installation space	Minimal	Moderate	Requires more space for piping

Advantages and Disadvantages

2-Pipe System

- **Advantages**:
  - Simple design and easy to install.
  - Lower installation and maintenance costs.
  - Energy-efficient for buildings with uniform thermal requirements.
- **Disadvantages**:
  - Cannot provide simultaneous heating and cooling.
  - Less suitable for buildings with varying thermal loads across zones.
  - Requires seasonal switchover, causing potential downtime.

3-Pipe System

- **Advantages**:
  - Can provide simultaneous heating and cooling.
  - Suitable for moderate mixed-use buildings.
- **Disadvantages**:
  - Lower energy efficiency due to mixing of hot and cold water in the return line.
  - Higher operational costs compared to 2-pipe systems.

4-Pipe System

- **Advantages**:
  - Provides independent heating and cooling to all zones.
  - Excellent for large buildings with diverse thermal loads.
  - Maximizes occupant comfort and operational flexibility.
- **Disadvantages**:
  - Highest installation and maintenance costs.
  - Requires more space for piping infrastructure.
  - More complex controls and equipment.

Example Applications

2-Pipe System

- **Hotels**: Where all rooms typically require the same mode (heating in winter, cooling in summer).
- **Small Offices**: With consistent thermal loads across zones.

3-Pipe System

- **Mixed-Use Retail Spaces**: Where some areas may require cooling (e.g., interior zones) while others need heating (e.g., perimeter zones).
- **School Buildings**: Where classrooms and auditoriums may have differing temperature requirements.

4-Pipe System

- **Hospitals**: Operating rooms, patient rooms, and offices may have simultaneous heating and cooling demands.
- **Luxury Hotels**: Suites with high expectations for individualized thermal comfort.
- **Corporate Offices**: Large buildings with varying solar exposures and occupant densities.

Summary of Selection Criteria

Criterion	Best system	Reason
Budget constraints	2-Pipe System	Lower installation and operational costs
Thermal comfort	4-Pipe System	True independent zone control
Energy efficiency	2-Pipe or 4-Pipe System	Avoids mixing of water streams
Building type	Depends: Hotels (2-pipe), Hospitals (4-pipe), Mixed Use (3-pipe)	Matches application needs
Space availability	2-Pipe System	Requires the least piping infrastructure

7.10 Comparison of Compressor-Based Chillers, Absorption Chillers, and Adsorption Chillers

Feature	Compressor-based chillers	Absorption chillers	Adsorption chillers
Operating principle	Use mechanical compression (via a compressor) to circulate refrigerant for cooling	Use heat (e.g., from steam, hot water, or natural gas) to drive a chemical reaction for cooling	Use heat to drive a physical adsorption/desorption cycle on a solid sorbent
Energy source	Electricity (mechanical energy to run the compressor)	Heat energy (low-grade waste heat or direct heat source) with minimal electricity	Low-grade heat energy (e.g., waste heat, solar heat) with minimal electricity
Typical working fluids	Refrigerants (e.g., R-134a, R-410a, R-1234yf)	Lithium bromide-water ($LiBr$-H_2O) or ammonia-water (NH_3-H_2O)	Water (as refrigerant) and silica gel or zeolite (as adsorbent)
Efficiency (COP)	High (COP ~3–6 for electric chillers)	Moderate (COP ~0.6–1.2)	Low to moderate (COP ~0.5–0.8)
Cooling capacity range	Wide range, from small (few kW) to large (several MW)	Medium to large systems (typically >100 kW)	Small to medium systems (typically 5–500 kW)
Start-up time	Quick start-up (seconds to minutes)	Slow start-up (minutes to hours)	Moderate start-up time (minutes)
Operating conditions	Performs well across varying load conditions	Most efficient with steady loads; less flexible	Effective for steady loads; may struggle with fluctuating loads
Maintenance requirements	Regular maintenance (e.g., compressor, refrigerant leaks)	Requires maintenance of heat exchangers and LiBr solution. Risk of crystallization	Minimal maintenance; no moving parts in the main cycle
Environmental impact	Dependent on refrigerant type; newer systems use low-GWP refrigerants	Uses eco-friendly refrigerants, but LiBr is corrosive and requires careful handling	Environmentally friendly; uses water as refrigerant
Installation cost	Moderate to high, depending on capacity	High initial cost due to large system size and complexity	High initial cost, but simpler design than absorption chillers
Operating cost	Higher due to electricity use, unless using renewable electricity	Lower if waste heat or low-cost heat is available	Very low if powered by waste heat or solar thermal energy
Applications	Universal: residential, commercial, industrial	Best for industrial or large-scale buildings with access to waste heat	Niche applications: waste heat recovery, solar cooling, remote locations
Size and weight	Compact and lightweight relative to cooling capacity	Large and heavy due to heat exchangers and solution	Moderate; smaller than absorption chillers

Feature	Compressor-based chillers	Absorption chillers	Adsorption chillers
Noise levels	Can be noisy due to the compressor	Very quiet; no mechanical compression	Very quiet; no mechanical compression
Lifespan	Long (15–25 years)	Long (20–25 years) if well-maintained	Long (20 + years) with minimal degradation

Practical Examples

- **Compressor-Based Chillers:**
  - **Application:** Office buildings, malls, hospitals, and industrial cooling.
  - **Example:** A 500-ton centrifugal chiller for central air conditioning in a large hotel.
- **Absorption Chillers:**
  - **Application:** Industrial plants with waste heat (e.g., chemical plants, power plants) or district cooling systems.
  - **Example:** A 300-ton LiBr absorption chiller using waste heat from a cogeneration plant.
- **Adsorption Chillers:**
  - **Application:** Buildings with solar thermal systems, remote cooling needs, or industries with low-grade waste heat.
  - **Example:** A 50-kW adsorption chiller powered by solar collectors in an off-grid eco-resort.

7.11 Comparison of Direct Expansion (DX) HVAC Systems and Evaporative Coolers

Feature	Direct expansion (DX) HVAC systems	Evaporative coolers
Operating principle	Use refrigerant to absorb heat from indoor air and reject it outdoors using a vapour-compression cycle	Use water evaporation to cool and humidify the air
Energy source	Electricity (compressors, fans, and controls)	Electricity (fans, water pump); no compressors required
Efficiency (energy use)	Moderately efficient; COP ranges from 2.5 to 6, depending on system type	Highly energy-efficient; consumes only 10–20% of the energy of DX systems
Cooling capacity	Suitable for a wide range of capacities, from small units to large commercial systems	Limited cooling capacity; best for small to medium applications
Humidity control	Can dehumidify the air, maintaining indoor comfort in humid climates	Increases humidity, unsuitable for already humid environments

Feature	Direct expansion (DX) HVAC systems	Evaporative coolers
Air quality	Filters dust and particles; some systems include HEPA or UV filters	Provides fresh air; filters dust, but less effective than advanced air filtration
Climatic suitability	Effective in all climates, including hot and humid conditions	Effective only in hot and dry climates; struggles in high humidity
Initial cost	High, especially for VRF or rooftop packaged systems	Low to moderate initial cost
Operating cost	High, due to electricity consumption of compressors and fans	Very low, due to minimal electricity usage
Maintenance	Moderate to high; requires regular refrigerant checks, filter cleaning, and compressor servicing	Low; mainly involves cleaning filters and replacing water
Environmental impact	Refrigerants may contribute to global warming if not managed properly; newer systems use low-GWP refrigerants	Eco-friendly; no refrigerants involved, only water
Lifespan	Long (15–20 years with proper maintenance)	Shorter (10–15 years), depending on water quality and usage
Noise levels	Moderate to low, depending on design and application	Low; mainly fan noise
Scalability	Highly scalable; VRF systems can manage multiple zones independently	Limited scalability; best for single or small-zone applications

Comparison of Climatic Suitability

Climatic factor	DX HVAC systems	Evaporative coolers
Hot and dry climates	Effective, though higher energy consumption may be a concern	Extremely effective; air is cooled significantly due to high evaporation rates
Hot and humid climates	Very effective; dehumidifies air to maintain comfort	Not suitable; increases humidity, leading to discomfort
Moderate climates (temperate)	Effective, though energy efficiency depends on proper sizing	Suitable in moderately dry conditions, but less efficient in humid seasons
Cold climates	Not typically needed; may function in heating mode (heat pumps)	Ineffective; cooling is unnecessary, and adding humidity may not be desirable

Practical Examples

- **DX HVAC Systems:**
  - **Application:** Office buildings, retail stores, data centres, and hospitals in any climate.
  - **Example:** A rooftop packaged unit providing consistent cooling in a mixed-use commercial building in a humid city like Miami.

- **Evaporative Coolers:**
  - **Application:** Warehouses, workshops, or homes in arid regions.
  - **Example:** An evaporative cooler installed in a warehouse, where humidity is consistently low.

Key Considerations for Selection

- **Climate Suitability:**
  - Use **DX systems** in regions with high humidity or variable climates.
  - Use **evaporative coolers** in arid or semi-arid climates.
- **Energy Costs and Sustainability:**
  - Opt for **evaporative coolers** for low operational costs and minimal environmental impact.
  - Opt for **DX systems** where precise temperature and humidity control are critical.
- **Application Size and Scope:**
  - Choose **DX systems** for large-scale, multi-zone buildings or regions with stringent comfort requirements.
  - Choose **evaporative coolers** for localized cooling needs or where natural ventilation is prioritized.

Chapter 8
HVAC Equipment Selection Cariteria

8.1 Overview

The chapter on sizing and selection of HVAC equipment and components using rules of thumb provides a practical approach to designing efficient HVAC systems. It highlights simplified guidelines and rules of thumbs that assist engineers and technicians in estimating the size and capacity of equipment like cooling and heating coils, dampers, silencers, filters, expansion and buffer tanks, valves, pumps and fans. Emphasizing the importance of balancing efficiency, and performance, the chapter explains key factors to size and select various HVAC components. While rules of thumb are valuable for quick estimates and preliminary designs, the chapter underscores the importance of verifying these estimates with detailed calculations or software modelling for accuracy and compliance with industry standards.

8.2 Cooling and Heating Coils

Sizing and selecting cooling and heating coils in HVAC systems is critical for achieving efficient heat transfer, dehumidification, and temperature control. This section provides the rules of thumb for cooling and heating coil sizing and selection in HVAC and building services.

Understand the Application

- **Purpose**: Determine if the coil is for cooling, heating, or both.
- **System Type**: Single-zone systems, multi-zone systems, or dedicated outdoor air systems.
- **Refrigerant**: Ensure compatibility with the refrigerant type (e.g., R-410A, R-32, etc.).

V. Vakiloroaya, *HVAC and Building Services Rules of Thumb*, Springer Tracts in Civil Engineering, https://doi.org/10.1007/978-3-032-10285-0_8

Cooling Capacity

- Use the formula to estimate cooling capacity:

$$Q = m{\cdot}Cp{\cdot}\Delta T$$

- where:
  - Q: Cooling capacity (kW)
  - m: Air mass flow rate (kg/s)
  - C_p: Specific heat of air (~1.005 kJ/kg C)
  - ΔT: Temperature difference (supply air temp – return air temp, in °C)

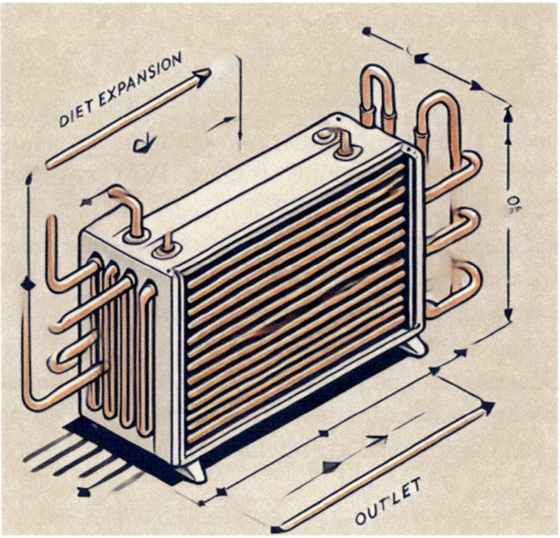

An illustration of a Direct Expansion (DX) Cooling Coil

Face Velocity

- Maintain appropriate face velocities to ensure efficient heat transfer and prevent water carryover:
  - **Cooling Coils**: 2–2.5 m/s.
  - **Heating Coils**: Up to 3 m/s.
- High velocities can lead to coil inefficiency and condensate blow-off.

Rows and Fins

- **Number of Rows**: More rows increase cooling/heating capacity but also increase pressure drop:
  - Cooling coils: Typically **4–8 rows**.
  - Heating coils: Typically **1–4 rows**.

- **Fin Density**: Typical fin spacing ranges from **8–16 fins per inch** (FPI). Higher density increases heat transfer but also pressure drop.

Refrigerant Circuiting

- Select the proper refrigerant circuiting based on load and flow requirements:
  - **Single-Circuit Coils**: For small to medium systems.
  - **Multi-Circuit Coils**: For larger systems or when precise load control is needed.

Pressure Drop

- Maintain a reasonable airside pressure drop for efficient operation:
  - **Cooling Coils**: 50–100 Pa.
  - **Heating Coils**: 30–70 Pa.
- For direct expansion (DX) coils check the refrigerant-side pressure drop as well, ensuring the system can handle it.

Following table shows the **typical pressure drop for the air side** and **water side** of a chilled water cooling coil, based on a range of **rows (4–10)** and **fins per inch (FPI)**:

Rows	FPI	Air side pressure drop (Pa)	Water side pressure drop (kPa)
4	8	50–75	10–15
4	10	60–85	10–15
4	12	80–100	10–15
6	8	70–100	15–20
6	10	90–120	15–20
6	12	110–140	15–20
8	8	90–130	20–25
8	10	110–160	20–25
8	12	140–180	20–25
10	8	120–180	25–30
10	10	150–200	25–30
10	12	180–240	25–30

Coil Selection

- Match coil capacity to the system design conditions:
  - **Entering Air Temperature**: Typical range for cooling: 24–28 °C dry bulb, 17–21 °C wet bulb.
  - **Leaving Air Temperature**: Typical range: 12–14 °C for cooling.
  - **For DX coils, Refrigerant Saturation Temperature**: Ensure the coil can handle the system's evaporating temperature (typically ~5–10 °C).

Dehumidification

- Ensure the coil's cooling capacity includes latent heat removal for dehumidification:
  - Typical humidity targets: 50–60% RH.

Construction Material

- Choose appropriate materials for the coil and tubes:
  - **Copper Tubes** with aluminium fins: Common and cost-effective.
  - **All-Aluminium Coils**: Lightweight and corrosion-resistant and suitable for environments where copper corrosion is a concern.
  - **Coated Coils**: For corrosive or coastal environments.

Common Pitfalls to Avoid

- **Oversizing**: Leads to short cycling and poor humidity control.
- **Undersizing**: Results in inadequate cooling or heating.
- **High Face Velocity**: Causes inefficiency and water carryover.
- **Improper Refrigerant Circuiting**: Reduces coil performance.

8.3 Filters

By following these rules of thumb, HVAC filters can be selected and sized effectively to meet the specific needs of various buildings and applications while optimizing performance, air quality, and energy efficiency. Followings are the guide for selecting and sizing filters in HVAC systems, including the types of filters and their applications.

An illustration of a HEPA Filter

Types of HVAC Filters

Filter type	Description	Applications	Advantages	Disadvantages
Panel filters	Flat or pleated filters made of fiberglass or polyester	Residential, light commercial spaces	Low cost, easy replacement	Limited efficiency (MERV 1–4)
Pleated filters	Accordion-like design to increase surface area	Residential, commercial spaces	Higher efficiency than panel filters	Higher pressure drop than basic filters
HEPA filters (high-efficiency particulate air)	Captures 99.97% of particles ≥0.3 μm	Hospitals, cleanrooms, laboratories	Very high efficiency	High pressure drop, requires fan upgrades
ULPA filters (ultra-low penetration air)	Captures 99.99% of particles ≥0.12 μm	Pharmaceutical, semiconductor facilities	Superior filtration	Very expensive, high maintenance
Electrostatic filters	Use static electricity to attract particles	Residential, commercial spaces	Reusable, efficient for smaller particles	Efficiency decreases over time
Activated carbon filters	Removes odours and gases using activated charcoal	Restaurants, industrial, residential	Controls odours, removes VOCs	Limited particulate filtration
Bag (pocket) filters	Multi-pocket design for large surface area	Large commercial, industrial spaces	Handles high airflows, high capacity	Bulky, moderate pressure drop
Washable filters	Reusable filters that can be cleaned and reused	Residential, light commercial spaces	Cost-effective over time	Lower efficiency, requires maintenance

Filter Efficiency Standards

- **MERV Rating (Minimum Efficiency Reporting Value):**
  - **MERV 1–4**: Basic dust, lint, and pollen removal.
  - **MERV 5–8**: Better filtration for residential and light commercial spaces (removes mold, dust mites).
  - **MERV 9–12**: Captures finer particles, suitable for higher air quality standards.
  - **MERV 13–16**: Removes bacteria, smoke, and fine particulates; used in hospitals and labs.
  - **MERV 17–20**: Equivalent to HEPA filters for critical environments.

Sizing Guidelines

- **Filter Area**:
  - Use larger filter surface areas to reduce pressure drop and increase efficiency.
  - Oversizing filters can prolong their life and improve system performance.

- **Face Velocity**:
  - Recommended face velocity: **1.5–2.5 m/s (300–500 fpm)**.
  - Higher velocities can cause increased pressure drops and reduce filter efficiency.
- **Pressure Drop**:
  - **Rule of Thumb**: Keep pressure drop below **250 Pa (1 in. WC)** for energy efficiency.
  - Monitor and replace filters when pressure drop increases significantly (e.g., double the initial value).

Following table show the typical **pressure drop** of different types of air filters used in HVAC and building services:

Filter type	Initial pressure drop (Pa)	Final pressure drop (Pa)	Efficiency (typical MERV rating)
Panel filter (fiberglass)	25–75	100–125	MERV 1–4
Pleated filter	50–150	200–250	MERV 6–12
Bag filter	150–300	300–400	MERV 13–16
Electrostatic filter	50–100	150–200	MERV 10–15
HEPA filter	250–500	500–750	MERV 17–20 (99.97% @ 0.3 μm)
ULPA filter	500–750	750–1000	99.999% @ 0.1–0.2 μm
Activated carbon filter	50–150	200–300	N/A (for odour and gas removal)
Metal mesh filter	25–75	75–100	Low efficiency (pre-filters)

Notes

- **Initial Pressure Drop** is measured when the filter is clean, while **Final Pressure Drop** is at the recommended point for replacement or cleaning.
- HEPA and ULPA filters have much higher pressure drops and are often used in applications requiring stringent air quality (e.g., cleanrooms, hospitals).
- **Efficiency** is typically represented by the Minimum Efficiency Reporting Value (MERV) rating or particle filtration efficiency.
- Filters with higher efficiency ratings generally have higher pressure drops, so balancing airflow requirements and filtration needs is crucial.

Selection Guidelines

- **Application**:
  - Residential: **MERV 5–8** pleated filters for general air quality.
  - Commercial: **MERV 9–13** filters for higher particulate removal.
  - Industrial: **Bag filters**, **HEPA filters**, or **ULPA filters** for clean air and specific processes.
  - Odour/VOC Control: **Activated carbon filters**.

- **Pollutant Types**:
  - Dust and large particles: Basic filters (MERV 1–4).
  - Allergens, mold, finer dust: Pleated filters (MERV 9–12).
  - Smoke, bacteria, viruses: HEPA or MERV 13–16 filters.
- **Maintenance**:
  - Ensure filters are easily accessible for replacement or cleaning.
  - Use washable or reusable filters in areas where frequent replacement is impractical.
- **Energy Efficiency**:
  - High-efficiency filters (e.g., HEPA) can increase energy consumption due to higher pressure drops; ensure system compatibility.
  - Opt for filters with a balance of efficiency and pressure drop.
- **Environmental Conditions**:
  - Humid or corrosive environments may require special filter materials.
  - Use filters designed for high dust-load environments in industrial settings.

Rules of Thumb for Different Applications

Application	Recommended filter type	Key considerations
Residential	Pleated filters (MERV 5–8)	Affordable, moderate efficiency, suitable for common pollutants
Offices/commercial buildings	MERV 9–13 filters, Bag filters	Handles higher occupant loads, improves indoor air quality (IAQ)
Hospitals	HEPA filters (MERV 17–20)	Critical air quality, bacteria and virus filtration, infection control
Industrial facilities	Bag filters, HEPA filters	High airflow, large particulate capacity, durable for harsh environments
Restaurants	Activated carbon + MERV 8 filters	Controls odours, grease particles, and basic pollutants
Data centres	Electrostatic or HEPA filters	Protects sensitive equipment from dust and fine particulates
Cleanrooms	ULPA filters	Extremely high air cleanliness requirements for sensitive processes

Maintenance and Replacement

- **Inspection Frequency**: Inspect filters monthly in high-use environments or quarterly in low-use areas.
- **Replacement Frequency**:
  - Basic filters: Every 1–3 months.
  - High-efficiency filters: Every 6–12 months.
  - Monitor pressure drops to determine when replacement is necessary.

- **Seal Integrity**: Ensure proper sealing to prevent bypass air from compromising filtration efficiency.

8.4 Fans

The rules of thumb for HVAC fan laws, sizing, and selection are essential for designing efficient air movement systems.

Fan Laws
Fan performance is governed by the following proportional relationships:

Fan Law 1: Flow Rate (Q)
- **$Q \propto RPM$**
- Airflow is directly proportional to the fan speed (RPM). Doubling the RPM doubles the airflow.

Fan Law 2: Pressure (P)
- **$P \propto (RPM)^2$**
- Static or total pressure increases with the square of the fan speed. Doubling the RPM increases the pressure by four times.

Fan Law 3: Power (W)
- **$W \propto (RPM)^3$**
- Power consumption increases with the cube of the fan speed. Doubling the RPM increases the power consumption by eight times.

Static Pressure

- Estimate the static pressure in the system, including:
  - Duct friction losses.
  - Fittings (elbows, dampers).
  - Terminal units and filters.
  - Typical ranges:
    - Low-pressure systems: 250–500 Pa.
    - High-pressure systems: 500–1500 Pa.

Fan Type

- Select a fan type based on application:
  - **Centrifugal Fans**: High-pressure applications, good for ducted systems.
  - **Axial Fans**: High flow rates, low pressure, suitable for ventilation or cooling towers.
  - **Mixed-Flow Fans**: Balance of pressure and flow rate.

Fan Efficiency

- Choose fans with high efficiency (70–90%) to reduce energy consumption.
- Use fans with performance aligned with the system's operating point.

Selection Guidelines

Fan Selection Criteria

1. **Operating Point**: Match the required flow rate and pressure to the fan performance curve.
2. **System Resistance**: Account for all losses in the duct system.
3. **Fan Size**: Select a fan slightly larger than required to allow for operational adjustments.
4. **Noise Levels**: Ensure fan noise is within acceptable levels (e.g., ≤50 dB for offices).

Motor Selection

- Use a motor with sufficient power to handle the fan's requirements, considering power $\propto$ (RPM)^3.
- Consider using **variable frequency drives (VFDs)** to adjust fan speed and optimize energy consumption.

8.5 Attenuators

Attenuators (or silencers) in HVAC systems reduce noise generated by equipment like fans, ductwork, and air terminals. Proper selection and sizing ensure efficient noise control without significantly impacting airflow and pressure. Below are the key rules of thumb for attenuator types, sizing, and selection. By adhering to these rules of thumb and design principles, HVAC attenuators can be sized and selected to control noise while maintaining system efficiency and occupant comfort.

An illustration of an Attenuator

Attenuator Types

- **Straight Duct Silencers**:
  - Inline silencers installed directly in ducts.
  - **Best For**: General noise reduction in straight duct runs.
  - **Rule of Thumb**: Use where space is limited and moderate noise reduction is needed.
- **Elbow Silencers**:
  - Installed at duct bends to manage directional noise and turbulence.
  - **Best For**: Duct systems with noise generated at bends or turns.
  - **Rule of Thumb**: Place at 90-degree bends where significant noise attenuation is required.
- **Circular Silencers**:
  - Designed for round ductwork.
  - **Best For**: Spiral or round ducts in compact installations.
  - **Rule of Thumb**: Use for smaller systems where space constraints exist.
- **Pod or Splitter Silencers**:
  - Feature sound-absorbing baffles to reduce noise.
  - **Best For**: Applications requiring high noise attenuation.
  - **Rule of Thumb**: Ideal for areas with high airflow rates and significant noise challenges.
- **Cross-Talk Attenuators**:
  - Prevent noise transfer between rooms via shared ducts.
  - **Best For**: Office spaces, conference rooms, and hospital settings.
  - **Rule of Thumb**: Place where privacy or quiet operation is essential.

Sizing Guidelines

- **Airflow Velocity**:
  - Maintain airflow velocity below **7–10 m/s** through the attenuator to avoid generating additional noise.
  - **Rule of Thumb**: Oversized attenuators are preferable to reduce velocity and noise.
- **Length**:
  - Longer attenuators provide better noise reduction.
  - Typical lengths: **900–1800 mm** depending on the required attenuation.
  - **Rule of Thumb**: Aim for longer silencers in critical noise-sensitive areas but balance with space constraints.

- **Pressure Drop**:
  - Keep pressure drop through the attenuator below **25–50 Pa**.
  - **Rule of Thumb**: Select attenuators with low resistance to airflow to minimize fan energy consumption.

Following table shows typical pressure drops for attenuators (duct silencers) based on their dimensions, airflow rates, and lengths. The values are approximate and will vary depending on the specific design and manufacturer.

Attenuator dimensions (H × W) (mm)	Length (m)	Airflow rate (m^3/s)	Pressure drop (Pa)
300 × 300	1.0	0.5	20–40
300 × 300	1.5	0.5	30–60
300 × 300	2.0	0.5	40–80
300 × 300	2.5	0.5	50–100
600 × 600	1.0	1.5	25–45
600 × 600	1.5	1.5	40–70
600 × 600	2.0	1.5	55–90
600 × 600	2.5	1.5	65–115
900 × 600	1.0	2.5	35–55
900 × 600	1.5	2.5	50–85
900 × 600	2.0	2.5	70–115
900 × 600	2.5	2.5	85–140
1200 × 600	1.0	3.5	40–70
1200 × 600	1.5	3.5	60–105
1200 × 600	2.0	3.5	80–140
1200 × 600	2.5	3.5	100–175
1500 × 900	1.0	5.0	50–80
1500 × 900	1.5	5.0	75–120
1500 × 900	2.0	5.0	100–160
1500 × 900	2.5	5.0	125–200

Notes

Pressure drop increases with the length of the attenuator due to the additional internal resistance.

Higher airflow rates result in greater pressure drops, as the air velocity through the attenuator increases.

Larger attenuators (greater cross-sectional dimensions) generally have **lower pressure drops** at the same airflow rate compared to smaller ones.

Internal baffle configuration, packing density, and material can also significantly influence the pressure drop.

Cross-Sectional Area

- Match the attenuator's cross-sectional area to the duct size.
- **Rule of Thumb**: Avoid reducing duct size at the attenuator to prevent increased turbulence and noise.

Selection Guidelines

- **Noise Reduction Requirements**:
  - Determine the required insertion loss (attenuation in dB) for the system.
  - **Rule of Thumb**: Typical attenuation targets:
    - Fan noise: **10–20 dB**.
    - Mechanical room: **15–30 dB**.
    - Office areas: **10–15 dB**.
- **Frequency-Specific Performance**:
  - Noise reduction effectiveness varies by frequency.
  - **Rule of Thumb**: Use attenuators designed to address low-frequency noise (e.g., <250 Hz) for fans and equipment with large rotating components.
- **Material Selection**:
  - Use perforated metal linings with sound-absorbing materials (e.g., fiberglass, foam).
  - For corrosive environments, use attenuators with stainless steel or coated interiors.
- **Placement**:
  - Position as close to the noise source as possible (e.g., fan discharge or intake).
  - **Rule of Thumb**: Allow a straight duct section of at least **2–3 duct diameters** before and after the attenuator for optimal performance.

Installation Rules of Thumb

- **Avoid Sharp Transitions**:
  - Avoid abrupt changes in duct size near the attenuator to reduce turbulence.
  - Use smooth transitions if a change in size is necessary.
- **Support and Accessibility**:
  - Properly support attenuators to avoid sagging or damage.
  - Ensure they are accessible for inspection and cleaning.
- **Fire Safety**:
  - Use attenuators with non-combustible or fire-rated materials in critical areas.

Applications and Typical Noise Levels

Application	Typical noise level (dB)	Recommended attenuation (dB)
Residential HVAC	35–45	10–15
Commercial offices	40–50	10–20
Hospitals (operating rooms)	30–40	15–25
Data centres	50–60	20–30
Industrial facilities	60–80	30–40

Performance Metrics

- **Insertion Loss (IL)**:
  - Measure of noise reduction provided by the attenuator (in dB).
  - **Rule of Thumb**: Higher IL is required for noisy environments or sensitive spaces.
- **Dynamic Insertion Loss**:
  - Performance under operating conditions (with airflow).
  - **Rule of Thumb**: Select attenuators tested for both static and dynamic conditions.
- **Regenerated Noise**:
  - Noise generated by the attenuator itself due to turbulence.
  - **Rule of Thumb**: Ensure the attenuator's regenerated noise is below the required background noise level.

Practical Considerations

- **Acoustic Modelling**:
  - Use acoustic analysis software to predict noise levels and select appropriate attenuators.
  - **Rule of Thumb**: Include all noise sources (fans, terminals, equipment) in the model.
- **System Balancing**:
  - Adjust dampers and airflow rates after attenuator installation to maintain design performance.
- **Maintenance**:
  - Inspect and clean attenuators periodically to maintain airflow and noise attenuation performance.

8.6 Fire Dampers

Sizing and selecting fire dampers for HVAC and building services is critical to ensure safety, compliance with fire codes, and effective system operation. Followings are the rules of thumb for proper selection and sizing of fire dampers.

An illustration of a Fire Damper

Types of Fire Dampers

- **Curtain-Type Fire Dampers**
  - **Design**: Use interlocking blades that drop into place when the fusible link melts.
  - **Application**: Common in horizontal ducts where airflow resistance is minimal.
- **Multi-Blade Fire Dampers**
  - **Design**: Consist of multiple blades that close when the fusible link melts.
  - **Application**: Suitable for larger ducts, providing better airflow control when open.
- **Combination Fire/Smoke Dampers**
  - **Design**: Serve as both fire and smoke dampers, equipped with actuators and smoke sensors.
  - **Application**: Used in areas requiring control of both fire and smoke spread, such as mixed-use spaces.
- **Static Fire Dampers**
  - **Design**: Operate in systems where airflow is stopped during a fire event.
  - **Application**: Typically installed in systems that shut down upon fire detection.
- **Dynamic Fire Dampers**
  - **Design**: Rated for operation under airflow conditions during a fire.
  - **Application**: Used in systems designed to remain operational during fire events.

- **Circular Fire Dampers**
  - **Design**: Specifically for round ducts, often with a spring-loaded blade.
  - **Application**: Ideal for compact or space-limited duct designs.

Determine Application Type

- Identify the application to select the appropriate type of fire damper:
  - **Static Fire Dampers**: For systems where airflow stops during a fire.
  - **Dynamic Fire Dampers**: For systems designed to operate during a fire.

Understand Fire Resistance Rating

- Match the fire damper's rating to the fire-resistance rating of the barrier (wall, floor, or partition).
  - Typical fire damper ratings: **1.5** or **3 h**.
  - Ensure compliance with local fire codes.

Airflow Velocity and Pressure

- Select dampers that can handle the system's maximum airflow velocity and pressure.
- Standard fire dampers handle velocities up to **15 m/s** and static pressure up to **2.5 kPa**.
- For higher velocities or pressures, choose heavy-duty dampers.

Fire Dampers PD

Below is a typical table showing the pressure drop of fire dampers for various airflow rates and dimensions commonly used in HVAC systems. The pressure drop values are estimated and should be refined for actual design by consulting the manufacturer's data.

Damper size (mm)	Airflow rate (L/s)	Pressure drop (Pa)
200 × 200	100	10
200 × 200	300	30
200 × 200	500	60
300 × 300	500	10
300 × 300	1000	20
300 × 300	1500	40
400 × 400	1000	10
400 × 400	2000	30
400 × 400	3000	60
500 × 500	2000	10
500 × 500	4000	50
500 × 500	6000	100
600 × 600	3000	10
600 × 600	5000	30
600 × 600	8000	90

Explanation

- **Damper Size (mm)**: Dimensions of the fire damper.
- **Airflow Rate (L/s)**: Volume of air passing through the damper.
- **Pressure Drop (Pa)**: Resistance to airflow caused by the damper, measured in Pascals.

Duct Size and Damper Dimensions

- Fire damper dimensions must match the duct size to avoid airflow restrictions.
- Standard dampers are available in various rectangular and circular sizes.
- For non-standard sizes, custom dampers may be required.

Installation Location

- Install fire dampers within the fire-rated barrier.
- Ensure proper alignment with the duct system to maintain fire resistance integrity.
- Use fire-rated caulking or sealants to seal gaps.

Following table specifies the location and reason for installation of a fire damper.

Location	Reason for installation	Purpose
Fire-rated walls or partitions	To maintain the integrity of fire-rated barriers, preventing fire and smoke from passing through ductwork	To compartmentalize a building and restrict the spread of fire between different areas
Ceilings of fire-resistant compartments	To prevent fire and smoke from breaching compartmentalized spaces through duct penetrations	To protect upper floors or adjacent compartments
Vertical ducts passing between floors	To prevent the vertical spread of fire and smoke through ductwork	To comply with fire compartmentation regulations and protect occupants on different floors
Air transfer openings in walls	To limit the spread of fire and smoke where ducts or grilles allow air transfer between fire-separated areas	To ensure fire barriers maintain their designed level of resistance
Escape routes and staircases	To prevent fire and smoke from spreading into evacuation routes, compromising safety	To ensure the safe evacuation of building occupants
Mechanical or HVAC Shafts	To prevent fire and smoke from using duct shafts as conduits between floors or compartments	To safeguard the building's structural integrity and occupant safety
High-risk areas (e.g., kitchens, labs)	To isolate areas with a higher risk of fire, preventing flames and smoke from entering ducts connected to other parts of the building	To protect other areas from high-risk zones where fire is more likely to originate
Return air systems	To prevent fire and smoke from being drawn into return air systems and distributed throughout the building	To limit the recirculation of fire-contaminated air

(continued)

Location	Reason for installation	Purpose
Supply air systems serving multiple zones	To avoid the spread of fire and smoke to other zones through shared supply ducts	To maintain compartmentalization and ensure air distribution systems don't compromise safety
Ducts serving pressurized stairwells	To ensure pressurization systems work effectively to keep smoke out of stairwells and safe escape routes	To maintain safe egress routes during a fire

Closure Mechanism

- **Fusible Link**: Most fire dampers close when the fusible link melts at a specific temperature (e.g., 74 °C or 100 °C).
- Ensure the link temperature matches the application.

Accessibility

- Provide accessible locations for inspection, maintenance, and testing of fire dampers.
  - Install access doors near fire dampers.
  - Check local building regulations and fire safety codes.

Maintenance and Testing Requirements

- Fire dampers should be tested and maintained as per local code in each country (e.g., annually or every 4 years, depending on the jurisdiction).
- Select dampers with easy-to-access components for periodic testing.

Combination Fire/Smoke Dampers

- For spaces requiring both fire and smoke protection, use combination fire/smoke dampers.

Consider Airflow Restriction

- Fire dampers add resistance to the airflow (pressure drop).
  - Check the pressure drop in the damper specification and adjust fan selection if necessary.

Example: Sizing and Selecting a Fire Damper

Scenario

- A rectangular duct passes through a 2-h rated firewall.
- Duct dimensions: **600 mm × 300 mm**.
- Air velocity: **9 m/s**.
- System pressure: **750 Pa**.

Steps

- **Determine Fire Rating**:

A 1.5-h rated damper meets the requirement for a 2-h wall.

Select Type:
Since the system operates during a fire, a **dynamic curtain-type fire damper** is chosen for its suitability in airflow conditions.

8.7 Volume Control Dampers (VCD) and Outside Air Damper (OAD)

An illustration of a Volume Control Damper

Types of Volume Control Dampers (VCDs)
Volume Control Dampers are used to regulate or isolate airflow within a duct system. Below are common types:

Opposed Blade Dampers

- **Description**: Blades rotate in opposite directions, providing a more linear and precise control of airflow.
- **Applications**: Used in balancing and fine control situations in HVAC systems.
- **Advantages**:
  - Precise airflow control.
  - Even distribution of pressure drop across the damper.
- **Limitations**:
  - Higher cost compared to parallel blade dampers.

Parallel Blade Dampers

- **Description**: Blades rotate in the same direction, commonly used for isolation or large-scale modulation.
- **Applications**: Mixing dampers and systems where precise control is less critical.
- **Advantages**:
  - Simple design, lower cost.
- **Limitations**:
  - Less precise than opposed blade dampers.

Single-Blade Dampers

- **Description**: A single rotating blade controls the airflow in small ducts.
- **Applications**: Smaller ducts or simple systems with limited modulation needs.
- **Advantages**:
  - Compact and cost-effective.
- **Limitations**:
  - Limited control range.

Butterfly Dampers

- **Description**: A single circular blade rotates within the duct to control airflow.
- **Applications**: Suitable for round duct systems or small ventilation units.
- **Advantages**:
  - Easy to operate.
  - Low pressure drop when fully open.
- **Limitations**:
  - Not ideal for precise control.

Motorized Dampers

- **Description**: Equipped with an actuator for automated control via a Building Management System (BMS).
- **Applications**: Systems requiring frequent or remote adjustments.
- **Advantages**:
  - Automated control.
  - Integration with HVAC controls.
- **Limitations**:
  - Higher cost and maintenance.

Installation Location

Location	Reason for installation	Purpose
Branch ducts in HVAC systems	To regulate airflow to individual zones or rooms by adjusting the volume of air supplied through the branch ducts	To balance the system, ensuring even distribution of air and maintaining thermal comfort
Main supply or return ducts	To control the total airflow delivered to or extracted from the system	To optimize HVAC system performance and accommodate changing load demands
Zones with varying load requirements	To adjust airflow based on occupancy, usage, or environmental conditions in different zones	To enhance energy efficiency and occupant comfort by meeting specific zone requirements
Mixing boxes or variable air volume (VAV) units	To fine-tune the amount of air being mixed or supplied to specific areas of the building	To allow precise control over temperature and air distribution in VAV systems
Duct runs to long or distant zones	To adjust airflow for zones farther from the air handling unit to compensate for pressure loss over long duct runs	To maintain consistent airflow and comfort across all areas of the building
Bypass ducts in HVAC systems	To redirect excess airflow when certain zones do not require air or when systems operate under varying loads	To maintain system efficiency and prevent overpressurization of ducts
Large open spaces (e.g., auditoriums)	To regulate airflow for large spaces with variable occupancy and environmental needs	To adapt to fluctuating airflow demands and reduce energy consumption
Air supply or extraction grilles	To control the airflow rate delivered through or extracted from individual grilles	To allow fine-tuning of airflow in specific locations for comfort or system balance
Exhaust or ventilation systems	To regulate airflow in exhaust systems, particularly in areas such as restrooms, kitchens, or laboratories	To ensure proper removal of contaminants, odors, or heat while maintaining system balance
Pressure control in HVAC zones	To maintain desired pressure levels between adjacent zones or rooms, especially in controlled environments like cleanrooms or operating rooms	To ensure proper containment or exclusion of contaminants and compliance with design specifications

Types of Outside Air Dampers (OADs)

Outside Air Dampers regulate the amount of fresh air entering the HVAC system and are often used in economizer cycles.

Economizer Dampers

- **Description**: Part of an economizer system to regulate outdoor air intake for free cooling.
- **Applications**: Large commercial HVAC systems for energy savings.
- **Advantages**:
  - Reduces cooling load by using outside air.

- **Limitations**:
  - Requires proper calibration and maintenance.

Barometric Relief Dampers

- **Description**: Automatically open based on pressure differential, allowing air to exit.
- **Applications**: Relief air systems, often paired with economizers.
- **Advantages**:
  - Passive operation, no actuator needed.
- **Limitations**:
  - Less control precision.

Gravity Dampers

- **Description**: Non-motorized dampers that open with air pressure and close by gravity.
- **Applications**: Simple exhaust systems.
- **Advantages**:
  - No external power or actuator required.
- **Limitations**:
  - Limited control and prone to leakage.

Sizing and Selection

Sizing Dampers

- **Face Velocity**:
  - Recommended face velocity: **8–12 m/s** for VCDs and OADs.
  - Lower velocities reduce noise and pressure drop.
- **Pressure Drop**:
  - Ensure the pressure drop across the damper does not exceed **25–50 Pa**.

Following table shows pressure drop for volume control dampers (VCD) for typical airflow rates and dimensions used in HVAC systems. These values are general estimates and should be cross-checked with specific manufacturer data for precise design.

Damper size (mm)	Airflow rate (L/s)	Pressure drop (Pa)
200 × 200	100	10
200 × 200	300	90
200 × 200	500	240
300 × 300	500	25
300 × 300	1000	90
300 × 300	1500	200
400 × 400	1000	20
400 × 400	2000	75

(continued)

Damper size (mm)	Airflow rate (L/s)	Pressure drop (Pa)
400 × 400	3000	160
500 × 500	2000	25
500 × 500	4000	90
500 × 500	6000	200
600 × 600	3000	25
600 × 600	5000	70
600 × 600	8000	120

Explanation

- **Damper Size (mm)**: Dimensions of the volume control damper.
- **Airflow Rate (L/s)**: Volume of air passing through the damper.
- **Pressure Drop (Pa)**: Resistance to airflow caused by the damper.
- **Duct Size**:
  - Match damper size to duct dimensions to minimize airflow resistance.

Damper Material

- **Galvanized Steel**: Standard material for general HVAC use.
- **Aluminium**: Lightweight and corrosion-resistant for humid environments.
- **Stainless Steel**: High durability, used in corrosive or industrial applications.

Control and Actuation

- **Manual Dampers**: Use in areas where adjustment is infrequent.
- **Motorized Dampers**: Essential for automated systems, integrated with BMS.

Leakage Class

- Select dampers based on leakage class requirements:
  - **Class 1**: Low leakage, for high-precision control.
  - **Class 2**: Medium leakage, for general HVAC systems.

Placement Rules

- **Upstream Placement**: Place dampers upstream of critical equipment (e.g., fans, coils) for better control.
- **Accessibility**: Ensure dampers are accessible for maintenance and adjustment.

Common Applications

VAV Systems: Volume control dampers balance airflow to zones.
Economizer Systems: Outside air dampers regulate fresh air intake.
Exhaust Systems: Gravity or motorized dampers control exhaust airflow.
Isolation: Dampers isolate sections of the system during maintenance or non-use.

Maintenance Considerations

- **Cleaning**: Remove dust and debris regularly to prevent operational issues.

- **Actuator Check**: Ensure motorized actuators respond correctly to control signals.
- **Seal Inspection**: Inspect seals to prevent leakage and maintain efficiency.

8.8 Waterproof Louvers

Waterproof louvers are critical components in HVAC systems, designed to protect air intake and exhaust openings from rainwater, snow, and wind-driven debris while allowing adequate airflow. Proper selection and sizing of waterproof louvers ensure efficient operation and prevent damage to HVAC equipment.

Types of Waterproof Louvers

- **Fixed Blade Louvers**:
  - Features stationary blades arranged at specific angles.
  - Designed to direct rain and water away from the opening while maintaining airflow.
  - Common in general HVAC applications with moderate weather protection needs.

An illustration of a Waterproof Louver

- **Drainable Blade Louvers**:
  - Includes a drainage channel on the blades or frame to collect and divert rainwater away from the system.

- Provides superior rain rejection compared to fixed blade louvers.
- Ideal for areas with frequent or heavy rainfall.

- **Combination Louvers**:
  - Combines the functions of an air intake/exhaust louver and a damper in a single unit.
  - Often includes motorized or manual control to regulate airflow and seal off openings when not in use.
  - Suitable for environments requiring both airflow control and weather protection.

- **Sand Trap Louvers**:
  - Incorporates features to filter out sand and dust particles in addition to rainwater protection.
  - Used in desert or high-dust environments.
  - Typically includes a dual-layer design for effective filtration.

- **Acoustic Louvers**:
  - Includes sound-attenuating materials in the louver assembly.
  - Provides noise control while offering rain protection and airflow management.
  - Common in noise-sensitive applications like hospitals and data centers.

- **High-Performance (Storm-Resistant) Louvers**:
  - Designed to handle extreme weather conditions, including wind-driven rain and hurricane-force winds.
  - Often tested for compliance with AMCA (Air Movement and Control Association) or similar standards.
  - Used in coastal or storm-prone areas.

Sizing and Selection

Airflow Requirements

- **Free Area**: Select a louver with adequate free area (the portion of the louver that allows air to pass through).
  - Rule of Thumb: The free area should be **50–60% of the louver's face area** for optimal performance.
  - Ensure that the selected louver meets the system's airflow (CFM or m^3/h) without causing excessive pressure drop.

Rain Rejection Performance

- Choose louvers with high water penetration resistance (often rated as Class A, B, or C based on AMCA standards).
  - **Class A**: Excellent water rejection (99% or higher).
  - **Class B**: Good water rejection (up to 95%).
  - **Class C**: Moderate water rejection (up to 80%).

Pressure Drop

- The pressure drop across the louver should not exceed **75–100 Pa (0.3–0.4 inches of water gauge)** for typical HVAC systems.
- Use manufacturer data to verify pressure drop for the required airflow.

Following table shows typical pressure drops for waterproof louvers, based on their dimensions, airflow rates, and free area percentage. These values are approximate and vary depending on the louver's design and manufacturer specifications.

Louver dimensions (H × W) (mm)	Free area (%)	Airflow rate (m^3/s)	Face velocity (m/s)	Pressure drop (Pa)
500 × 500	50	0.5	2.0	25–40
500 × 500	50	0.8	3.0	40–70
500 × 500	50	1.0	4.0	70–100
1000 × 1000	60	1.5	2.5	20–35
1000 × 1000	60	2.5	4.0	35–60
1000 × 1000	60	3.5	5.5	60–90
1500 × 1000	70	3.0	2.0	15–25
1500 × 1000	70	4.5	3.0	25–45
1500 × 1000	70	6.0	4.0	45–75
2000 × 1000	70	6.0	2.5	20–35
2000 × 1000	70	8.0	3.5	35–60
2000 × 1000	70	10.0	4.5	60–90

Key Factors

Free Area Percentage: Louvers with a higher free area (e.g., 70%) have **lower pressure drops**, while those with smaller free areas (e.g., 50%) show **higher pressure drops** due to restricted airflow.

Face Velocity: Pressure drop increases as the face velocity (airflow divided by the free area) increases.

Louver Size: Larger louvers typically have lower pressure drops at the same airflow rate because of a larger free area for air to pass through.

Design Variations: Waterproof louvers are designed to resist rain penetration, which may slightly increase pressure drop compared to standard louvers due to additional obstructions.

Face Velocity

- Maintain appropriate face velocity to balance airflow and rain rejection.
  - Rule of Thumb: Face velocity should typically be **2.0–3.5 m/s (400–700 FPM)**.
  - For storm-resistant louvers, limit face velocity to **2.5 m/s (500 FPM)**.

Material Selection

- Choose materials based on environmental conditions:
  - **Aluminium**: Lightweight, corrosion-resistant, suitable for most applications.

  - **Stainless Steel**: Durable and corrosion-resistant, ideal for coastal or industrial environments.
  - **Galvanized Steel**: Cost-effective, suitable for mild environments.

Mounting and Installation

- Ensure proper installation to maintain performance:
  - Include a recessed or sloped sill to direct water away.
  - Provide adequate clearance for maintenance and cleaning.
  - Consider bird screens or insect screens for additional protection.

Size Selection

- Rule of Thumb: The louver size should be **20–25% larger than the duct or opening size** to account for free area loss and ensure adequate airflow.

Acoustic Considerations

- If noise control is required, use acoustic louvers or add sound attenuation features.
- Rule of Thumb: Ensure at least **15–25 dB noise reduction** for applications in noise-sensitive areas.

Example Applications

- **Office Building**:
  - **Type**: Drainable blade louver.
  - **Requirements**: Moderate airflow with good rain protection.
  - **Sizing**: 1.2 m × 1.5 m louver for a 1.0 m × 1.2 m duct opening.
- **Industrial Facility in a Coastal Area**:
  - **Type**: Storm-resistant louver (Class A rain rejection).
  - **Material**: Stainless steel.
  - **Sizing**: 25% oversized to handle high wind-driven rain.
- **Data Centre**:
  - **Type**: Acoustic louver with drainable blades.
  - **Requirements**: Noise attenuation and water rejection.
  - **Performance**: 20 dB noise reduction and Class A rain rejection.

8.9 Grilles and Diffusers

Grilles and diffusers are essential components of HVAC systems, used to distribute and return air effectively while maintaining indoor comfort. Choosing the correct type, size, and placement is vital for achieving optimal air distribution and system efficiency.

Types of Grilles and Diffusers

Supply Grilles

- **Purpose**: Distribute conditioned air into the room.
- **Design**: Fixed or adjustable blades to control airflow direction.
- **Applications**: Offices, residential spaces, and retail environments.
- **Examples**:
  - Single deflection grille (adjustable blades in one direction).
  - Double deflection grille (adjustable blades in two directions).

Return Grilles

- **Purpose**: Allow air to flow back into the HVAC system for reconditioning.
- **Design**: Typically fixed blades or perforated panels.
- **Applications**: Any return air system.
- **Examples**:
  - Fixed-blade return grille.
  - Perforated return grille.

Ceiling Diffusers

- **Purpose**: Distribute supply air evenly in all directions.
- **Design**: Circular, square, or rectangular with multiple airflow patterns.
- **Applications**: Offices, classrooms, and conference rooms.
- **Examples**:
  - Circular ceiling diffuser.
  - Square plaque diffuser.

Slot Diffusers

- **Purpose**: Provide a linear air distribution with sleek aesthetics.
- **Design**: Long, narrow slots with adjustable airflow direction.
- **Applications**: Modern offices, retail stores, and airports.
- **Examples**:
  - Single-slot diffuser.
  - Multi-slot diffuser.

Linear Bar Grilles and Diffusers

- **Purpose**: For both supply and return air in modern architectural spaces.
- **Design**: Long, continuous appearance with minimal visual impact.
- **Applications**: High-end commercial and residential spaces.
- **Examples**:
  - Linear bar grille.
  - Continuous slot diffuser.

Egg Crate Grilles

- **Purpose**: Typically used for return air or exhaust applications.
- **Design**: Grid-like appearance for high free area.

- **Applications**: Commercial buildings and industrial spaces.
- **Examples**:
  - Standard egg crate grille.

Floor Grilles and Diffusers
- **Purpose**: Provide supply air from floor-mounted systems.
- **Design**: Heavy-duty construction to withstand foot traffic.
- **Applications**: Raised floor systems in data centers, theaters, and offices.
- **Examples**:
  - Heavy-duty floor grille.

High-Throw Jet Nozzles
- **Purpose**: Deliver air over long distances.
- **Design**: Circular nozzles with adjustable throw angles.
- **Applications**: Large spaces like stadiums, warehouses, and auditoriums.

Comparison of Grilles and Diffusers

Feature	Supply grilles	Return grilles	Ceiling diffusers	Slot diffusers	Egg crate grilles	Linear bar diffusers
Primary function	Supply air distribution	Return air intake	Even air distribution	Linear air distribution	High free area exhaust	Aesthetic air distribution
Design	Adjustable/ fixed blades	Fixed blades or perforated	Multi-directional	Narrow slots	Grid-like appearance	Continuous sleek design
Airflow direction	Adjustable	Fixed	Multi-directional	Adjustable	Fixed	Adjustable
Aesthetic impact	Moderate	Moderate	Low	High	Low	High
Typical placement	Walls or ceilings	Walls or ceilings	Ceilings	Ceilings	Walls or ceilings	Walls, ceilings, or floors
Best use case	General air supply	General return/ exhaust	General HVAC spaces	Modern and aesthetic spaces	High volume return	Architectural spaces

Sizing and Selection

Airflow Requirements
- Calculate required **CFM (Cubic Feet per Minute)** or **m^3/h** for the space.
  - Use manufacturer-**provided air throw and pressure** drop data.
  - Ensure total grille/diffuser airflow matches the calculated space requirements.

Velocity Guidelines
- Recommended velocities:
  - Supply air: **250–400 FPM (1.2–2.0 m/s)** at the diffuser.
  - Return air: **300–500 FPM (1.5–2.5 m/s)** at the grille.

Noise Levels

- Select grilles/diffusers with noise levels below **30–35 NC** (Noise Criteria) for most occupied spaces.
- Use larger diffusers or grilles to reduce velocity and noise.

Throw and Spread

- Ensure the air throw covers the entire room without causing drafts.
  - Rule of Thumb: Air throw should be **75–150% of the room's longest dimension.**

Aesthetic Considerations

- Use slot or linear diffusers in high-end spaces for a modern look.
- Use egg crate grilles for return air where appearance is less critical.

Mounting Location

- Ceiling-mounted diffusers are ideal for uniform airflow.
- Wall-mounted grilles work well for smaller spaces with directional airflow needs.

Free Area

- Ensure sufficient free area to minimize pressure drop:
  - Supply grilles: **50–60% free area.**
  - Return grilles: **60–80% free area.**

Material Selection

- Use aluminium or steel for durability in commercial applications.
- Use corrosion-resistant materials in humid or corrosive environments.

Specialty Applications

- Use jet nozzles for long-distance air throws in large spaces.
- Use heavy-duty floor grilles for areas with foot traffic.

Sizing and Selection

Office Building (Meeting Room, 20 m^2)

- **Airflow Requirement**: 500 m^3/h (300 CFM).
- **Type**: Square ceiling diffuser with 4-way air distribution.
- **Size**: 600 mm × 600 mm (standard tile size).
- **Face Velocity**: ~2 m/s.

Retail Store (Large Open Space, 500 m^2)

- **Airflow Requirement**: 15,000 m^3/h (8800 CFM).
- **Type**: Slot diffusers along perimeter for aesthetic design.
- **Size**: 6-slot diffuser, each 1.2 m long.

Auditorium (Long Distance Air Throw, 1000 m^2)

- **Airflow Requirement**: 30,000 m^3/h (17,600 CFM).
- **Type**: High-throw jet nozzles.
- **Placement**: Mounted at 6 m height, directed at audience seating.

8.10 Humidifiers and Dehumidifiers

An illustration of Ultrasonic Humidifier

Types of Humidifiers

Type	Description	Applications	Advantages	Disadvantages
Steam humidifiers	Produces steam using electricity or gas	Hospitals, industrial facilities	Precise control, high capacity	High energy consumption, initial cost
Evaporative humidifiers	Uses airflow over a water-soaked medium	Homes, offices, retail spaces	Energy efficient, simple operation	Less precise, limited capacity
Ultrasonic humidifiers	Uses ultrasonic vibrations to create mist	Data centres, museums, offices	Low energy use, precise control	Requires high-quality water to avoid scaling
Adiabatic humidifiers	Water is atomized into fine droplets	Industrial spaces, greenhouses	Energy efficient, good cooling effect	Limited to specific applications
Portable humidifiers	Standalone units for individual spaces	Residential, small offices	Easy to use, affordable	Limited to small areas

Types of Dehumidifiers

Type	Description	Applications	Advantages	Disadvantages
Refrigerant dehumidifiers	Removes moisture by cooling air below dew point	Homes, offices, industrial spaces	Efficient for high humidity, affordable	Less effective in low temperatures
Desiccant dehumidifiers	Uses a desiccant material to absorb moisture	Warehouses, cold climates	Effective in low temperatures and humidity	Higher energy cost for regeneration
Ventilation dehumidifiers	Reduces humidity by exchanging indoor and outdoor air	Homes, offices, humid climates	Improves ventilation and IAQ	Limited control over indoor conditions
Portable dehumidifiers	Standalone units for small areas	Residences, small offices	Easy to use, affordable	Limited to small spaces, lower capacity

Sizing Guidelines

For Humidifiers

- **Determine the Desired Humidity Level**: Typical indoor relative humidity (RH) ranges from 40–60% for comfort and health.
- **Calculate Humidity Load**:
  - **Formula**: Humidity Load (kg/h) = Volume of Space (m^3) × Air Change Rate (ACH) × ΔH humidity Ratio (kg/kg)*1.2
  - ΔH is the difference in humidity ratio (**kg of water per kg of dry air**) between the incoming air and the desired air condition. Humidity ratio is typically obtained from a psychrometric chart or calculated using the specific moisture content of the air at given temperature and relative humidity conditions.
  - Consider **outside air infiltration**, ventilation rates, and initial versus target RH levels.
- **Adjust for Outdoor Conditions**: Humidifier capacity may need to increase in colder climates to counteract drier outdoor air.

For Dehumidifiers

- **Determine Moisture Removal Needs**:
  - Calculate latent heat load (moisture content in the air).
  - Use psychrometric charts to find moisture removal requirements based on air conditions.
  - **Rule of Thumb**:
    - Residential: 10–30 L/day for small spaces.
    - Commercial/Industrial: 2–5 L/h/100 m^2, depending on activity levels.

- **Consider Space Conditions**:
  - High ceilings, higher occupant density, or moisture-generating processes (e.g., cooking, cleaning) require larger capacity.
- **Climate Factors**: Select systems based on climate—desiccant dehumidifiers for cooler climates, refrigerant systems for warm, humid regions.

Selection Guidelines

Humidifiers
- **Application-Specific**:
  - **Healthcare**: Steam humidifiers for precise control.
  - **Residential/Offices**: Evaporative or ultrasonic humidifiers for comfort.
- **Water Quality**: Ensure water treatment (deionization, filtration) for ultrasonic or steam systems to prevent scaling.
- **Energy Efficiency**: Choose adiabatic humidifiers for cooling and low-energy applications.

Dehumidifiers
- **Climate and Temperature**:
  - Refrigerant systems for warm and humid climates.
  - Desiccant systems for cooler and less humid climates.
- **Capacity Needs**:
  - Match capacity with the calculated moisture load to prevent under-sizing or over-sizing.
- **Special Applications**:
  - Use desiccant dehumidifiers for spaces requiring low dew points (e.g., pharmaceutical facilities, museums).

Rules of Thumb

Humidifiers
Residential: 5–10 L/day for a single room, depending on size and outdoor air conditions.
Office/Commercial: 10–50 L/h/1000 m^2 of conditioned space.
Industrial: 50–200 L/h/1000 m^2, depending on process requirements.

For example, a 1000 m^2 office with 3 m ceilings has a volume of 3000 m^3. At a humidity addition rate of 10–50 L/h, this corresponds to a humidification rate of ~0.0033–0.0167 L/m^3/h, which aligns with typical office requirements. For another example, a 1000 m^2 space with 6 m ceilings has a volume of 6000 m^3. At 50–200 L/h, this corresponds to 0.008–0.033 L/m^3/h, which is typical for industrial processes.

Dehumidifiers
- **Residential**: 0.5–1 L/day/10 m^2 of floor area.

Office/Commercial: 1–3 L/h/100 m^2.
Industrial: 5–10 L/h/100 m^2, adjusted for high moisture processes.

Rules of Thumb for Humidifiers and Dehumidifiers

Application type	Humidifiers (capacity)	Dehumidifiers (capacity)	Equivalent rate (L/ m^3 h)–3 m Ceiling
Residential	5–10 L/day for a single room	0.5–1 L/day/10 m^2	~0.00007–0.00014
Office/commercial	10–50 L/h/1000 m^2	1–3 L/h/100 m^2	0.0033–0.0167
Industrial	50–200 L/h/1000 m^2 (process-dependent)	5–10 L/h/100 m^2 (high-moisture)	0.0167–0.0667

Always adjust moisture capacity for actual air volume when ceilings exceed the standard 2.7–3.0 m height. If the ceiling height doubles, the humidifier or dehumidifier capacity must also roughly double to maintain the same $L/m^3\ h$ rate.

8.11 Electrical Heating Coils

Electrical heating coils are commonly used in HVAC systems to heat air, either for space heating or for conditioning in air-handling units (AHUs). Below is a detailed guide to the types, sizing, and selection rules of thumb for electrical heating coils.

An illustration of Electrical Coil

Types of Electrical Heating Coils

- **Duct-Mounted Electric Heaters**
  - Installed directly in air ducts.
  - Common for zone heating or reheating applications.
  - Available in single-phase or three-phase configurations.

- **Open Coil Heaters**
  - Exposed resistance wires (e.g., nickel-chromium).
  - High heat transfer efficiency.
  - Suitable for clean environments with controlled airflow.
- **Finned Tubular Heaters**
  - Resistance wires embedded in metal tubes with fins for increased heat transfer.
  - Durable and suitable for harsh or contaminated environments.
- **Strip Heaters**
  - Thin, flat heating elements mounted in ducts or on surfaces.
  - Ideal for small-scale applications or supplemental heating.

Rules of Thumb for Sizing and Selection

Capacity Calculation

- To calculate the required heating capacity:

$$Q = 1206 \times V \times \Delta T$$

- **Q**: Heating capacity (W)
- **V**: Airflow (m^3/s)
- **ΔT**: Temperature rise (°C)

Note

- The constant **1206** is derived from the specific heat of air (1.006 kJ/kg °C), the density of air at standard conditions (~1.2 kg/m^3), and the unit conversion from kJ/s to W:

$$Q = \rho \times cp \times 1000$$

- At standard air conditions (1.2 × 1.006 × 1000 ≈ 1206).

This formula assumes **dry air at sea level**. For high altitude, humid air, or significant temperature extremes, adjust density accordingly.

Example

- Airflow: 2 m^3/s
- Temperature Rise: 30 °C

$$Q = 1206 \times 2 \times 30 = 72{,}360\,\text{W} = 72.36\,\text{kW}$$

Power Supply and Voltage

- Check available power supply: **Single-phase** or **Three-phase**.
- Common voltages:
  - Single-phase: 120 V, 208 V, 240 V.
  - Three-phase: 208 V, 400 V, 480 V.

- Ensure the heater's capacity matches the supply voltage and current.

Airflow Considerations

- **Minimum airflow requirement**: Maintain a minimum airflow to prevent overheating and element burnout.
- Typical minimum: **2 m/s (400 ft/min)** across the coil.

Safety Features

- Include safety devices such as:
  - **Over-**temperature cutout switches.
  - Airflow switch**es** to ensure proper operation.

Placement and Clearance

- Ensure adequate clearance around the heater for safe operation and maintenance.
- Avoid locations where the heater may be exposed to water, dust, or debris.

Temperature Rise

- Typical temperature rise range: **10–40 °C (18–72 °F)**.
- For reheating applications, aim for a smaller temperature rise (e.g., **10–15 °C**).

Efficiency

- Electric heating is typically **100% efficient** in converting electrical energy to heat.

Rules of Thumb for Selection

- **Heating Load Requirements**
  - Select a coil with a capacity that matches or slightly exceeds the calculated load.
- **Control Options**
  - Use **step controls** or **modulating controls** for precise temperature regulation.
- **Material**
  - Open coil heaters: Use in clean, low-contaminant environments.
  - Finned tubular heaters: Use in humid or contaminated airflows.
- **Installation Environment**
  - For outdoor applications or corrosive environments, use **stainless steel or corrosion-resistant materials**.
- **System Integration**
  - Ensure compatibility with the HVAC control system for proper sequencing and safety interlocks.

Common Applications

- **Space Heating**: Supplement heating for rooms or zones.
- **Reheat Coils**: Installed downstream of cooling coils to control humidity and air temperature.
- **Duct Heating**: Heating for air supplied to large spaces or processes.
- **Makeup Air Units**: Preheat incoming ventilation air.

8.12 Expansion Tanks

By following these guidelines, HVAC expansion tanks can be properly sized and selected to ensure system stability, pressure control, and reliable operation. Here are the key rules of thumb for selecting and sizing expansion tanks for HVAC systems in cooling and heating applications.

An illustration of a Expansion Tank

Types of Expansion Tanks

Type	Description	Applications	Advantages	Disadvantages
Diaphragm/ bladder tanks	Features a flexible diaphragm or bladder that separates air and water	Most modern HVAC systems, both heating and cooling	Compact, low maintenance, no air-water contact	Limited capacity for very large systems
Plain steel tanks	Non-separated design where air and water are in direct contact	Older systems, larger systems with low-pressure needs	Simpler design, suitable for large volumes	Requires regular air replenishment
Sealed tanks	Fully enclosed tanks with pre-charged air	High-pressure systems, space-constrained applications	Compact, no evaporation or air leakage	Higher cost, requires proper pre-charging
Thermal expansion tanks	Specifically designed for domestic water systems	Domestic hot water systems	Controls pressure due to thermal expansion	Not used for hydronic HVAC systems

Sizing Guidelines

Expansion tanks are sized based on system volume, temperature range, and operating pressure.

Sizing Formula:

$$V_t = \frac{V_s \times E}{1 - \frac{P_i}{P_f}}$$

Where:

- $\mathbf{V_t}$: Required volume of the expansion tank (in litters or cubic meters).
- $\mathbf{V_s}$: Volume of the fluid in the system (in litters or cubic meters).
- **E**: Expansion coefficient of the fluid, which depends on the type of fluid (e.g., water or glycol) and its temperature range.
- $\mathbf{P_i}$: Initial system pressure (in absolute pressure, e.g., kPa or bar).
- $\mathbf{P_f}$: Final system pressure, which is typically the maximum allowable pressure (in absolute pressure, e.g., kPa or bar).

Explanation

Fluid Expansion: The formula calculates the volume of the expansion tank required to accommodate the increase in fluid volume as it heats up and expands.

Pressure Effect: The term $1 - (P_i/P_f)$ accounts for how much the system's pressure changes as the fluid expands, affecting the expansion tank's size.

Simplification: Ensure that P_i and P_f are in **absolute pressure** (not gauge pressure). If you have gauge pressure, convert it by adding atmospheric pressure (e.g., 101.3 kPa or 1.013 bar).

Key Notes

- If $\mathbf{P_i}$ is close to $\mathbf{P_f}$, the denominator 1 − (Pi/Pf) becomes small, which increases V_t, indicating a larger expansion tank is needed.
- Ensure the fluid type and temperature range are considered to calculate the correct **E** value.

Key Parameters

- **Volume of the fluid in the system (V_s):**
  - Includes the total volume of water in the pipes, equipment, and coils.
  - Approximation:
    - Residential systems: **10–20 L/kW** (cooling or heating load).
    - Commercial systems: **20–30 L/kW**.

Required Volume of the Expansion Tank in Cooling Systems

- **Tank Sizing Approximation:**
  - Vt ≈ 3–5% of the total system volume (cooling loop volume).
  - Use 3% if the system has a relatively stable temperature range (e.g., chilled water systems).

  - Use **5%** if the system operates with more significant temperature fluctuations or uses a glycol-water mix (glycol has a higher expansion coefficient than water).

- **Key Considerations for Cooling Systems:**
  - Account for the specific heat transfer fluid. A higher glycol concentration increases the expansion coefficient.
  - Ensure that the system's minimum and maximum operating pressures are known to refine the estimate.

Required Volume of the Expansion Tank in Heating Systems

- **Tank Sizing Approximation:**
  $V_t \approx$ 8–12% of the total system volume (heating loop volume).
  - Use 8% if the system operates within a low-temperature range, such as radiant floor heating (30–50 °C).
  - Use 12% for high-temperature systems, such as boiler-fed systems (up to 90–100 °C).

- **Key Considerations for Heating Systems:**
  - Higher temperatures lead to more significant water expansion, necessitating larger tanks.
  - Pressure differences between the initial cold fill pressure and maximum system pressure must also be factored in for precise sizing.

Example 1: Cooling System

Scenario

- A chilled water system with a total system volume (Vs) of **10,000 L**.
- The system operates at a relatively stable temperature range 5–12 °C.
- No glycol is used (pure water).

Rule of Thumb

- $V_t \approx$ 3–5% of V_s.

Calculation

- Lower estimate: $V_t = 0.03 \times 10{,}000 = 300$ L
- Upper estimate: $V_t = 0.05 \times 10{,}000 = 500$ L

Result The expansion tank size should be between **300 and 500 L**, depending on safety margins and pressure limits.

Example 2: Heating System

Scenario

- A boiler-fed heating system with a total system volume (V_s) of **5000 L**.
- The system operates between 20 °C (cold fill) and 90 °C (maximum operating temperature).
- Pure water is used.

Rule of Thumb

- $V_t \approx$ 8–12% of V_s

Calculation

- Lower estimate: $V_t = 0.08 \times 5000 = 400$ L
- Upper estimate: $V_t = 0.12 \times 5000 = 600$ L

Result The expansion tank size should be between **400 and 600 L**.

Total System Volume Based on the Circulation Time

If the flow rate of chilled water is known, one can estimate the system volume based on the time it takes to circulate the fluid.

Formula:

$$V_s = V \times t$$

Where:

- **Vs**: Total system volume (in litters or cubic meters).
- **V**: Flow rate (in m^3/s or L/min).
- **t**: Circulation time, typically assumed as 3 min for HVAC systems.

Estimating Flow Rate (V)

- One can calculate the flow rate from the cooling capacity and temperature difference:

$$V = \frac{Q}{\Delta T \times \rho \times c_p}$$

Where:

- **Q**: Cooling capacity (in kW).
- **ΔT**: Temperature difference across the system (typically 5 °C for chilled water systems).
- **ρ**: Density of water (1000 kg/m^3).
- $\mathbf{c_p}$: Specific heat capacity of water (4.18 kJ/kg °C).

- **Temperature Range**:
  - Determine the temperature increase for the system.
  - Typical ranges:
    - Heating systems: **20–90 °C**.
    - Cooling systems: **4–15 °C**.
- **Expansion Coefficient**:
  - The expansion of water is a function of temperature.
  - Approximate values:
    - 4–60 °C: **2.5%**.
    - 4–90 °C: **3.8%**.

- **Operating Pressure**:
  - Consider system working pressure and height.
  - Typical pressures:
    - Residential systems: **1–2 bar (100–200 kPa)**.
    - Commercial systems: **2–4 bar (200–400 kPa)**.
- **Safety Margin**:
  - Add a **10–20% safety margin** to calculated tank volume.

Selection Guidelines

For Heating Systems
- Use a diaphragm or bladder tank for most modern systems.
- Ensure the tank can handle high temperatures (e.g., up to 90 °C).
- Select larger tanks for systems with long pipe runs or large water volumes.

For Cooling Systems
- Diaphragm tanks are ideal for chilled water systems.
- Consider the lower temperature range (e.g., 4–15 °C) for sizing.
- Account for glycol mixture if used (adjust expansion coefficient).

For Pressurized Systems
- Use sealed tanks with high-pressure ratings for closed-loop systems.
- Ensure proper pre-charge pressure to match system requirements.

Rules of Thumb for Sizing

- **Residential Systems**:
  - Heating: **10–20 L/kW** of heating load.
  - Cooling: **5–15 L/kW** of cooling load.
  - Tank size: **8–12% of system volume**.
- **Commercial Systems**:
  - Heating: **15–30 L/kW** of heating load.
  - Cooling: **10–25 L/kW** of cooling load.
  - Tank size: **8–10% of system volume** for heating, **3–5%** for cooling.
- **High-Temperature Applications**:
  - For systems above 90 °C, increase the tank size by **10–20%** to account for higher expansion.
- **Glycol Systems**:
  - Adjust expansion coefficient based on the glycol concentration (e.g., **30% glycol increases expansion factor by ~15%**).

Common Considerations

- **Pre-Charge Pressure**:
  - Match the tank's pre-charge pressure to the system's static pressure (based on building height and pump location).
  - **Rule of Thumb**: 10 kPa per meter of system height +20 kPa safety margin
- **Location**:
  - Install near the pump suction side for optimal operation.
  - Ensure the tank is accessible for maintenance.
- **System Compatibility**:
  - Ensure the tank's materials are compatible with the fluid (e.g., glycol-water mixtures).
- **Maintenance**:
  - For diaphragm tanks, check air pressure annually.
  - For plain steel tanks, ensure regular air replenishment to prevent waterlogging.

Practical Applications

Building type	System type	Recommended tank type	Sizing considerations
Residential Homes	Heating, Cooling	Diaphragm/Bladder Tanks	Small to medium tanks; typically **8–15% of system volume**
Office buildings	Chilled Water	Sealed Tanks	Larger tanks; account for glycol and long pipe runs
Hospitals and labs	Heating, Cooling	High-Pressure Tanks	High capacity; account for redundancy in critical systems
Industrial facilities	Process Cooling	Steel or Sealed Tanks	Oversized tanks to handle large temperature swings

8.13 Buffer Tanks

Buffer tanks are used in HVAC systems to enhance system stability, manage thermal storage, and reduce equipment cycling. They are often included in chilled water, hot water, and heat pump systems. Proper sizing and selection ensure efficiency and longevity of the HVAC system.

An illustration of a Buffer Tank

Types of Buffer Tanks

- **Chilled Water Buffer Tanks**:
  - Used in chilled water systems to increase system water volume and reduce short-cycling of chillers.
  - Common in systems with variable load conditions, such as air-cooled chillers or VRF systems with water-cooled condensers.
- **Hot Water Buffer Tanks**:
  - Applied in hydronic heating systems to store excess thermal energy from boilers or heat pumps.
  - Helps maintain consistent hot water supply and reduces boiler short-cycling.
- **Thermal Storage Tanks**:
  - Designed to store excess heating or cooling energy during off-peak periods for use during peak demand.
  - Common in systems using renewable energy or operating in regions with time-of-use energy pricing.
- **Combination Buffer Tanks**:
  - Tanks designed to handle both chilled and hot water systems by maintaining separate compartments or loops.
  - Useful in systems requiring seasonal changeover between heating and cooling.

- **Solar Thermal Buffer Tanks**:
  - Used in solar thermal systems to store excess solar energy for domestic hot water or space heating.

Sizing Buffer Tanks

Chilled Water Buffer Tanks

- **General Sizing Formula**:

$$\text{Volume (Liters)} = 10 * \text{Chiller Capacity (kW)}$$

  This formula is a **rule of thumb** and works for many typical HVAC applications, but it might need adjustment for specific system requirements:

- **Systems with Variable Flow Rates:** The formula assumes constant flow. If your system has variable flow rates (e.g., with VFD pumps), a larger buffer tank may be needed.
- **High Thermal Stability Requirements:** For systems requiring precise temperature control, larger buffer tanks might be recommended.
- **Delta-T Variations:** If ΔT is different from the standard 5 °C the flow rate will change, affecting the tank size.
- **Rules of Thumb**:
  - **Small Systems (<300 kW)**: Provide at least 20–30 L of buffer per kW of chiller capacity.
  - **Large Systems (>300 kW)**: Provide at least 10 L/kW.
  - Ensure at least 1 min of system water volume turnover to prevent short-cycling.

Buffer Tank Volume Formula

$$Tank\ Volume\left(Liters\right) = \frac{t_r \times Q}{4.187 \times \Delta T}$$

where:

t_r: Desired Minutes of Run Time (s).
Q: System Load (kW).
ΔT: Temperature Differential (°C).
4.187 = Specific heat of water (kJ/kg °C), with density ≈ 1 kg/L.

Key Parameters Explained

- **Desired Minutes of Run Time:**
  - The minimum time the system should run continuously without cycling off.

- This is crucial to prevent frequent chiller cycling, which can lead to wear and tear and inefficient operation.
- Typical values range from **5 to 10 min** for most applications.

- **System Load (kW):**
  - The cooling or heating load the chiller or boiler is designed to handle.
- **Temperature Differential ΔT (°C):**
  - The difference in temperature between the supply and return water in the system.

This formula calculates the tank volume needed to store enough water to maintain the system's operation for the specified time without cycling, based on the energy balance between the load and the available temperature difference.

Thermal Storage Tanks

- **Rules of Thumb**:
  - Size based on the maximum thermal demand of the building during peak hours.
  - Typically, 40–60 L/kW of peak thermal load.
  - Account for charging/discharging cycles and energy storage requirements.

Solar Thermal Buffer Tanks

- **Rules of Thumb**:
  - Residential systems: 50–80 L/m^2 of solar collector area.
  - Commercial systems: 30–50 L/m^2, depending on system demand.

Selection Criteria

- **System Capacity**:
  - Match the buffer tank volume to the thermal output of the chiller, boiler, or heat pump.
- **System Flow Rate**:
  - Ensure the tank supports the system's flow rate without causing excessive pressure drop.
- **Tank Material**:
  - **Steel Tanks**: Durable, suitable for high-pressure systems.
  - **Stainless Steel Tanks**: Corrosion-resistant, ideal for potable water systems or corrosive environments.

- **Insulation**:
  - High-quality insulation reduces heat loss or gain, improving system efficiency.
- **Connection Ports**:
  - Ensure adequate and properly sized inlet/outlet ports to connect with system piping.

- **Placement**:
  - Vertical tanks save floor space; horizontal tanks may be required for height-constrained spaces.
- **Control Features**:
  - Look for tanks with integrated temperature sensors, sight glasses, or ports for external controls.

Examples of Buffer Tank Applications

- **Residential Chilled Water System**:
  - Chiller capacity: 20 kW.
  - Buffer tank volume: 400 L (20 L/kW).
- **Commercial Hot Water Heating System**:
  - Boiler capacity: 200 kW.
  - Buffer tank volume: 2500 L (12.5 L/kW).
- **Solar Thermal System for an Office Building**:
  - Collector area: 50 m^2.
  - Buffer tank volume: 2500 L (50 L/m^2).

Difference Between Buffer Tank and Expansion Tanks

Buffer Tank

- **Purpose**
  - To increase the thermal inertia of a system by storing thermal energy (either chilled or hot water).
  - To minimize equipment cycling (e.g., chillers, boilers, or heat pumps) and improve system stability and efficiency.
  - Used in systems with variable loads to prevent rapid temperature fluctuations.
- **Applications**:
  - Chilled water systems.
  - Hot water heating systems.
  - Thermal storage systems.
  - Solar heating systems.
- **Key Features**:
  - Stores water at the operating pressure of the system.
  - Does not compensate for pressure changes caused by thermal expansion.
  - Insulated to minimize heat loss or gain.
- **Sizing**:
  - Based on system capacity, flow rates, and desired temperature stability.

Expansion Tank

- **Purpose**:
  - To manage pressure fluctuations in a closed-loop system caused by thermal expansion of water when it is heated.
  - Prevents overpressurization that could damage the system's components.
- **Applications**:
  - Hydronic heating systems (hot water loops).
  - Chilled water cooling systems (cold water loops).
- **Key Features**:
  - Divided into two sections: one for system water and the other for air or gas (usually nitrogen), separated by a diaphragm or bladder.
  - Absorbs the expansion of water to maintain consistent system pressure.
  - Not insulated, as its purpose is not thermal energy storage.
- **Sizing**:
  - Based on the total system water volume, maximum temperature, and system pressure.

Comparison

Aspect	Buffer tank	Expansion tank
Primary function	Stores thermal energy for system stability	Manages pressure fluctuations due to water expansion
System type	Used in thermal systems (heating/ cooling)	Used in closed-loop systems to control pressure
Key design	Insulated tank with large volume capacity	Divided sections (water/air) with a diaphragm or bladder
Impact on pressure	Does not directly affect system pressure	Directly impacts and controls system pressure
Sizing basis	System capacity, thermal load, and flow rate	Total system volume, temperature rise, and pressure
Material	Typically steel with insulation	Steel with a flexible diaphragm or bladder

Example in an HVAC System

- A **chilled water system** may have:
  - **Buffer Tank**: To ensure smooth chiller operation by adding thermal inertia.
  - **Expansion Tank**: To accommodate water expansion when the temperature changes in the closed-loop.

8.14 HVAC Valves

HVAC valve sizing and selection are critical to ensuring proper system performance, energy efficiency, and longevity. Followings are some key rules of thumb for valve sizing and selection.

Understand the Application

- Determine the purpose of the valve: controlling flow, balancing, isolating, or relieving pressure.
- Know the type of system: chilled water, hot water, steam, or air.

Flow Coefficient (Kv)

- The **Kv value** of a valve represents its flow capacity at a pressure drop of 1 bar for water at 15 °C.
- Select a valve with a **Kv value** that matches the calculated system flow requirements.
- Avoid oversizing or undersizing. **Oversizing** can lead to poor control and hunting, while **under sizing** restricts flow and reduces efficiency.

Pressure Drop

- Target a pressure drop across the valve of **20–35 kPa** for most water systems.
- For steam systems, ensure the valve can handle higher pressure drops and temperatures without excessive wear.

Valve Authority

- Valve authority = ΔP (pressure drop across the valve) $\div$ ΔP (total system pressure drop, including valve).
- Aim for a valve authority between **0.3 and 0.5** for optimal control.

Valve Type
Select the appropriate valve type based on application:

- **Globe Valves**: Precise flow control, ideal for modulating applications.
- **Ball Valves**: Quick shutoff, minimal pressure drop, suitable for isolation and some control applications.
- **Butterfly Valves**: Large systems, lower cost, less precise control.
- **Gate Valves**: On/off isolation, not suitable for throttling.
- **Pressure-Independent Control Valves (PICVs)**: Maintain constant flow regardless of system pressure variations.

Material Compatibility

- Match valve materials to the fluid type and operating conditions.
- For water: Bronze, brass, or stainless steel.
- For steam: Cast iron, steel, or high-temperature alloys.
- Ensure materials are corrosion-resistant for the system medium.

Actuation

- Choose the appropriate actuator type:
  - **Manual**: For basic on/off control.
  - **Electric/Pneumatic**: For modulating or automated systems.
  - Ensure the actuator torque matches the valve's operational requirements.

System Design Considerations

- Account for pipe size but avoid sizing valves purely by pipe diameter.
- Ensure sufficient clearance for installation, maintenance, and operation.
- Confirm that the valve meets the required pressure and temperature ratings.

Noise and Cavitation

- For high-pressure systems, choose valves designed to reduce noise and cavitation, which can damage equipment and reduce lifespan.

Regulations and Standards

- Ensure compliance with relevant codes and standards (e.g., ISO, EN, or local regulations).

Common Pitfalls to Avoid

- **Oversizing**: Leads to unstable operation and inefficient energy use.
- **Undersizing**: Causes flow restrictions and system performance issues.
- **Ignoring Kv**: Always match the valve Kv to the design requirements.
- **Incorrect Pressure Ratings**: Can lead to valve failure.

Globe Valves

- **Chilled Water Systems**:
  - To reg**ulate the flow in secondary chilled water loops.**
  - **Installed on bypass lines around equipment (e.g., chillers or AHUs) to control flow rates.**
  - **Balan**cing valves in fan coil unit loops or terminal units.
- **Hot Water Systems**:
  - Used to control flow in heating circuits.
  - Often paired with temperature sensors for maintaining precise temperatures.

An illustration of HVAC globe valve, commonly used in building services

- **Steam Systems**:
  - In systems requiring modulation, such as steam humidifiers or steam distribution.
- **Cooling Tower Systems**:
  - Regulate the flow of water in bypass or bleed-off lines.
- **Pressure Reducing Stations**:
  - Control the downstream pressure by throttling flow.
- **Process Systems**:
  - Used in industrial processes requiring precise flow or pressure control.

Sizing of Globe Valves

- **Flow Coefficient (Cv)**:
  - The valve is selected based on the required flow coefficient:

    $$C_v = Q\sqrt{\frac{SG}{\Delta P}}$$

    - Q: Flow rate (m^3/h or L/s).
    - ΔP: Pressure drop across the valve (kPa or bar).
    - SG: Specific gravity (water = 1.0).
  - For chilled water and hot water systems, the pressure drop (ΔP) across the globe valve is typically between **10 and 30 kPa**.

- **Flow Range**:
  - Select a globe valve that operates within 60–80% of its maximum Cv value to ensure accurate control.
- **Pipe Size Matching**:
  - Use a valve with the same nominal diameter as the pipe unless calculations indicate the need for a smaller valve for better control.
- **Pressure Ratings**:
  - Select valves rated for the system's operating pressure, typically **PN16** or **PN25** in HVAC applications.
- **Actuation**:
  - For automated systems, ensure compatibility with actuators (electric or pneumatic) based on torque and signal requirements.

Selection Rules of Thumb

- **Material:**
  - Choose based on the fluid and operating conditions:
    - Cast iron or carbon steel for water systems.
    - Stainless steel for corrosive environments.
    - Bronze or brass for smaller systems.
- **End Connections**:
  - Flanged connections for larger systems or higher pressures.
  - Threaded or soldered for smaller pipes.
- **Flow Characteristics**:
  - Equal percentage or linear characteristics depending on the application:
    - Equal percentage for variable flow systems.
    - Linear for constant flow systems.

Practical Examples

Example 1: Globe Valve in a Chilled Water System

- **System Details**:
  - Chilled water flow rate: 1.5 L/s
  - Pressure drop across valve: 7 kPa
  - Specific gravity: 1
- **Sizing**:

$$C_v = Q\sqrt{\frac{SG}{\Delta P}} = 5.4 \times \sqrt{\frac{1}{7 \times 0.01}} = 20.4$$

- **Selection**:
  - Select a globe valve with Cv of 20. (1.5L/s is equal to 5.4 m3/h).
  - Material: Cast iron.
  - End connection: Flanged.

Example 2: Globe Valve for Steam Humidifier

- **System Details**:
  - Steam flow rate: 0.25 kg/s (1.5 m3/h)
  - Pressure drop: 0.2 bar
  - Specific gravity: 0.6
- **Sizing**:

$$C_v = Q\sqrt{\frac{SG}{\Delta P}} = 1.5 \times \sqrt{\frac{0.6}{0.2}} = 2.6$$

- **Selection**:
  - Select a small stainless-steel globe valve with Cv of 3–4.
  - Actuation: Electric modulating actuator for precise flow control.

Example 3: Cooling Tower Bypass Valve

- **System Details**:
- Flow rate: 10 L/s (36 m3/h)
- Pressure drop: 10 kPa
- **Sizing**:

$$C_v = Q\sqrt{\frac{SG}{\Delta P}} = 36 \times \sqrt{\frac{1}{10 \times 0.01}} = 113.8$$

- **Selection**:
  - Select a flanged cast iron globe valve with Cv of 115.
  - Actuation: Pneumatic actuator for easy integration into the control system.

Advantages of Globe Valves

1. Precise flow regulation.
2. Can handle high-pressure drops.
3. Suitable for throttling applications.
4. Durable and reliable in demanding systems.

Disadvantages of Globe Valves

1. Higher pressure drop compared to other valves (e.g., gate or ball valves).
2. Bulkier design can require more space.
3. More expensive than simpler valve types.

Ball Valves

Ball valves are primarily used in HVAC and building services for on/off control, isolation, and bypass applications. They are not typically designed for precise throttling but are excellent for applications requiring quick shutoff and minimal pressure drop.

An illustration of HVAC Ball valve, commonly used in building services

Where Ball Valves Are Installed

- **Isolation Applications**:
  - Installed **upstream and downstream of major equipment (e.g., chillers, pumps, cooling towers, AHUs) for maintenance.**
  - **Isolation for** fan coil units or terminal units in chilled or hot water systems.
- **Control Bypass Applications**:
  - Used in bypass lines to enable manual flow adjustments.
- **System Draining and Filling**:
  - Installed at system low points for draining.
  - Installed at high points or inline for filling and air venting.
- **Cooling Tower Systems**:
  - For isolation of piping to/from cooling towers.
- **Chemical Treatment and Sampling**:
  - Used in chemical treatment loops or sampling points in cooling water systems.
- **Firefighting Systems**:
  - For isolation of branches in sprinkler systems.

Sizing of Ball Valves

1. **Flow Capacity**:
   - Ball valves typically have a high flow coefficient (Cv) and very low pressure drop.
   - Match the valve size to the pipe size unless specific flow throttling is required.
2. **Pressure Ratings**:
   - Select based on system pressure, commonly **PN16** or **PN25** for HVAC applications.
3. **Actuation**:
   - For automated systems, ensure compatibility with actuators (electric, pneumatic) based on torque requirements.
4. **Material Selection**:
   - Use materials based on fluid type and operating conditions:
     - Bronze or brass for smaller systems or potable water.
     - Stainless steel for corrosive environments.
     - Carbon steel for chilled or hot water systems.
     - PVC for low-temperature, non-pressurized applications.

Selection

- **Port Type**:
  - **Full Port: For minimal pressure drop and high flow capacity.**
  - **Reduced Port**: For applications where throttling is occasionally required or smaller flow rates are acceptable.
- **Connection Type**:
  - **Threaded**: For small-diameter pipes in low-pressure systems.
  - **Flanged**: For larger pipes and higher pressures.
  - **Welded or Grooved**: For permanent connections in industrial applications.
- **Temperature and Fluid Compatibility**:
  - Ensure the valve material is compatible with the system fluid and temperature range.

Practical Examples

Example 1: Isolation Valve for a Chiller

- **System Details**:
  - Pipe size: 100 mm
  - Flow rate: 20 L/s
  - Pressure rating: PN16.
- **Selection**:
  - Material: Cast steel or stainless steel.
  - Type: Full port ball valve.
  - Connection: Flanged.

Example 2: Cooling Tower Make-Up Water Valve

- **System Details**:
  - Pipe size: 50 mm
  - Pressure: 6 bar
  - Fluid: Potable water.
- **Selection**:
  - Material: Brass or bronze.
  - Type: Reduced port ball valve.
  - Connection: Threaded.

Example 3: Fan Coil Unit Isolation Valve

- **System Details**:
  - Pipe size: 25 mm
  - Flow rate: 0.2 L/s
- **Selection**:
  - Material: Brass.
  - Type: Full port.
  - Actuation: Manual lever.

Example 4: Firefighting Branch Isolation

- **System Details**:
  - Pipe size: 150 mm
  - Pressure: 12 bar
- **Selection**:
  - Material: Stainless steel or carbon steel.
  - Type: Full port ball valve.
  - Connection: Grooved.

Advantages of Ball Valves

1. Quick and easy operation.
2. Tight shutoff, suitable for isolation.
3. Minimal pressure drop due to full-port design.
4. Compact and lightweight.

Disadvantages of Ball Valves

1. Not suitable for precise throttling or flow regulation.
2. Seals can degrade in high-temperature applications.
3. Limited to specific ranges of pressure and temperature.

Butterfly Valves

Butterfly valves are widely used in HVAC and building services for on/off control and throttling of fluid flow. They are preferred for their compact size, lightweight, and cost-effectiveness, especially in large-diameter piping systems.

Where Butterfly Valves Are Installed in HVAC and Building Services

- **Chilled Water Systems**:
  - Isolation valves in main headers or branch lines.
  - Balancing and throttling valves in primary and secondary loops.
- **Hot Water Systems**:
  - Isolation and flow control in large-diameter heating systems.
- **Cooling Tower Systems**:
  - Isolation and control for cooling water supply and return lines.
  - Throttling in cooling tower bypass lines.
- **Air Handling Systems**:
  - Dampers in large air ducts, acting as airflow control valves.
- **Firefighting Systems**:
  - Used as isolation valves in sprinkler system mains.
- **Process and Utility Water Systems**:

Isolation valves for non-critical applications.

An illustration of an Butterfly valve, commonly used in building services

Sizing of Butterfly Valves

- **Flow Coefficient (Cv)**:
  - Butterfly valves generally have a high flow capacity, but their Cv decreases significantly at partially open positions.
  - Ensure the valve is properly sized for the desired flow rate and pressure drop.

- **Pipe Size Matching**:
  - The valve size is typically matched to the pipe size unless a smaller size is needed for throttling purposes.
- **Pressure Drop**:
  - Designed for low-pressure drops, with minimal resistance when fully open.
- **Valve Material**:
  - Select based on the fluid and system operating conditions:
    - Cast iron or ductile iron for chilled and hot water systems.
    - Stainless steel for corrosive or high-temperature environments.
- **Actuation**:
  - Manual (lever or gear-operated) for simple systems.
  - Electric or pneumatic actuators for automated systems.
- **Pressure and Temperature Ratings**:
  - Typically rated for **PN10, PN16, or PN25**, depending on the application.
  - Temperature range varies by material, from −20 °C to 200 °C for HVAC applications.

Selection

- **Disc Material**:
  - Aluminum bronze, stainless steel, or nylon-coated for corrosive environments.
- **Seat Material**:
  - EPDM for water systems up to 120 °C.
  - PTFE for higher temperatures or chemical resistance.
- **Type**:
  - **Wafer**: Compact and suitable for lower-pressure applications.
  - **Lug**: For systems where the valve must remain in place during downstream maintenance.
  - **Double Offset**: For higher pressure and reduced wear.
  - **Triple Offset**: For critical applications requiring tight shutoff and high durability.
- **Leakage Class**:
  - Class VI (bubble-tight) for HVAC isolation applications.

Practical Examples

Example 1: Butterfly Valve in a Chilled Water Main

- **System Details**:
  - Pipe diameter: 300 mm
  - Flow rate: 100 L/s
  - Pressure rating: PN16.

- **Selection**:
  - Valve size: 300 mm
  - Material: Ductile iron body with aluminium bronze disc.
  - Type: Lug type for ease of maintenance.
  - Seat: EPDM.

Example 2: Cooling Tower Isolation Valve

- **System Details**:
  - Pipe diameter: 200 mm
  - Flow rate: 50 L/s
  - Fluid: Cooling water with potential corrosion.
- **Selection**:
  - Valve size: 200 mm
  - Material: Stainless steel disc and nylon-coated body.
  - Type: Wafer type.
  - Actuation: Manual gear-operated.

Example 3: Firefighting System Isolation Valve

- **System Details**:
  - Pipe diameter: 150 mm
  - Pressure: 12 bar
- **Selection**:
  - Valve size: 150 mm
  - Material: Cast iron body with stainless steel disc.
  - Type: Lug type.
  - Seat: Nitrile for compatibility with firefighting water.

Example 4: AHU Balancing Valve

- **System Details**:
  - Pipe diameter: 80 mm
  - Flow rate: 15 L/s
- **Selection**:
  - Valve size: 80 mm
  - Material: Ductile iron body with stainless steel disc.
  - Type: Wafer type with manual lever.

Advantages of Butterfly Valves

1. Compact and lightweight.
2. Cost-effective for large-diameter applications.
3. Quick operation with 90° rotation.
4. Low pressure drop when fully open.
5. Suitable for automation.

Disadvantages of Butterfly Valves

1. Limited precision for throttling compared to globe valves.
2. Not suitable for high-pressure drops or tight shutoff in critical applications.
3. Can cause cavitation if used improperly in throttling.

Gate Valves

Gate valves are widely used in HVAC and building services for isolation purposes. They are not typically used for throttling or flow control because they are designed to be fully open or fully closed. Their ability to provide minimal pressure drop when fully open makes them ideal for applications requiring unobstructed flow.

An illustration of an Gate valve, commonly used in building services

Where Gate Valves Are Installed

- **Chilled Water Systems**:
  - Isolation of pipelines, **pumps, chillers, and cooling** towers.
  - Installed on branch lines for system isolation.
- **Hot Water Systems**:
  - Used for isolating boilers, heating units, and main headers.
- **Firefighting Systems**:
  - Installed as isolation valves in sprinkler mains and branches.
- **Cooling Tower Systems**:
  - Isolating large-diameter supply and return pipelines.

- **Drainage and Filling**:
  - Used in system drain lines or large system filling lines.

Sizing of Gate Valves

- **Match Valve to Pipe Size**:
  - Typically, the valve size matches **the pipe size unless specific system constraints** require otherwise.
- **Pressure Ratings**:
  - Standard pressure ratings for HVAC applications are PN16 or PN25.
  - For high-pressure applications, higher ratings like PN40 may be used.
- **Material Compatibility**:
  - **Cast iron** or **ductile iron** for standard HVAC water systems.
  - **Stainless steel** or **bronze** for corrosive environments or potable water.
- **Valve Configuration**:
  - **Rising Stem**: Easy to observe whether the valve is open or closed.
  - **Non-Rising Stem**: Suitable for areas with space constraints.
- **End Connections**:
  - **Flanged**: For large-diameter pipes and high-pressure systems.
  - **Threaded**: For smaller systems or low-pressure applications.
  - **Grooved**: For quick installation in HVAC systems.
- **Operation Type**:
  - Manual handwheel for simple systems.
  - Electric actuators for automation.

Selection

1. **Flow Considerations**:
   - Gate valves have minimal resistance to flow when fully open.
   - Ensure the valve is fully open during normal operation to avoid wear.
2. **Temperature Rating**:
   - Select materials that can handle the system's maximum operating temperature.
3. **Leakage Class**:
   - Gate valves are typically designed for zero leakage when closed.
4. **Size Limits**:
   - Gate valves are preferred for larger pipe sizes (e.g., >50 mm) due to their robust design.

Practical Examples

Example 1: Chilled Water Main Isolation Valve

- **System Details**:
  - Pipe size: 250 mm
  - Flow rate: 90 L/s
  - Pressure: PN16.
- **Selection**:
  - Valve type: Rising stem gate valve.
  - Material: Ductile iron body with stainless steel stem.
  - Connection: Flanged.

Example 2: Hot Water Boiler Isolation Valve

- **System Details**:
  - Pipe size: 100 mm
  - Pressure: PN16.
  - Temperature: 80 °C
- **Selection**:
  - Valve type: Non-rising stem gate valve.
  - Material: Cast iron body with brass wedge.
  - Connection: Threaded.

Example 3: Firefighting Sprinkler Main Isolation

- **System Details**:
  - Pipe size: 150 mm
  - Pressure: 12 bar
- **Selection**:
  - Valve type: Rising stem gate valve.
  - Material: Bronze or stainless steel.
  - Connection: Flanged.

Example 4: Cooling Tower Supply Line Valve

- **System Details**:
- Pipe size: 200 mm
- Flow rate: 50 L/s
- **Selection**:
  - Valve type: Non-rising stem gate valve.
  - Material: Ductile iron with stainless steel internals.
  - Connection: Grooved.

Advantages of Gate Valves

1. Minimal pressure drop when fully open.
2. Suitable for large-diameter piping systems.
3. Durable and reliable for isolation.
4. Tight shutoff capability.

Disadvantages of Gate Valves

1. Not suitable for throttling or flow regulation.

2. Requires significant space for operation (especially rising stem types).
3. Susceptible to wear if operated partially open.
4. Slower to operate compared to ball or butterfly valves.

Pressure-Independent Control Valves

Pressure-Independent Control Valves (PICVs) combine the functions of a balancing valve and a control valve, maintaining a constant flow regardless of pressure fluctuations in the system. They are widely used in modern HVAC systems to improve energy efficiency, provide precise control, and reduce maintenance.

Where PICVs Are Installed

- **Chilled Water Systems**:
  - At the inlet of terminal units such as fan coil units (FCUs), air handling units (AHUs), or chilled beams.
  - In primary and secondary loops to control flow to specific zones.
- **Hot Water Systems**:
  - At the inlet of radiators, heating coils, or other heating terminals.
- **Variable Flow Systems**:
  - To optimize performance in systems with variable speed pumps or fluctuating demand.
- **Zone Control Applications**:
  - For precise temperature and flow regulation in building zones.
- **District Cooling and Heating Systems**:
  - At the interface between the district system and the building's internal distribution network.

Illustration of a Pressure-Independent Control Valves, commonly used in building services

Sizing of PICVs

- **Flow Rate**:
  - Size based on the design flow rate required for the terminal unit or zone.
  - Ensure the valve can handle the maximum flow rate needed without exceeding its range.
- **Pressure Range**:
  - Check the valve's operating pressure range to ensure compatibility with the system.
  - Typical PICVs operate effectively between 16 and 400 kPa differential pressure.
- **Pipe Size**:
  - Match the valve size to the terminal connection or pipe size unless specific design constraints dictate otherwise.
- **Valve Authority**:
  - PICVs inherently maintain a high valve authority (close to 1) due to their pressure-independent operation.
- **Flow Characteristics**:
  - Ensure the valve's flow characteristics match the system requirements, typically linear or equal percentage.
- **Control Signal**:
  - Verify compatibility with the Building Management System (BMS), as most PICVs use modulating actuators (e.g., 0–10 V or 4–20 mA).

Selection

- **Material:**
  - **Brass or bronze** for smaller sizes and standard water systems.
  - **Stainless steel** for corrosive or high-temperature systems.
- **Connection Type**:
  - **Threaded**: For smaller systems, such as FCUs.
  - **Flanged**: For larger units like AHUs.
- **Actuator**:
  - Select electric or pneumatic actuators based on the control system requirements.
  - Modulating actuators are standard for precise temperature control.
- **Pressure Rating**:
  - Select valves rated for the system's maximum operating pressure (e.g., PN16, PN25).
- **System Compatibility**:
  - Ensure the valve is compatible with water or other fluids used in the system.

Practical Examples

Example 1: PICV for Fan Coil Unit

- **System Details**:
  - FCU flow rate: 0.1 L/s
  - Pressure differential: 50 kPa
- **Selection**:
  - Valve size: 15 mm
  - Material: Brass.
  - Flow range: 0.05–0.2 L/s
  - Actuator: Electric modulating (0–10 V).

Example 2: PICV for Air Handling Unit

- **System Details**:
  - AHU flow rate: 5 L/s
  - Pressure differential: 200 kPa
- **Selection**:
  - Valve size: 50 mm
  - Matcrial: Stainless steel.
  - Flow range: 4–6 L/s
  - Actuator: Electric modulating (4–20 mA).

Example 3: PICV for Chilled Beam System

- **System Details**:
  - Flow rate per chilled beam: 0.03 L/s
  - Number of beams: 20.
- **Selection**:
  - Valve size: 10 mm for each beam.
  - Material: Brass.
  - Flow range: 0.02–0.05 L/s
  - Actuator: Modulating with BMS compatibility.

Advantages of PICVs

- **Energy Efficiency**:
  - Reduces pump energy consumption by maintaining a constant flow.
- **Precise Control**:
  - Provides accurate temperature regulation at terminal units.
- **Reduced Commissioning Time**:
  - Eliminates the need for separate balancing valves.
- **Improved System Stability**:
  - Minimizes pressure-related fluctuations across the system.

Disadvantages of PICVs

1. **Higher Initial Cost**:
   - More expensive than traditional balancing and control valve combinations.
2. **Limited Applications**:
   - Not suitable for systems with fixed flows or very low pressure differentials.
3. **Complexity**:
   - Requires proper installation and integration with control systems.

8.15 HVAC Isolators

HVAC isolators are critical components used to minimize vibrations and noise transmitted from equipment to the building structure. Proper sizing and selection depend on the equipment type, weight, operating conditions, and building structure.

Types of HVAC Isolators

- **Spring Isolators**:
  - **Application**: Large, heavy equipment like chillers, air handlers, and pumps.
  - **Characteristics**: Metal springs with or without neoprene bases to provide high deflection and vibration reduction.
  - **Advantages**: High deflection, suitable for low-frequency vibrations.
- **Rubber-in-Shear Isolators**:
  - **Application**: Light to medium-weight equipment like fans, small compressors, and terminal units.
  - **Characteristics**: Neoprene or natural rubber material designed to absorb vibrations.
  - **Advantages**: Cost-effective, easy to install, and resistant to wear and tear.
- **Elastomeric Pads**:
  - **Application**: Small equipment and lightweight systems like mini-splits, ductless units, and fan coil units.
  - **Characteristics**: Made of layered rubber or neoprene for simple, static isolation.
  - **Advantages**: Economical and suitable for equipment with minimal vibration.
- **Seismic Isolators**:
  - **Application**: Equipment in seismic zones or areas prone to vibrations from external forces.
  - **Characteristics**: Combines spring isolators with restraints to limit movement during seismic events.

- **Advantages**: Provides both vibration isolation and stability.

- **Floating Floors**:
  - **Application**: Acoustic-sensitive areas like studios, hospitals, or labs.
  - **Characteristics**: Specialized isolators beneath a suspended floor system.
  - **Advantages**: Superior sound and vibration isolation.

Sizing and Selection

- **Equipment Weight**:
  - Determine the total operating weight of the equipment (including fluid, piping, etc.).
  - Select isolators that can support 1.2 to 2 times the equipment weight to ensure safety.
- **Frequency of Vibration**:
  - **Natural Frequency of Isolator**: Should be at least 1.4 to 3 times lower than the excitation frequency of the equipment for effective isolation.
  - Use **spring isolators** for low-frequency applications (below 10 Hz).
  - Use **rubber isolators** for high-frequency applications (above 10 Hz).
- **Deflection**:
  - **Rubber Pads**: Provide minimal deflection (up to 3 mm).
  - **Spring Isolators**: Typically offer deflection of 12–50 mm.
  - Select deflection based on vibration isolation requirements; higher deflection offers better isolation for lower frequencies.
- **Static Load per Isolator**:
  - Divide total equipment weight by the number of isolators.
  - Ensure the isolator's rated load matches or exceeds the calculated static load.
- **Environmental Conditions**:
  - For outdoor or corrosive environments, use galvanized, stainless steel, or neoprene-coated isolators.
- **Space Constraints**:
  - Low-profile isolators, like elastomeric pads, are suitable for confined spaces.
  - Spring isolators require more vertical space.
- **Seismic and Wind Load Considerations**:
  - Use seismic isolators where seismic events or high wind loads are a factor.
  - Include restraints to limit excessive motion during such events.
- **Acoustic Sensitivity**:
  - For sound-critical areas, combine isolators with acoustic treatments.
  - Floating floors or specialized isolators may be required for high-end acoustic performance.

Practical Examples

- **Chiller System (2000 kg)**:
  - **Solution**: Spring isolators with 25 mm deflection.
  - **Reason**: Heavy equipment with low-frequency vibration.
- **Ductless Mini-Split (50 kg)**:
  - **Solution**: Neoprene pads or rubber-in-shear isolators.
  - **Reason**: Lightweight equipment with minimal vibration.
- **Seismic Zone Application**:
  - **Solution**: Seismic spring isolators with restraints.
  - **Reason**: Combines vibration isolation and seismic safety.
- **Fan Coil Unit in a Residential Building**:
  - **Solution**: Elastomeric pads.
  - **Reason**: Economical, simple, and effective for light equipment.

8.16 HVAC Control Systems

Types of Control Systems for HVAC and Building Services

- **Thermostats:**
  - **Description:** Basic temperature control devices, often used in residential or small-scale commercial buildings.
  - **Example:** A programmable thermostat in a home that maintains desired temperature ranges.
- **Direct Digital Control (DDC) Systems:**
  - **Description:** Use digital processors to monitor and control HVAC systems.
  - **Example:** A building automation system (BAS) in an office complex that controls temperature, ventilation, and lighting.
- **Pneumatic Control Systems:**
  - **Description:** Utilize compressed air to operate HVAC components like dampers or valves.
  - **Example:** Older commercial buildings where air compressors drive valve actuators.
- **Proportional-Integral-Derivative (PID) Controllers:**
  - **Description:** Provide precise control by adjusting HVAC outputs based on feedback.
  - **Example:** A PID controller in a data center maintaining server room temperatures.

- **Building Automation Systems (BAS):**
  - **Description:** Centralized systems for managing HVAC, lighting, and security.
  - **Example:** An intelligent BAS in a hospital that integrates HVAC with emergency systems for consistent performance.
- **Programmable Logic Controllers (PLC):**
  - **Description:** Used in industrial applications for robust and customized HVAC control.
  - **Example:** A PLC system controlling air circulation in a pharmaceutical manufacturing facility.

Rules of Thumb for Sizing and Selecting Control Systems

- **Match System Complexity with Building Needs:**
  - **Rule:** Choose a simpler system (like thermostats) for small-scale applications and advanced systems (like BAS) for large buildings.
  - **Example:** Use a single thermostat for a small retail store and a BAS for a university campus.
- **Account for Building Size and Zones:**
  - **Rule:** Larger buildings or those with many zones require more sophisticated systems.
  - **Example:** A DDC system with zoning capabilities for a multi-story office building.
- **Energy Efficiency Goals:**
  - **Rule:** Select systems that support energy management and monitoring for sustainability goals.
  - **Example:** Implement a BAS that integrates occupancy sensors for efficient energy use.
- **Scalability:**
  - **Rule:** Ensure the control system can be expanded to meet future needs.
  - **Example:** A modular BAS in a growing tech campus.
- **Integration Requirements:**
  - **Rule:** Opt for systems compatible with other building services like fire alarms or security.
  - **Example:** An HVAC system linked to a smoke extraction system in a high-rise building.
- **Cost vs. Functionality:**
  - **Rule:** Balance budget constraints with the desired level of control and automation.
  - **Example:** A mid-tier thermostat system for a medium-sized hotel.

- **Environmental Conditions:**
  - **Rule:** Select systems suited to the building's operating environment.
  - **Example:** A weather-resistant pneumatic system for outdoor installations in harsh climates.

Practical Examples

- **Retail Store:**
  - **Control System:** Programmable thermostat.
  - **Sizing Rule:** Ensure the thermostat can handle the HVAC load for the building size, typically 1 thermostat/1000 ft^2.
- **Hospital:**
  - **Control System:** Building Automation System (BAS).
  - **Sizing Rule:** Use redundancy in critical zones like operating rooms. Ensure systems comply with ASHRAE standards for healthcare.
- **Data Centre:**
  - **Control System:** DDC with PID control for precise temperature and humidity management.
  - **Sizing Rule:** Design cooling to handle heat loads, usually 200–500 W/ft^2.
- **High-rise Office:**
  - **Control System:** BAS integrated with lighting and fire systems.
  - **Sizing Rule:** Divide into zones for each floor, with independent controls for each zone to manage diverse usage patterns.

Types of HVAC Control Systems

Control type	Description	Applications	Advantages	Disadvantages
Thermostats	Simple on/off control based on temperature set points	Residential, small commercial systems	Easy to install and use, cost-effective	Limited control, not energy-efficient
Programmable thermostats	Allows scheduling for temperature changes at specific times	Residential, light commercial spaces	Improves efficiency, reduces energy costs	Requires user programming, limited flexibility
Direct digital control (DDC)	Uses sensors and software to precisely manage HVAC operations	Large commercial, industrial systems	Highly accurate, scalable, energy-efficient	Higher cost, requires skilled setup
Building management systems (BMS)	Integrated systems that control HVAC, lighting, and other building functions	Large buildings, campuses	Centralized control, detailed monitoring	High cost, complex to implement

Control type	Description	Applications	Advantages	Disadvantages
Zoning systems	Divides building into zones, each controlled separately	Residential, commercial, and mixed-use	Improves occupant comfort, saves energy	Initial setup cost, may require ductwork modifications
Variable air volume (VAV) controls	Adjusts airflow based on zone demand	Office buildings, hospitals, schools	Energy-efficient, enhances comfort	Requires VAV boxes and compatible equipment
Proportional-integral-derivative (PID) controllers	Advanced control algorithm for precise temperature and pressure regulation	Laboratories, critical environments	High precision, adapts to varying conditions	Complex setup and tuning
IoT-enabled systems	Wireless, cloud-connected controls with remote monitoring	Smart homes, advanced commercial systems	Real-time data, remote access, adaptive control	Dependent on internet connectivity

Sizing Guidelines

- **System Size and Complexity**:
  - Match control system complexity to the building size and HVAC configuration.
    - Small buildings: Simple thermostats or programmable thermostats.
    - Medium-sized buildings: DDC systems or zoning controls.
    - Large buildings or campuses: BMS or IoT-enabled systems.
- **Zone Count**:
  - Divide the building into zones based on usage, occupancy, and thermal loads.
  - **Rule of Thumb**: 1 thermostat or sensor/**500–1000 ft^2 (46–93 m^2)** of occupied space.
  - Ensure independent controls for areas with distinct schedules or environmental requirements.
- **Sensor Placement**:
  - Place sensors at the centre of zones, away from direct sunlight, drafts, or heat sources.
  - **Rule of Thumb**: 1 temperature sensor for every **15–25 m^2 (160–270 ft^2)** in critical zones.
- **Control Point Capacity**:
  - For DDC or BMS systems, ensure enough capacity to handle all inputs and outputs (e.g., sensors, actuators, dampers).
  - Add **20–30% spare capacity** for future expansion.

Selection Guidelines

- **Application**:
  - **Residential**: Use simple or programmable thermostats with zoning if necessary.
  - **Commercial**: Employ DDC or VAV systems for efficiency and control.
  - **Critical Environments**: Use PID controllers or BMS for high precision and redundancy.
- **Energy Efficiency Goals**:
  - Include energy-saving features such as setback schedules, demand ventilation, and load optimization.
  - Use systems compatible with energy management certifications (e.g., LEED, ENERGY STAR).
- **Integration Needs**:
  - Choose a BMS if HVAC needs to integrate with lighting, security, or fire safety systems.
  - Ensure compatibility with existing equipment and protocols (e.g., BACnet, Modbus).
- **Scalability**:
  - For future-proofing, select systems with modular components or cloud-based software for expansion.
- **Maintenance and Monitoring**:
  - Systems with remote diagnostics and predictive maintenance reduce downtime.
  - Ensure local availability of replacement parts and technical support.

Thermostat
The best location for installing an air conditioning thermostat is crucial for ensuring accurate temperature readings and optimal system performance. The thermostat should be installed in a location that provides an accurate representation of the room's average temperature while avoiding external influences like heat, drafts, or direct airflow. A carefully chosen thermostat location can enhance comfort, improve energy efficiency, and prolong the lifespan of the HVAC system. Here are guidelines for selecting the ideal location:

Key Considerations for Thermostat Placement

- **Height from the Floor**:
  - Install the thermostat at **approximately 1.5 m (5 ft)** above the floor.
  - This height ensures the sensor detects the average room temperature, avoiding stratification effects (cooler air near the floor or warmer air near the ceiling).

- **Avoid Heat Sources**:
  - Do not place the thermostat near **heat-producing devices** (e.g., ovens, stoves, computers, or televisions).
  - Avoid locations exposed to **direct sunlight**, as it can cause inaccurate temperature readings.
- **Distance from Air Supply**:
  - Install the thermostat away from **air supply vents or return air ducts** to prevent it from being influenced by the immediate airflow.
  - Ideally, place it **at least 1–2 m (3–6 ft)** from supply vents.
- **Avoid Drafty Areas**:
  - Do not install near **windows, doors, or other drafty areas** where external air can affect the reading.
- **Central Location**:
  - Position the thermostat in a **central location** within the zone or area it controls.
  - Place it in the room where people spend the most time, such as a **living room** or **main office area**.
- **Wall Selection**:
  - Choose an **interior wall** for installation, as exterior walls may be affected by outdoor temperature fluctuations.
  - The wall should be stable and free from vibrations that could interfere with the device.
- **Accessibility**:
  - Ensure the thermostat is easily accessible for programming, adjustments, and maintenance.
- **Avoid Humid Areas**:
  - Do not install in areas with high humidity or potential moisture exposure, such as **bathrooms, kitchens**, or near humidifiers.

Special Considerations

- **Multi-Zone Systems**:
  - If the HVAC system controls multiple zones, install thermostats in the **primary area** of each zone.
  - Use zoning controls to ensure even comfort throughout the building.
- **Smart Thermostats**:
  - If using a **smart thermostat**, ensure it has a strong Wi-Fi signal at its location.
  - Many smart thermostats come with remote sensors to monitor temperatures in multiple areas.

- **Commercial Applications**:
  - In commercial buildings, place the thermostat in the **occupied zones** and avoid direct installation in corridors or utility spaces.

Examples of Good and Bad Locations

Good locations	Bad locations
Central wall in the living room	Near a window or exterior wall
Interior wall of a main office	Above a heat-emitting appliance
Hallway away from air vents	Near an air vent or return air grille
Bedroom interior wall	Direct sunlight exposure

8.17 Access Panels

Following table summarizing location for Access Panels installation.

Location	Reason for installation	Purpose
Ductwork for Fire Dampers	To provide access for inspecting, testing, and maintaining fire dampers as required by fire codes and standards	To ensure fire dampers remain functional and compliant with fire safety regulations
Ductwork for Volume Control Dampers (VCDs)	To allow adjustment, repair, or replacement of VCDs	To facilitate proper airflow balancing and system maintenance
Ductwork for Smoke Dampers	To enable inspection, testing, and servicing of smoke dampers	To ensure smoke dampers operate effectively during a fire emergency
Ductwork with Airflow Sensors	To provide access for cleaning, calibration, or replacement of airflow measurement devices	To maintain accurate airflow monitoring and HVAC system performance
Fan Coil Units and AHUs	To allow access to internal components such as filters, coils, fans, and electrical connections for maintenance and cleaning	To ensure efficient operation and prolong equipment life
Grease Ducts in Commercial Kitchens	To enable cleaning and inspection of grease buildup in compliance with fire safety regulations	To reduce fire risks and ensure safe operation of kitchen exhaust systems
Chilled Water or Heating Pipes	To allow access to valves, strainers, and insulation for maintenance and troubleshooting	To simplify repair and ensure efficient operation of heating and cooling systems
Electrical or Control Panels in HVAC Systems	To facilitate servicing or upgrading of electrical connections, relays, or controllers	To maintain operational reliability and allow system adjustments

Location	Reason for installation	Purpose
Plenums and Air Duct Junctions	To inspect or clean areas prone to dust accumulation or where duct leaks may occur	To improve air quality and ensure system efficiency
Concealed Valves and Shut-Offs	To provide access to hidden valves, such as isolation valves or balancing valves, within walls or ceilings	To enable quick and easy access for emergency shut-off or system adjustments
Hydronic Systems and Pumps	To enable access for servicing circulating pumps, pressure relief valves, and expansion tanks	To maintain effective hydronic system performance and safety
VAV Boxes and Dampers in Ceilings	To allow maintenance or replacement of VAV box components or damper actuators	To ensure consistent temperature control and energy efficiency in HVAC zones

Key Considerations

Ease of Access: Panels should be positioned to allow unobstructed access to equipment without compromising structural integrity or aesthetics.

Size and Labelling: Panels must be adequately sized and labelled for easy identification of the serviceable components behind them.

Integration with Building Design: Panels should be installed discreetly to maintain the visual appearance of walls, ceilings, or floors.

Chapter 9
Practical Example Using ROT

9.1 Overview

This chapter provides a practical guide to applying Rules of Thumb (ROT) for quick and effective problem-solving in various scenarios. Through a series of solved examples, it demonstrates how to use these simplified guidelines to estimate, design, and analyse common challenges encountered in engineering, construction, and building services. Each example is structured to highlight the assumptions, methodologies, and step-by-step calculations involved, ensuring clarity and ease of application. The goal is to equip readers with a hands-on understanding of how rules of thumb can serve as reliable tools for preliminary assessments and decision-making, saving time while maintaining reasonable accuracy.

9.2 Example 1: Designing a Complete Chilled-Water Cycle

Design a chilled-water system for a **2000 m² office building** located in a region with a design ambient **dry bulb temperature of 35 °C** and **relative humidity of 30%**. Use standard rules of thumb for estimating key parameters, such as cooling load, chiller size, and pump flow rates. Provide a practical and scalable framework for system design.

Cooling Load Estimation

- **Rule of Thumb**: Office cooling load = **100–150 W/m²**, depending on insulation, occupancy, and equipment.
- **Calculation**: Assume **120 W/m²** for a well-insulated office:

V. Vakiloroaya, *HVAC and Building Services Rules of Thumb*, Springer Tracts in Civil Engineering, https://doi.org/10.1007/978-3-032-10285-0_9

$$\text{Cooling Load} = 2000\,\text{m}^2 \times 120\,\text{W}/\text{m}^2 = 240{,}000\,\text{W}\,(240\,\text{kW})$$

Chiller Selection

- **Rule of Thumb**: Add a **10–15% safety factor** to the cooling load for equipment sizing.
- **Calculation**: Chiller Capacity = 240 kW × 1.1 = 264 kW
- **Selection**: Choose a water-cooled chiller with a capacity of **~270 kW**.

Cooling Tower Selection

- **Rule of Thumb**: Cooling towers are sized to reject **1.25–1.3 times the chiller capacity**.
- **Calculation**: Cooling Tower Capacity = 264 kW × 1.25 = 330 kW
- **Selection**: A cooling tower with a capacity of **~330 kW** is suitable. Ensure it is rated for the local wet bulb temperature (~24–26 °C).

Air Handling Units (AHUs)

- **Rule of Thumb**: AHUs handle **50–70 W/m^2**. Divide the building into zones.
- **Calculation**: Assume **3 zones of ~667 m^2 each**:

$$\text{Capacity per AHU} = 667\,\text{m}^2 \times 70\,\text{W}/\text{m}^2 = 46{,}690\text{W}\,(47\,\text{kW})$$

- **Selection**: Use **3 AHUs**, each rated at **~50 kW**.

Expansion Tank

- **Rule of Thumb**: Expansion tank volume = **10% of the total system water volume**.
- **System Water Volume Estimate**: Assume **3 L/m^2 of floor area**: Total Volume = 2000 m^2 × 3 L/m^2 = 6000 L
- **Expansion Tank Volume**: 6000 L × 0.1 = 600 L

Buffer Tank

- **Rule of Thumb**: Buffer tank volume = **15 L/kW of chiller capacity**.
- **Calculation**: Buffer Volume = 264 kW × 15 L/kW = 3960 L

Piping System

- **Rule of Thumb**: Piping velocity for chilled water = **1.2–2.0 m/s**.
- **Flow Rate Calculation**:

$$\text{Flow rate} = \text{kW}/(\text{Cp} \times \text{Temperature Difference}) = 264/(4.186 \times 5) = 12.61/\text{s}$$

- **Pipe Size**: A **65–80 mm** diameter pipe typically handles this flow rate.

Insulation

- **Rule of Thumb**: Use insulation thickness based on pipe diameter and location:
  - Pipes up to **80 mm diameter**: **25 mm thick insulation**.
  - Larger pipes: **40 mm thick insulation**.
- Material: Use closed-cell insulation material, such as **elastomeric foam**.

Valves and Controls

- **Isolation Valves**: Use butterfly valves for large-diameter pipes and ball valves for smaller pipes.
- **Balancing Valves**: Install on branches to balance flow between zones.
- **Control Valves**: Use 2-way or 3-way valves on AHUs for chilled water flow control.
- **System Control**: Implement a **Building Management System (BMS)** for monitoring and control.

Make-Up Water System for Cooling Tower

- **Rule of Thumb**: Cooling towers lose **3–5% of circulation flow** to evaporation, drift, and blowdown.
- **Calculation**: Assume a circulation rate of **12.6 L/s**:

$$\text{Evaporation Loss} = 12.6\,\text{L/s} \times 0.03 = 0.378\,\text{L/s}\left(\approx 1.36\,\text{m}^3/\text{h}\right)$$

- Include an **automatic make-up water system** to replenish losses.

Pumps

- **Rule of Thumb**: Pump head = **15–25 m (150–250 kPa)** for medium-sized buildings.
- **Flow Rate**: **12.6 L/s**.
- Use variable frequency drives (VFDs) for energy savings.

Summary Table for 2000 m² Office

Component	Selected size/capacity
Chiller	~270 kW water-cooled
Cooling tower	~330 kW
AHUs	3 units, each ~50 kW
Expansion tank	600 L
Buffer tank	~4000 L
Piping	65–80 mm diameter
Insulation	25 mm thick for small pipes
Pumps	12.6 L/s at 20 m head
Make-up water	~1.36 m^3/h
Valves & controls	Isolation, balancing, and control

9.3 Example 2: Comparison of Evaporative Cooling System and Direct Expansion (DX) Water-Cooled Rooftop Package for a 300 m^2 Office for the Following Data

Ambient Design Conditions

- **Dry-bulb temperature**: 35 °C.
- **Relative humidity**: 40%.
- **Building area**: 300 m^2.
- **Occupancy Type**: Office space.
- **Cooling Load**: Based on **rule of thumb** for office cooling: **120–150 W/m^2**.
  - Assume **150 W/m^2** for high-performance systems.

$$\text{Cooling Load} = 300\,\text{m}^2 \times 150\,\text{W} / \text{m}^2 = 45{,}000\,\text{W} = 45\,\text{kW}$$

System Calculations and Comparisons

Evaporative Cooling System

- **Principle**: Uses evaporation to cool air, achieving comfort by reducing air temperature while increasing humidity.
- **Performance Assumption**: Typical evaporative coolers lower air temperature to **85–90% of the wet-bulb temperature difference**.
  - Wet-bulb temperature at 35 °C DBT and 40% RH ≈ **24 °C**.
  - Final supply air temperature = 35 – (35–24) × 0.85 ≈ 25.65 °C
- **Sizing**: Airflow required to meet the cooling load:
  - Assume temperature drop: 35–25.65 ≈ 9.35 °C
  - Formula:

$$\text{Airflow} = \frac{\text{kW}}{1.2 \times \Delta\text{T}} = \frac{45}{1.2 \times 9.35} - 4\text{m}^3 / \text{s}$$

  - Required airflow: **4 m^3/s** or **14,400 m^3/h**.
- **Electricity Usage**:
  - Fan power = **0.4–0.8 kW/m^3/s**. Assume **0.6 kW/m^3/s**: Fan Power = 4 × 0.6 = 2.4 kW.
  - Water pump power = **~0.2 kW**.
  - Total power: **2.6 kW**.
- **Thermal Comfort**:
  - Supply air at **25.65 °C** with increased humidity.
  - May feel less comfortable in high-humidity zones.

DX Water-Cooled Rooftop Package

- **Principle**: Uses refrigerant-based cooling with a water-cooled condenser. Provides precise temperature and humidity control.
- **Sizing**:
  - Cooling capacity = 45 kW.
  - Typical efficiency (EER): **3.5–4.5**. Assume **4.0**.
  - Compressor power:

$$\text{Power} = \frac{\text{kW}}{\text{EER}} = \frac{45}{4} = 11.25\text{kW}$$

  - Additional power for water pump: **0.5–1.0 kW**. Assume **0.8 kW**.
  - Total power: **12.05 kW**.
- **Thermal Comfort**:
  - Supply air at **16–18 °C**.
  - Maintains target indoor conditions of **24 °C, 50% RH**.
  - Superior thermal comfort compared to evaporative cooling, especially in humid conditions.

Comparison Table

Parameter	Evaporative cooling	DX water-cooled rooftop package
Cooling capacity	45 kW	45 kW
Supply air temperature	25.65 °C	16–18 °C
Humidity control	No	Yes
Electricity usage	2.6 kW	12.05 kW
Initial cost	Low	Moderate
Maintenance cost	Low	Moderate
Thermal comfort	Moderate (humid zones)	High (precise control)
Water usage	High (evaporation)	Moderate (condenser cooling)
Suitability	Dry climates	All climates

Advantages and Disadvantages

Evaporative Cooling

- **Advantages:**
- Low energy consumption (operates mainly on fans and pumps).
- Simple and cost-effective.
- Environmentally friendly (no refrigerants).

Disadvantages:

- Limited cooling potential; depends on ambient humidity.
- Increased indoor humidity can reduce comfort.
- Ineffective in humid climates.

DX Water-Cooled Rooftop Package

- **Advantages:**
- Provides precise temperature and humidity control.
- Suitable for all climates, including humid conditions.
- Better thermal comfort and consistent performance.

Disadvantages:

- Higher electricity consumption.
- More complex and costly to install and maintain.

Electricity Usage Comparison

For **8 h of daily operation** over **30 days**:

- **Evaporative Cooling**:

$$\text{Energy Consumption} = 2.6\,\text{kW} \times 8\,\text{h} \times 30\,\text{days} = 624\,\text{kWh}$$

- **DX Rooftop Package**:

$$\text{Energy Consumption} = 12.05\,\text{kW} \times 8\,\text{h} \times 30\,\text{days} = 2892\,\text{kWh}$$

Conclusion

- **Evaporative Cooling** is a better option in dry climates with lower electricity usage and costs but is less effective in maintaining thermal comfort, especially in humid environments.
- **DX Water-Cooled Rooftop Package** provides superior comfort, precise humidity control, and consistent performance, making it more suitable for offices in mixed or humid climates, despite higher energy usage and costs.

9.4 Example 3: Sizing and Selecting a Heat Recovery Direct Expansion (DX) System for a 300 m^2 Gym for Following Assumptions

Assumptions

- **Building type**: Gym (higher occupant density and ventilation needs).
- **Area**: 300 m^2.
- **Ambient design conditions**:
  - Dry-bulb temperature: **35 °C**.
  - Relative humidity: **40%**.
  - Humidity ratio at 35 °C, 40% RH ≈ **0.014 kg water/kg dry air**.

- **Indoor set-points**:
  - Temperature: **23 °C**.
  - Relative humidity: **45%**.
  - Humidity ratio at 23 °C, 45% RH ≈ **0.008 kg water/kg dry air**.
- **Ventilation requirement**: **30 air changes/h (ACH)** for gyms, based on ASHRAE standards.
- **Heat recovery efficiency**: Assume **70% sensible heat recovery efficiency**.

Ventilation Airflow Requirement
Using the **30 ACH rule of thumb**:

$$\text{Airflow}\left(\text{m}^3/\text{h}\right) = \text{Room Volume}\left(\text{m}^3\right) \times \text{ACH}$$

- Room volume: Assume ceiling height = **4 m**.

$$\text{Room Volume} = 300\,\text{m}^2 \times 4\,\text{m} = 1200\,\text{m}^3$$

$$\text{Airflow}\left(\text{m}^3/\text{h}\right) = 1200\,\text{m}^3 \times 30 = 36{,}000\,\text{m}^3/\text{h}$$

- Convert to m^3/s:

$$\text{Airflow}\left(\text{m}^3/\text{s}\right) = 36{,}000 = 10\,\text{m}^3/\text{s}$$

Cooling Load Calculation

Sensible Cooling Load

$$\text{Sensible Heat} = 1.2\,\text{kJ}/\text{m}^3{}^\circ\text{C} \times \text{Airflow}\left(\text{m}^3/\text{s}\right) \times \Delta T$$

- **Temperature difference**:
- Outdoor T_{out} = 35 °C, Indoor T_{in} = 23 °C.

$$\Delta T = 35 - 23 = 12^\circ\text{C}$$

$$\text{Sensible Heat} = 1.2 \times 10\,\text{m}^3/\text{s} \times 12^\circ\text{C} = 144\,\text{kW}$$

Latent Cooling Load

$$\text{Latent Heat} = 2500\,\text{kJ}/\text{kg} \times \text{Airflow}\left(\text{kg}/\text{s}\right) \times \Delta W$$

- Airflow in kg/s: 1.2 kg/m^3 × 10 m^3/s = 12 kg/s
- Humidity ratio difference:

$$\Delta W = 0.014 - 0.008 = 0.006\,\text{kg water / kg dry air}$$

$$\text{Latent Heat} = 2500 \times 12\,\text{kg / s} \times 0.006 = 180\,\text{kW}$$

Total Cooling Load

$$\text{Total Load} = \text{Sensible Heat} + \text{Latent Heat} = 144\,\text{kW} + 180\,\text{kW} = 324\,\text{kW}$$

Heat Recovery Effect
With **70% sensible heat recovery efficiency**, the load on the cooling system reduces significantly:

- Sensible load recovery:

$$\text{Recovered Sensible Heat} = 144\,\text{kW} \times 0.7 = 100.8\,\text{kW}$$

$$\text{Remaining Sensible Load} = 144\,\text{kW} - 100.8\,\text{kW} = 43.2\,\text{kW}$$

- Latent load remains unchanged as heat recovery devices typically do not handle latent loads effectively:

$$\text{Remaining Latent Load} = 180\,\text{kW}$$

- Total adjusted load:

$$\text{Adjusted Load} = 43.2\,\text{kW} + 180\,\text{kW} = 223.2\,\text{kW}$$

System Sizing

- Select a **DX system** capable of handling **223.2 kW** total load.
- Divide the system into multiple units for redundancy and load distribution.
  - Example: 3 units of **75 kW each**.

Electricity Usage
Assuming an average **EER of 4.0** for the DX system:

$$\text{Compressor Power} = 223.2 / 4 = 55.8\,\text{kW}$$

Advantages of Heat Recovery DX System

Energy Efficiency: Heat recovery significantly reduces the sensible cooling load.
Thermal Comfort: Provides precise temperature and humidity control.
Compact Design: No water requirements, ideal for areas with water scarcity.
Flexibility: Can be modularized for future expansion.

Disadvantages

Higher Initial Cost: Heat recovery systems are more expensive to install.
Limited Latent Heat Recovery: Latent load must still be handled by the DX system.
Maintenance: Requires skilled maintenance for refrigerant and recovery components.

Equipment Selection Example

- **Heat Recovery Unit**: Choose a unit with **70% efficiency** capable of handling 10 m^3/s airflow.
- **DX Cooling System**: Select **3 × 75 kW** packaged rooftop DX units.
- **Ductwork**: Design for **10 m^3/s** airflow, insulated to prevent condensation and heat gain.
- **Controls**: Advanced controls for temperature and humidity regulation.

9.5 Example 4: Sizing and Selecting a Heat Recovery Direct Expansion (DX) System for Above Example (Example 3) Using 70% Outdoor Air and 30% Indoor Return Air

If the heat recovery system uses **70% outdoor air** and **30% indoor return air**, the recalculated cooling load and system sizing are as follows:

Key Adjustments

1. **Mixed Air Properties**
   - Outdoor air: 70% at 35 °C DBT, 40% RH (humidity ratio W_{out} = 0.014 kg water/kg dry air)
   - Indoor return air: 30% at 23 °C DBT, 45% RH (humidity ratio Win = 0.008 kg water/kg dry air)

Mixed Air Humidity Ratio

$$W_{mixed} = (0.7 \times W_{out}) + (0.3 \times W_{in})$$

$$W_{mixed} = (0.7 \times 0.014) + (0.3 \times 0.008) = 0.0122 \text{kg water / kg dry air}$$

Mixed Air Temperature

$$T_{mixed} = (0.7 \times T_{out}) + (0.3 \times T_{in})$$

$$T_{mixed} = (0.7 \times 35) + (0.3 \times 23) = 31.6°C$$

Ventilation Airflow Requirement
Airflow remains the same as previously calculated:

$$\text{Airflow}\left(m^3 / h\right) = 36{,}000\, m^3 / h$$

$$\text{Airflow}\left(m^3 / s\right) = 10\, m^3 / s$$

Cooling Load Calculation

Sensible Cooling Load

$$\text{Sensible Heat} = 1.2\, kJ / m^3 °C \times \text{Airflow}\left(m^3 / s\right) \times \Delta T$$

- Mixed air temperature = **31.6** °**C**, Indoor set point = **23** °**C**.

$$\Delta T = 31.6 - 23 = 8.6°C$$

$$\text{Sensible Heat} = 1.2 \times 10\, m^3 / s \times 8.6°C = 103.2\, kW$$

Latent Cooling Load

$$\text{Latent Heat} = 2500\, kJ / kg \times \text{Airflow}\left(kg / s\right) \times \Delta W$$

- Airflow in kg/s: 1.2 kg/m^3 × 10 m^3/s = 12 kg/s
- Humidity ratio difference:

$$\Delta W = 0.0122 - 0.008 = 0.0042\, \text{kg water} / \text{kg dry air}$$

$$\text{Latent Heat} = 2500 \times 12\, kg / s \times 0.0042 = 126 kW$$

Total Cooling Load

$$\text{Total Load} = \text{Sensible Heat} + \text{Latent Heat} = 103.2\,\text{kW} + 126\,\text{kW} = 229.2\,\text{kW}$$

Heat Recovery Effect
With **70% sensible heat recovery efficiency**, the load on the cooling system reduces further:

- Sensible load recovery:

$$\text{Recovered Sensible Heat} = 103.2\,\text{kW} \times 0.7 = 72.24\,\text{kW}$$

$$\text{Remaining Sensible Load} = 103.2\,\text{kW} - 72.24\,\text{kW} = 30.96\,\text{kW}$$

- Latent load remains unchanged:

$$\text{Remaining Latent Load} = 126\,\text{kW}$$

- Total adjusted load:

$$\text{Adjusted Load} = 30.96\,\text{kW} + 126\,\text{kW} = 156.96\,\text{kW}$$

System Sizing

- Select a **DX system** capable of handling **156.96 kW** total load.
- Divide the system into multiple units for redundancy and load distribution:
  – Example: **2 units of 80 kW each**.

Electricity Usage
Assuming an average **EER of 4.0** for the DX system:

$$\text{Compressor Power} = 156.96 / 4 = 39.2\text{kW}$$

Summary of Results

Parameter	100% outdoor air	70% outdoor air + 30% indoor return
Total cooling load	223.2 kW	156.9 kW
Sensible cooling load	43.2 kW	30.96 kW
Latent cooling load	180 kW	126 kW
Compressor power (EER = 4)	55.8 kW	39.2 kW
System size	3 × 75 kW units	2 × 80 kW units

Advantages of Using Mixed Air

Lower Cooling Load: By reusing return air, the cooling load is significantly reduced, saving energy and operational costs.
Smaller System Size: Requires fewer or smaller DX units, reducing upfront costs.
Improved Humidity Control: Mixed air helps to balance indoor conditions, reducing latent load.

9.6 Example 5: Design and Calculate a Stair Pressurization System for a 7-Level Residential Building Tower

Design and calculate a stair pressurization system for a 7-level residential building tower with the following conditions:

Each level contains six apartment units, and each apartment unit is 80 m^2.
Each level has one door to the stairwell with a door area of 1.8 m^2.
The assumed air velocity through the door is 1 m/s.
Calculate and size the air relief system for each level's corridor. Select all components, including volume control dampers, fans, and grilles, for both the stair pressurization system and the air relief system. Provide a working sequence of operations for when a fire occurs on Level 3 of the building.

Problem Summary
We are designing a **stair pressurization system** and an **air relief system** for a **7-level residential building**, with the following details:

1. **Building Configuration:**
   - **7 levels**, with **6 apartments per level.**
   - Each apartment is **80 m^2**.
   - **Stairwell door area per level:** 1.8 m^2
2. **Design Assumptions:**
   - **Air velocity through leakage:** 1 m/s
   - Pressurization target: **50 Pa** (sufficient to prevent smoke ingress).
   - Overpressurization limit: **60 Pa** (to ensure door operability).

Step 1: Stair Pressurization Airflow Calculation

Leakage Airflow per Door
- For each door (leakage area 1.8 m^2) and air velocity 1 m/s:

$$Q_{door} = A_{door} \times V$$

$$Q = 1.8\,m^2 \times 1\,m/s = 1.8\,m^3/s$$

Convert to hourly flow:

$$Q_{door} = 1.8\,m^3/s \times 3600 = 6480\,m^3/h$$

Total Airflow for Stairwell

- For all **7 levels**, total leakage area is 7 × 1.8 = 12.6 m^2.

Airflow required:

$$Q_{total} = 12.6\,m^2 \times 1\,m/s = 12.6\,m^3/s$$

Convert to hourly flow:

$$Q_{total} = 12.6\,m^3/s \times 3600 = 45{,}360\,m^3/h$$

Step 2: Fan Selection for Stair Pressurization

1. **Airflow:**
   - Select a fan capable of 46,000 m^3/h (rounding up for safety).
2. **Static Pressure:**
   - Account for duct losses and leakage resistance: **300 Pa** (typical for pressurization systems).

Fan Specification

- Centrifugal fan with:
- Airflow: 46,000 m^3/h
- Static pressure: 300 Pa

Step 3: Corridor Air Relief System

Objective

- To relieve overpressurization caused by stairwell pressurization airflow.

Relief Fan

- Install a central air relief fan capable of handling total exhaust airflow:

$$Q_{relief} = Q_{stair} = 45{,}360\,m^3/h$$

- Relief Dampers and Grills
- Motorized dampers installed on each level corridor for automatic operation during fire.
- Grills sized for 6480 m^3/h

Step 4: Volume Control Dampers

Volume control dampers are installed in the stairwell to balance airflow across levels and maintain uniform pressurization.

Step 5: Sequence of Operation During Fire (e.g., Fire on Level 3)

1. **Fire Detection:**
   - Fire alarm system detects fire on **Level 3**.
2. **System Activation:**
   - **Stair pressurization fan** activates and delivers 45,360 m^3/h
   - **Air relief system** activates, with motorized damper on **Level 3 corridor** opening to exhaust 6480 m^3/h
3. **Pressure Control:**
   - Volume control dampers maintain uniform stairwell pressure (target: 50 Pa)
4. **Evacuation:**
   - Pressurized stairwell remains smoke-free, ensuring safe evacuation.
   - Air relief system prevents overpressurization in Level 3 corridor and assists in smoke removal.

Equipment Summary

Component	Specification
Stair pressurization fan	46,000 m^3/h, 300 Pa
Air relief fan	46,000 m^3/h, 200 Pa
Volume control dampers	One per stairwell level
Motorized dampers	One per corridor level (linked to fire alarm system)
Relief grills	Sized for 6480 m^3/h

9.7 Example 6: Ventilation System Design for Car Park

Design the ventilation system for a **3-level car park** of a residential building with the following conditions:

- The car park has a total of **180 car spots**, with **60 spots per level**.
- Each level has an area of **800 m^2**.
- A **negative pressure of 12%** is required.

Tasks to Solve

1. Calculate the **required exhaust airflow rate**.
2. Calculate the **required supply airflow rate**.
3. Determine the **number of jet fans** required for each level.

Assumptions and Key Information

1. **Total Area per Level**: 800 m^2
2. **Levels**: 3 levels, each with 60 car park spaces.
3. **Total Spaces**: 180 spaces
4. **Negative Pressure**: 12% i.e., supply airflow is 88% of exhaust airflow.
5. **Air Changes per Hour (ACH)**: Typical for car parks is 6 ACH
6. **Jet Fan Coverage**: 20 m × 20 m = 400 m^2
7. **Jet Fan Capacity**: 5000 m^3/h

Step 1: Total Volume

Ceiling Height: Assume a typical 2.5 m^2.
Volume per Level:

$$\text{Volume per level} = \text{Area per level} \times \text{Ceiling height}$$

$$\text{Volume per level} = 800\,\text{m}^2 \times 2.5\,\text{m} = 2000\,\text{m}^3$$

Total Volume:

$$\text{Total volume} = 3 \times 2000 = 6000\,\text{m}^3$$

Step 2: Exhaust and Supply Airflow Rates

Exhaust Airflow Rate:

$$Q_{\text{exhaust}} = \frac{6000 \times 6}{60} = 600\,\text{m}^3/\text{min} = 36{,}000\,\text{m}^3/\text{h}$$

Supply Airflow Rate: Given 12% negative pressure the supply airflow is 88% of exhaust airflow:

$$Q_{\text{supply}} = Q_{\text{exhaust}} \times 0.88$$

$$Q_{\text{supply}} = 36{,}000 \times 0.88 = 31{,}680\,\text{m}^3/\text{h}$$

Step 3: Jet Fan Sizing and Quantity

Jet Fan Coverage per Level: Each jet fan covers 400 m^2. Given each level is 800 m^2, then the fan per level will be 2.

$$\text{Fan per level} = 800 / 400 = 2$$

Total Jet Fans:

$$\text{Total fans} = 3 \times 2 = 6\,\text{fans}$$

Summary of Results

Parameter	Value
Total car park volume	6000 m^3
Exhaust airflow (total)	36,000 m^3/h
Supply airflow (total)	31,680 m^3/h
Jet fans per level	2 fans
Total jet fans	6 fans

Sequence of Operation

1. **Normal Operation**:
    - Jet fans operate intermittently based on CO/NOx sensor readings.
    - Exhaust and supply fans maintain airflow to achieve 6 ACH and 12% negative pressure
2. **Emergency (Fire Scenario)**:
    - Jet fans operate at full speed to direct smoke toward exhaust points.
    - Exhaust fans run at full capacity to remove smoke.
    - Supply fans may shut down to avoid feeding oxygen to the fire.

9.8 Example 7: Design and Size a Complete Mechanical Room for an Office Space of 400 m²

Design and size a complete mechanical room for an **office space of 400 m²**. Include the following components:

Water-Cooled Scroll Chiller: Select and size the appropriate capacity.
Cooling Tower: Determine the suitable type and size.
Expansion Tank: Size appropriately based on the system's needs.
Buffer Tank: Calculate the required size for the system.
Cooling Tower Make-Up Water and Chemical Treatment System: Estimate the water usage and design the treatment system.

Water Circulation Pumps: Size two pumps (each with full capacity, with one as a changeover).
Piping System: Calculate pipe sizes for the water distribution.
Fan Coil Units: Use 2-pipe fan coil systems and size accordingly.
Valves:

- Specify required valves between the chiller and fan coils.
- Specify required valves between the chiller and cooling tower.

Equipment Placement: Determine the space required for each component within the mechanical room.
Additional Equipment: Identify any other necessary components for the system.

Ambient Design Conditions

- Dry Bulb Temperature: **35 °C**
- Relative Humidity: **50%**

Rules of Thumb

- Cooling load for office spaces: **0.15 kW/m^2**

Step 1: Calculate Cooling Load

Cooling Load

$$\text{Cooling Load} = \text{Area} \times \text{Cooling Load per Area}$$

$$\text{Cooling Load} = 400\,\text{m}^2 \times 0.15\,\text{kW}/\text{m}^2 = 60\,\text{kW}$$

Step 2: Water-Cooled Scroll Chiller Selection

Chiller Sizing
- Chiller capacity = 60 kW
- Typical scroll chiller efficiency (COP): 4.5
- **Chiller Input Power**: Input Power = Cooling Load/COP = 60/4.5 = 13.33 kW

Space Requirement
- Approximate dimensions:
- Length: 1.8–2.0 m (1800–2000 mm).
- Width: 0.8–1.2 m (800–1200 mm).
- Height: 1.2–1.6 m (1200–1600 mm)
- Approximate weight: 600 kg

  Dry Weight: 400–800 kg.
  Operating Weight (with water): 500–900 kg.

- Always check the manufacturer's specifications for precise dimensions and weight as these can vary significantly by brand and model.

- The weight difference between dry and operating conditions depends on the volume of water within the chiller.

Step 3: Cooling Tower Selection

Heat Rejection

- Cooling towers reject heat from both the cooling load and the chiller input power:

$$\text{Total Heat Rejection} = \text{Cooling Load} + \text{Chiller Input Power}$$

$$\text{Total Heat Rejection} = 60 + 13.33 = 73.33\,\text{kW}$$

Cooling Tower Flow Rate

- Cooling tower flow rate (thumb rule): 0.05 L/s/kW of heat rejection

$$\text{Flow Rate} = 73.33 \times 0.05 = 3.6\,\text{L / s}$$

- Cooling Tower Type
- Select a cross-flow or induced draft cooling tower for compact space and efficiency.
- Cooling tower size: Approx. 2 m × 1.2 m × 1.2 m

Make-Up Water

- Cooling tower make-up water requirement:

$$\text{Make} - \text{Up Water} = \text{Evaporation Loss} + \text{Drift Loss} + \text{Blowdown}$$

- Typical for 60 kW: 160 L/h (approximately 3% of total water flow rate for cooling tower)

Chemical Treatment

- Include a chemical dosing system for scaling, corrosion, and biological control:
- Capacity: 0.1–0.2 L/h

Step 4: Expansion Tank

Expansion Tank Sizing

- Rule of thumb: 3% of system volume
- System volume (approx.): 20 L/kW

$$\text{System Volume} = 60 \times 20 = 1200\,\text{L}$$

$$\text{Expansion Tank Volume} = 0.03 \times 1200 = 36\,\text{L}$$

Step 5: Buffer Tank

Buffer Tank Sizing

- Rule of thumb: 8 L/kW of cooling capacity

$$\text{Buffer Tank Volume} = 8 \times 60 = 480\,\text{L}$$

- Space requirement: Approx. 1.2 m × 1.2 m

Step 6: Pumps

Pump Sizing

- Flow rate (primary pump): Cooling Load/(4.2 × ΔT)

$$\text{Flow Rate} = 60\,\text{kW} / \left(4.2 \times 5\text{C}\right) = 2.86\,\text{L} / \text{s}$$

- Total dynamic head: Approx. 20–25 m

Pump Selection

- Flow rate: 2.86 L/s
- Head: 20 m
- Select two pumps (one standby), each with full capacity.

Step 7: Piping

Pipe Sizing

- Use a velocity range of 1.5–2 m/s for main pipes.
- Flow rate: 2.86 L/s.

$$D = \sqrt{\frac{4 \times m}{\pi \times v}}$$

- Where

m: water flow rate in m^3/s
v: water velocity in m/s
For 1.5 m/s
Pipe Diameter ≈ 65 mm

Valves

- Between chiller and fan coils:
- Isolation valves for each fan coil.
- Balancing valves for flow control.
- Between chiller and cooling tower:
  - Isolation and balancing valves.

Step 8: 2-Pipe Fan Coil Units

Fan Coil Selection

- Cooling load per unit: Divide 60 kW by the number of fan coils.
- Assume 10 fan coils, each with 6 kW.
- Space requirement per fan coil: 1.0 m × 0.3 m

Step 9: Mechanical Room Layout

Total Space Requirements

1. Chiller: 1.8 m × 1.2 m
2. Cooling Tower: 1.8 m × 1.2 m
3. Buffer Tank: 1.2 m × 1.2 m
4. Expansion Tank: 0.5 m × 0.5 m
5. Pumps: 0.8 m × 0.5 m (each, two total).

Summary Table

Equipment	Size/capacity	Space requirement
Chiller	60 kW	1.8 m × 1.2 m
Cooling tower	73.33 kW	1.8 m × 1.2 m
Buffer tank	480 L	1.2 m × 1.2 m
Expansion tank	36 L	0.5 m × 0.5 m
Pumps (2)	2.86 L/s, 20 m	0.8 m × 0.5 m

Chapter 10
Design of Air Distribution Systems

10.1 Overview

Ductwork system is the lifelines of HVAC systems, responsible for the efficient distribution of air throughout a building. Proper design and optimisation of these systems are essential for achieving energy efficiency, thermal comfort, and reliable performance. In this chapter, we will discuss the principles, best practices, and considerations involved in designing ductwork and air distribution systems. These rules of thumb provide quick estimates for ductwork design and car park ventilation, but final designs should be verified with local building codes, fire safety standards, and computational simulations where applicable.

10.2 Air Distribution System

Primary Equipment

- **Air Handling Units (AHUs)**
  - Main system for heating, cooling, and circulating air.
  - Includes components like filters, fans, cooling/heating coils, and dampers.
- **Fans**
  - Types: Centrifugal, axial, mixed flow, inline, or plug fans.
  - Function: Move air through the ducts and distribute it to the occupied spaces.
- **Ductwork**
  - Materials: Galvanized steel, aluminium, or flexible ducting.
  - Types:

V. Vakiloroaya, *HVAC and Building Services Rules of Thumb*, Springer Tracts in Civil Engineering, https://doi.org/10.1007/978-3-032-10285-0_10

- Supply ducts: Deliver conditioned air to spaces.
- Return ducts: Return air to the HVAC system.
- Exhaust ducts: Expel air from the building.

Air Distribution Devices

- **Diffusers**
  - Function: Distribute air evenly and reduce velocity for comfort.
  - Types: Square, round, linear slot, perforated face.
- **Grilles**
  - Function: Allow air to flow into or out of a space.
  - Types: Supply grilles, return grilles, exhaust grilles.
- **Registers**
  - Similar to grilles but include adjustable dampers for airflow control.
- **Louvers**
  - Installed on exterior walls to allow fresh air intake or exhaust while preventing water and debris infiltration.

Dampers

- **Volume Control Dampers**
  - Regulate airflow in the ducts.
  - Types: Manual or motorized.
- **Fire Dampers**
  - Automatically close during a fire to prevent the spread of flames through ducts.
- **Smoke Dampers**
  - Restrict airflow to prevent the spread of smoke during a fire.
- **Combination Fire/Smoke Dampers**
  - Provide both fire and smoke containment.
- **Backdraft Dampers**
  - Prevent reverse airflow in ducts.
- **Opposed Blade Dampers (OBDs)**
  - Used for precise airflow control in diffusers or grilles.

Filters

- **Panel Filters**
  - Primary filters for coarse dust and particles.
- **Pleated Filters**
  - Higher efficiency than panel filters.

- **Bag Filters**
  - Used for finer particle filtration.
- **HEPA Filters**
  - High-efficiency particulate air filters for cleanrooms and hospitals.
- **Carbon Filters**
  - Remove odours and gases.

Control Devices

- **Thermostats**
  - Control room temperature by regulating the HVAC system.
- **Variable Air Volume (VAV) Boxes**
  - Adjust airflow to maintain zone temperature.
- **Static Pressure Controllers**
  - Manage duct pressure to ensure optimal airflow.
- **Sensors**
  - Types: Temperature, humidity, CO_2, pressure, and airflow sensors.
  - Provide feedback to HVAC control systems.

Insulation

- **Duct Insulation**
  - Prevents heat loss or gain.
  - Materials: Fiberglass, foam, or reflective insulation.
- **Acoustic Insulation**
  - Reduces noise from airflow or duct vibration.

Accessories

- **Airflow Balancers**
  - Devices used to balance airflow across the system.
- **Turning Vanes**
  - Installed inside duct bends to reduce turbulence.
- **Flexible Duct Connectors**
  - Used to connect fans or AHUs to ducts, reducing vibration transmission.
- **Air Volume Control Valves**
  - Adjust and regulate airflow in ducts.
- **Sound Attenuators (Silencers)**
  - Reduce noise from fans or air movement.

- **Access Panels**
  - Allow for inspection and maintenance of duct interiors.
- **Air Curtains**
  - Installed at building entrances to minimize air exchange between indoor and outdoor spaces.
- **Plenums**
  - Enclosed spaces used to distribute air from AHUs to the ductwork.
- **Hoods**
  - Installed in kitchens or laboratories for exhaust purposes.

10.3 Principles of Ductwork Design

Ductwork systems transport conditioned air to various zones within a building. Key principles for efficient duct design include:

- **Minimizing Pressure Loss**:
  - Use smooth, straight ducts wherever possible.
  - Minimize the number of bends, transitions, and fittings.
- **Proper Sizing**:
  - Oversized ducts lead to higher material costs, while undersized ducts increase energy consumption due to higher fan power requirements.
  - Use tools such as the duct calculator or software like AutoCAD MEP for precise sizing.
- **Airflow Distribution**:
  - Ensure even airflow distribution by selecting appropriate diffuser locations and types.
  - Balance airflow with dampers to maintain consistent conditions across zones.
- **Acoustic Control**:
  - Prevent noise issues by avoiding abrupt transitions and selecting materials with good sound attenuation properties.
- **Leakage Prevention**:
  - Seal joints and connections with appropriate sealing materials to reduce air leakage.
  - Conduct duct leakage testing during commissioning.

Determine Airflow Requirements

Designing and sizing an air distribution system for HVAC and building services requires balancing airflow, pressure, noise, and cost. Below are key rules of thumb for sizing and selecting air distribution systems:

- **Rule of Thumb**:
  - Calculate required airflow based on the space cooling or heating load: $v = Q / (Cp \times \Delta T \times \rho)$
    - Q: cooling or heating load (kj/s)
    - v: Mass flow rate (m^3/s)
    - Cp: Specific heat capacity of air (1.005 kJ/kg°C)
    - ΔT: Temperature difference (supply vs. return air, in °C)
    - ρ: Density of air (1.2 kg/m^3)

Select Duct Sizing Method

- Use common methods for duct sizing are as followings:
  - **Equal Friction Method**: Maintains consistent pressure drop (Pa/m).
  - **Velocity Reduction Method**: Controls velocity to reduce noise and friction losses.
  - **Static Regain Method**: Balances static pressure for efficient airflow distribution.

Air Velocity

- Limit air velocity to reduce noise and pressure loss:
  - Main ducts: **5–8 m/s**
  - Branch ducts: **3–5 m/s**
  - Terminal outlets: **2–3 m/s**

Duct Sizing

- Limit duct aspect ratio to reduce pressure drop:
  - Use rectangular or circular ducts sized to balance airflow and space constraints.
  - Circular ducts are preferred for efficiency (less surface area, lower friction loss).
  - For space-saving, rectangular ducts can be used with an aspect ratio not exceeding **4:1**.

Pressure Drop

- Design for a total pressure drop of **0.5–1.0 Pa/m** for main ducts.

- For supply air systems, use approximate pressure losses of:
  - 0.66 Pa/m of duct [0.08 in. of water per 100 linear ft. of duct] for quiet areas
  - 0.82 Pa/m [0.10 in. of water per 100 linear ft. of duct] for ordinary areas

– 1.21 Pa/m [0.15 in. of water per 100 linear ft. of duct] for factory areas.
– Minimize bends, fittings, and abrupt transitions to reduce additional pressure loss.

Typical Pressure Drops of HVAC Duct Components

The pressure drop in duct components is typically expressed in **Pascals (Pa)** or **inches of water column (in. w.c.)** and depends on factors such as air velocity, component geometry, and duct size. Below is a general guide to typical pressure drops for common HVAC duct components:

Duct component	Typical pressure drop (Pa)	Air velocity (m/s)	Description
90° Elbow (with vanes)	0.15–0.3 per elbow	8 (Main Ducts)	Smooth bend with turning vanes
90° Elbow (no vanes)	0.3–0.6 per elbow	8 (Main Ducts)	Sharp bend without turning vanes
45° Elbow	0.1–0.2 per elbow	8 (Main Ducts)	Less resistance due to a smoother turn
T-junction (straight)	0.2–0.4 per T-junction	6 (Branch Ducts)	T-junction with air flowing straight
T-junction (branch)	0.3–0.7 per T-junction	6 (Branch Ducts)	T-junction with air flowing into the branch outlet
Dampers (open)	1.0–2.5	6 (Branch Ducts)	Fully open damper
Dampers (partially closed)	5.0–15.0	6 (Branch Ducts)	Dampers partially restricting flow
Grilles & registers	2.0–5.0	4 (Terminal Ducts)	Includes supply and return grilles/registers
Flexible duct (per meter)	0.8–2.0	4 (Terminal Ducts)	Dependent on length and compression
Duct-to-diffuser transitions	0.3–0.6	4 (Terminal Ducts)	Includes loss due to shape transitions

Key Considerations

Air Velocity Impact: Pressure drop increases with higher air velocities.
Smoothness of Components: Smooth transitions and well-designed components (e.g., vanes in elbows) minimize pressure drops.
Component Design: Sharp bends, abrupt transitions, and restrictive components cause significant pressure losses.
System Design: Proper duct sizing and layout reduce the cumulative pressure drop across the system.

The following table presents the possible width and height combinations for rectangular ducts designed to maintain a pressure drop of 0.8 Pa/m. Duct sizes can be adjusted to the nearest standard dimensions. For instance, for an airflow rate of 500 L/s, a suitable duct size would be W = 300 mm and H = 200 mm.

Airflow (L/s)	Width (mm)	Height (mm)	Duct area (m^2)	Air velocity (m/s)
100	122.5	81.6	0.009996	10.004
200	173.2	115.5	0.020005	9.998
300	212.1	141.4	0.029991	10.003
400	244.9	163.3	0.039992	10.002
500	273.9	182.6	0.050014	9.997
600	300.0	200.0	0.060000	10.000
700	324.0	216.0	0.069984	10.002
800	346.4	230.9	0.079984	10.002
900	367.4	244.9	0.089976	10.003
1000	387.3	258.2	0.100001	9.999
1100	406.2	270.8	0.109999	10.000
1200	424.3	282.8	0.119992	10.001
1300	441.6	294.4	0.130007	9.999
1400	458.3	305.5	0.140011	9.999
1500	474.3	316.2	0.149974	10.002
1600	489.9	326.6	0.160001	9.999
1700	505.0	336.7	0.170034	9.998
1800	519.6	346.4	0.179989	10.001
1900	533.9	355.9	0.190015	9.999
2000	547.7	365.1	0.199965	10.002

Flexible Duct Sizing Table

This table provides approximate sizes for **flexible ducts** based on airflow rates and acceptable air velocities commonly used in HVAC systems.

Airflow (L/s)	Airflow (CFM)	Duct diameter (mm)	Duct diameter (inches)	Air velocity (m/s)	Air velocity (fpm)
50	106	150	6	2.83	558
100	212	200	8	3.18	626
150	318	250	10	3.06	604
200	424	300	12	2.83	558
300	636	350	14	3.12	614
400	848	400	16	3.18	626
500	1060	450	18	3.27	645
600	1272	500	20	3.06	604
800	1696	600	24	2.83	558
1000	2120	700	28	2.59	510
1200	2544	800	32	2.65	523

Notes

- **Air Velocity**: The air velocity is calculated based on standard HVAC design practices, aiming for velocities between **2.5–3.5 m/s** (approximately **500–700 fpm**) for flexible ducts to minimize noise and pressure loss.
  Flexible ducts typically come in standard sizes such as **150 mm, 200 mm, 250 mm, etc.** Use this table as a guideline for selecting flexible duct sizes based on the desired airflow and system requirements.
- **Supply Air Temperature**
  - Cooling: Supply air temperature should be around **12–15 °C**.
  - Heating: Supply air temperature should be around **30–50 °C**.
- **Return and Exhaust Air**
  - Provide return air openings equal to or slightly larger than supply openings to prevent negative pressure.
  - Exhaust air volume should match the supply air volume in tightly controlled spaces.
- **Diffusers and Grilles**
  - Size diffusers and grilles to maintain a face velocity of **2–5 m/s**.
  - Select diffusers based on room layout and airflow pattern requirements (e.g., linear, ceiling, or slot diffusers).
- **Noise Control**
  - Ensure noise levels remain within acceptable limits:
    - Offices: **NC 30–35**
    - Residential: **NC 25–30**
    - Meeting rooms: **NC 20–25**
  - Use sound attenuators or insulation for noise-prone ducts.
- **Insulation**
  - Insulate ducts to prevent heat gain/loss and condensation.
  - Use insulation with an R-value sufficient for the temperature difference and local climate conditions.
- **Duct Material**
  - Use galvanized steel, aluminium, or flexible ducts based on application and budget.
  - Flexible ducts should be limited to short runs (less than **1.5 m**) to reduce resistance.

Technical Specification Table for Different Sheet Metals Used in HVAC Duct Fabrication

Material	Common grades	Thickness range (mm)	Key properties	Applications	Advantages	Disadvantages
Galvanized steel	G90, G60	0.6–2.0	Corrosion-resistant, durable, cost-effective	General HVAC ductwork, rectangular and round ducts	Corrosion resistance, strength, affordability	Heavier than aluminum
Stainless steel	304, 316	0.5–1.5	High corrosion resistance, durable, hygienic	Hospitals, labs, cleanrooms, corrosive environments	Excellent for hygienic applications	Expensive, harder to fabricate
Aluminium	3003, 5052	0.8–2.5	Lightweight, corrosion-resistant, flexible	Lightweight ductwork, outdoor applications	Lightweight, corrosion resistance	Less strength, higher cost than steel
Black steel	A36, ASTM A283	0.8–1.5	High strength, economical	Industrial ducting, high-temperature applications	High strength, economical	Requires painting or coating for corrosion
Copper	C110, C122	0.6–1.2	Excellent thermal conductivity, antimicrobial	Specialty HVAC systems, hospital/food industry	Antimicrobial, corrosion-resistant	Expensive, heavy, limited availability
PVC-coated steel	Custom Grades	0.8–2.0	Corrosion-resistant, sound-insulating	Ducts in corrosive or high-humidity environments	Corrosion resistance, sound insulation	Higher cost than galvanized steel
Zinc coated steel	Z275	0.6–2.0	Corrosion resistance	General duct fabrication	Affordable, corrosion-resistant	Limited to low-to-medium corrosion areas
Fiber-reinforced plastic (FRP)	Custom Grades	2.0–6.0	Highly corrosion-resistant, lightweight	Chemical plants, wastewater treatment facilities	Extreme corrosion resistance, lightweight	Expensive, limited in structural strength

Additional Notes

- **Thickness Standards**:
  - For **low-pressure systems**, 0.6–1.0 mm is typical.
  - For **medium/high-pressure systems**, 1.0–2.0 mm is used.
- **Insulated Ducts**: Sheet metals (typically galvanized steel or aluminium) are paired with insulation materials to reduce heat loss/gain and noise.
- **Coating Options**: Paints, PVC coatings, or epoxy coatings are applied to increase corrosion resistance.
- **Fabrication Considerations**:
  - Galvanized and stainless steels are more weldable.
  - Aluminium is easier to bend and form but requires special welding techniques.
  - Copper is often soldered rather than welded.
- **Balancing Dampers**
  - Install dampers at branch ducts for airflow balancing.
  - Use opposed-blade dampers for precise adjustment.
- **Fan Sizing**
  - Size the fan to overcome total pressure losses in the system, including duct friction, fittings, and equipment.
  - Add a **10–15% safety margin** to fan capacity for unforeseen resistance.
- **Energy Efficiency**
  - Keep duct leakage below **3% of total airflow**.
  - Use Variable Air Volume (VAV) systems where applicable to reduce energy consumption.
- **Space Constraints**
  - Account for space limitations in ceilings, walls, or floors when designing ductwork.
  - Coordinate with structural and electrical systems to avoid conflicts.
- **Maintenance Access**
  - Design the system with access panels or doors for easy inspection, cleaning, and maintenance.
- **Safety and Fire Protection**
  - Install fire dampers and smoke dampers at fire-rated barriers.
  - Use fire-resistant duct materials where required.

10.4 Car Parks Air Flow Rate Calculation

To estimate the required **supply** and **exhaust airflow rates** in a car park, one can use rules of thumb based on the number of cars or the area of the car park. Here are the guidelines:

Based on Number of Cars

- Allocate **150 m^3/h/car** for normal ventilation.
- Allocate **200–250 m^3/h/car** for smoke extraction or emergency ventilation.

Based on Car Park Area

- For normal ventilation:
  - **4–6 air changes per hour (ACH)** in the car park area.
- For emergency smoke extraction:
  - **10–12 ACH**.

Example Calculation

1. Calculate the car park volume: Volume (m^3) = Area (m^2) × Ceiling height (m)
2. Calculate the airflow rate: Airflow (m^3/h) = Volume (m^3) × ACH

Combined Approach

- Use a hybrid approach based on both cars and area when possible:
  - Calculate the airflow based on the number of cars to address pollutant load.
  - Verify with the area-based calculation to ensure sufficient overall air circulation.
- Design emergency exhaust systems to handle **10–12 ACH**, regardless of car count, to manage smoke effectively during a fire.

Carbon Monoxide (CO) and Pollutant Load

- Adjust airflow based on CO concentration:
  - Ensure the CO level remains below **30 ppm**.
  - Incorporate **CO sensors** to modulate fan speeds dynamically, reducing energy consumption when pollutant levels are low.

Ventilation per Car Park Level

- For multi-level car parks, calculate supply and exhaust rates for each level independently.
- Account for higher pollutant loads near ramps and entrances.

Typical Flow Rates (Summary)

- **Normal Ventilation**:
  - **150 m^3/h/car**, or
  - **4–6 ACH**.
- **Smoke Extraction**:
  - **200–250 m^3/h/car**, or
  - **10–12 ACH**.

Example Calculation

- **Given:**
  - Car park area: **1000 m^2**.
  - Ceiling height: **3 m**.
  - Number of cars: **50**.
- **Normal Ventilation:**
  1. Based on area: Volume = 1000 m^2 × 3 m = 3000 m^3
     Airflow = 3000 m^3 × 6 ACH = 18,000 m^3/h
  2. Based on cars: Airflow = 50 cars×150 m^3/h = 7500 m^3/h
- **Smoke Extraction:**
  1. Based on area: Airflow = 3000 m^3 × 12 ACH = 36,000 m^3/h
  2. Based on cars: Airflow = 50 cars×200 m^3/h = 10,000 m^3/h
- Use the higher of the two airflow rates to ensure proper ventilation.

10.5 Insulation Thickness for HVAC Ducts

Proper insulation of HVAC ducts and piping is essential to minimize heat gain or loss, control condensation, and improve energy efficiency. The required insulation thickness depends on factors like the application, operating temperature, ambient conditions, and the type of insulation material. Below is a detailed guide, including rules of thumb, materials, and common practices.

Air Ducts (Supply, Return, and Exhaust)

- **Conditioned Air Supply Ducts**:
  - **Rule of Thumb**: Thickness typically ranges from **25–50 mm**, depending on the temperature difference between the air inside the duct and the surrounding environment.
  - Use thicker insulation for ducts passing through unconditioned spaces.

- **Exhaust or Return Ducts**:
  - **Rule of Thumb**: Thickness of **12–25 mm** may suffice, as there is less concern about temperature loss.
- **Condensation Control**:
  - For ducts carrying cool air, ensure insulation prevents surface temperatures from dropping below the dew point to avoid condensation.
  - **Rule of Thumb**: Use insulation with vapor barriers, especially in humid climates.

Insulation Thickness Recommendations (Duct Applications)

Temperature difference (°C)	Insulation thickness (mm)	Remarks
<10	25	Minimal heat loss or gain
10–20	25–40	For moderately conditioned spaces
>20	40–50	High-temperature differences or humid climates

Common Insulation Materials

Materials for Ducts

- **Fiberglass**:
  - Excellent thermal and acoustic insulation properties.
  - Typically used with a vapour barrier.
- **Mineral Wool (Rock or Slag Wool)**:
  - Fire-resistant and provides good thermal performance.
- **Phenolic Foam**:
  - Lightweight and has high thermal resistance.
- **Polyisocyanurate (PIR) and Polyurethane Foam (PUF)**:
  - High thermal resistance; suitable for ducts in extreme environments.

10.6 Duct Leakage Rate

Below is a table showing approximate air leakage rates in a duct system under various conditions. Leakage depends on duct pressure, construction quality, sealing practices, and whether the ducts are located indoors or outdoors.

Condition	Pressure class	Leakage rate (as % of design airflow)	Notes
Well-sealed and insulated	Low (250 Pa)	1–2	Sealed with high-quality mastic or tape; indoor installation
Well-sealed and insulated	Medium (500 Pa)	3–5	Sealed properly, tested for leakage; outdoor ducts may have slightly higher rates
Poorly-sealed	Low (250 Pa)	10–15	Inadequate sealing (e.g., gaps, improper tape or no sealant); common in older systems
Poorly-sealed	Medium (500 Pa)	15–20	Typical for ducts with poor sealing and minimal testing
Unsealed (e.g., leaky ducts)	Low (250 Pa)	20–30	Airflows largely unregulated due to significant gaps or no sealing
Unsealed (e.g., leaky ducts)	High (750+ Pa)	30–40	Duct systems with high operating pressures and no attention to leakage control
Flex ducts (unsealed joints)	Medium (500 Pa)	10–25	Often used in residential systems, with high leakage at connections or punctures
Metal ducts (unsealed joints)	High (750+ Pa)	15–30	Leakage at seams, poorly assembled joints, or deteriorated seals over time

Key Points

- Leakage rates are commonly expressed as a percentage of total design airflow.
- Testing and balancing can quantify leakage rates in real systems.
- Leakage increases energy consumption, reduces system efficiency, and may compromise indoor air quality.
- Standards such as SMACNA or ASHRAE provide detailed guidelines on acceptable leakage rates based on duct classification.

Testing Procedures for Duct Air Leakage

Duct leakage testing quantifies the volume of air escaping from a duct system and helps ensure compliance with industry standards such as those by **ASHRAE**, **SMACNA**, or local codes.

Preparation for Testing

- Seal all outlets, inlets, and vents using plastic sheeting or inflatable bladders.
- Isolate the duct section to be tested from the rest of the HVAC system.
- Use a calibrated duct leakage tester to measure leakage rates under specified pressure conditions.

Test Methods

- **Total Leakage Test**: Measures leakage for the entire system, including intentional and unintentional gaps.
- **Leakage to Outdoors Test**: Measures only the air escaping to unconditioned spaces.
- **Positive Pressure Test**: Air is forced into the duct and leaks are detected by measuring pressure drops.

- **Negative Pressure Test**: Air is drawn out of the duct to simulate operating conditions.

Industry Standards

- **SMACNA Duct Leakage Classifications**: Specifies allowable leakage based on duct pressure and sealing quality (e.g., Class A is the tightest, while Class C allows more leakage).
- **ASHRAE Standard 90.1**: Limits duct leakage for energy efficiency (e.g., 4% of total airflow for high-pressure ducts).

Case-Specific Example

Scenario

- **Building Type**: Office
- **Duct System Size**: 500 m^2 conditioned area
- **Duct Type**: Medium-pressure metal duct, 500 Pa
- **Total Airflow**: 10,000 m^3/h (2777 L/s)

Test Results

- **Measured Leakage**: 400 m^3/h
- **Acceptable Leakage** (SMACNA Class A): 2% × 10,000 m^3/h = 200 m^3/h

Result The leakage rate exceeds allowable limits for Class A ducts. Recommendations include re-sealing joints with mastic and replacing deteriorated gaskets.

10.7 Car Park Supply and Exhaust Fans Sizing and Selection

Sizing and selecting supply and exhaust fans for car parks in HVAC and building services involves ensuring adequate ventilation to maintain air quality and control smoke during emergencies. Below are the **rules of thumb** to guide this process:

Air Distribution

- Ensure even distribution of supply and exhaust air across the car park.
- Use jet fans, ducted systems, or strategically placed supply/exhaust fans to avoid stagnant areas.

Fan Sizing

- Select fans with a total airflow capacity that meets or exceeds the calculated ventilation requirement.
- For larger car parks, divide airflow among multiple fans for better coverage and redundancy.
- **Centrifugal Fans**: For ducted systems requiring high static pressure.
- **Axial Fans**: For open systems with lower static pressure needs.
- **Jet Fans**: For ductless systems to move air directly.

Static Pressure Calculation

- Determine the total static pressure in the system, including losses from ductwork, grilles, and filters:
  - Typical static pressure range:
    - Ducted systems: **250–500 Pa**.
    - Jet fan systems: **50–150 Pa**.
  - Add a **10–15% margin** to account for unforeseen pressure drops.

Air Velocity

- Maintain optimal air velocities:
  - Ducted supply/exhaust systems: **5–8 m/s** in main ducts and **3–5 m/s** in branch ducts.
  - Exhaust air openings: **2–3 m/s** to minimize noise and pressure drop.

Noise Control

- Limit noise levels to **70–75 dB(A)** near occupied areas.
- Use silencers, vibration isolators, and acoustic treatments if necessary.

Fire and Smoke Control

- Ensure fans are certified for fire-rated operation, such as 300 °C for 2 h or 400 °C for 2 h.
- Fans must continue operating during fire emergencies to maintain smoke control.

Carbon Monoxide (CO) Monitoring

- Integrate CO sensors to modulate fan operation based on pollutant levels.
- Maintain CO concentrations below **30 ppm** in normal operation.

Exhaust and Supply Balance

- Exhaust airflow should match or slightly exceed supply airflow to prevent positive pressure and ensure proper ventilation.
- Consider natural ventilation if permissible, supplemented by mechanical systems.

Backup Power

- Provide backup power for fans to ensure continuous operation during emergencies.

Fan Placement

- Place supply fans near fresh air intakes to avoid recirculation.
- Locate exhaust fans near pollutant or heat sources (e.g., vehicle ramps) for effective removal.
- Ensure fans are easily accessible for inspection and maintenance.

- Provide adequate clearance around fans for servicing.
- For critical applications, design with redundancy (e.g., N + 1 configuration) to ensure continuous ventilation during failures.

Computational Fluid Dynamics (CFD)

- For large or complex car parks, use CFD simulations to optimize fan placement, airflow patterns, and system efficiency.

10.8 Jet-Fan Design and Selection

Jet fans are commonly used for ventilation in enclosed spaces such as car parks, tunnels, and large industrial areas. They move air by inducing a thrust force to ventilate pollutants or smoke during emergencies.

Application Type

- Identify the purpose of the jet fan system:
  - **Normal Ventilation**: Remove pollutants like CO and NOx in parking garages.
  - **Smoke Control**: Extract smoke during a fire for safe evacuation and firefighting.

Calculate Ventilation Requirement

- For car parks: Provide air changes per hour (ACH) based on the type of system:
  - **Normal operation**: 6 ACH.
  - **Smoke control**: 10–12 ACH.

Demonstration of a Jet-Fan

Air Thrust Requirements

- Calculate the thrust required to achieve adequate airflow and overcome system resistance: $F = m \times v$
  - F: Thrust force (N)
  - m: Mass flow rate (kg/s)
  - v: Air velocity (m/s)
  - Use computational fluid dynamics (CFD) simulations to refine thrust calculations for large or complex spaces.

Coverage Area per Jet Fan

- Each jet fan typically covers **25–35 m** of area in a linear arrangement.
- Overlap fan coverage zones to ensure no dead spots.

Fan Placement

- Install jet fans at regular intervals based on airflow patterns and structural constraints.
- Position fans to direct airflow toward exhaust points, avoiding obstacles that could disrupt the flow.

Jet Fan Capacity

- Select jet fans based on thrust (N) and airflow capacity (m^3/s):
  - Small systems: **50–100 N** thrust.
  - Medium systems: **100–300 N** thrust.
  - Large systems: **300–500 N** thrust.

Air Velocity

- Jet fans typically generate air velocities between **10 and 20 m/s**.
- Avoid velocities exceeding **25 m/s** to minimize noise and vibration.

Noise Levels

- Ensure noise levels are within acceptable limits:
  - Typical maximum noise level: **70–75 dB(A)**.
  - Use silencers or acoustic treatment if necessary.

Electrical Power and Efficiency

- Jet fans typically consume **1–10 kW** per unit depending on size and capacity.
- Use energy-efficient motors (e.g., EC or IE3/IE4 motors) to reduce operational costs.

Fire Safety

- Ensure jet fans are certified for fire resistance to operate during emergencies.
- Fans should function for at least **2 h** at specified fire ratings.

System Control

- Use a Variable Frequency Drive (VFD) for speed control and energy savings.
- Integrate with a Building Management System (BMS) for automatic operation based on CO sensors, NOx sensors, or fire alarms.

Duct-Free Operation

- Jet fans eliminate the need for ducting, reducing system cost and space requirements.
  - Use strategically placed jet fans to create airflow paths toward exhaust shafts.
- Install fans at accessible locations for regular inspection and maintenance.
  - Provide sufficient clearance around fans for servicing.

Computational Fluid Dynamics (CFD)

- For complex spaces, use CFD simulations to design optimal jet fan placement, airflow patterns, and coverage.

Practical Rules

- The jet fans shall be installed parallel to the beams, ensuring adequate vertical clearance from the driveway areas below. They must be positioned at least 1.6 m away from any slab or ceiling obstructions and oriented to avoid the placement of other services, such as warning signs, exit signs, or pipework, within their discharge area.
- The jet fans shall be positioned between rows of sprinklers, spaced approximately 3 m apart. Sprinkler heads should generally be located at least 1.5 m from the centreline of each jet fan. The fire sprinklers installed adjacent to the jet fan's airstream shall be placed no more than 300 mm below the ceiling slab.
- Coordination between mechanical and fire services engineers is essential. For instance, jet fans in car parking areas must NOT be placed directly in front of sprinkler heads. Instead, jet fans should generally be positioned centrally between two rows of sprinklers.
- The jet fans in car parking areas shall be configured to automatically shut down when the installed fire sprinkler system is activated.
- Upon activation of the fire sprinkler system: All jet fans shall automatically shut down. The car park exhaust system shall operate at 100% capacity.

Ventilation Systems for Garbage Rooms
Garbage rooms require dedicated ventilation systems to manage odours, maintain air quality, and prevent contaminants from spreading to other areas. Proper sizing and selection are crucial for functional and efficient operation.

Ventilation Requirements

Air Changes per Hour (ACH)

- The typical design guideline is **6–15 ACH**, depending on the size and usage of the garbage room.

- Higher ACH is recommended for larger garbage rooms or those with high waste turnover.

Minimum Exhaust Airflow

- **Room Area-Based Rule**: Provide **10–15 L/s/m^2** (2–3 CFM/ft^2) of garbage room area.
- **Volume-Based Rule**: Use the calculated air changes based on the volume of the room.

Outdoor Air Supply

- Provide a minimum of **0.5–1 ACH** of outdoor air for makeup ventilation to maintain negative pressure.

Negative Pressure

- Maintain the garbage room under **negative pressure** (approximately **−5 to −10 Pa**) relative to adjacent spaces.
- This prevents odours and contaminants from escaping into other areas of the building.

Ventilation System Components

Exhaust Fan

- Use a centrifugal or inline exhaust fan with a capacity matching the calculated airflow.
- Fans should be rated for **corrosive environments** due to exposure to waste gases.

Air Inlets

- Install low-level air inlets to provide fresh air. Ensure that air paths do not allow recirculation or direct exhaust to inhabited spaces.

Odour Control

- Include filters or odour control systems if required:
- **Carbon Filters**: To neutralize odours.
- **UV-C Systems**: To prevent bacterial growth and odours in ductwork.

Ductwork

- Use materials resistant to corrosion (e.g., galvanized steel or aluminium).
- Ensure proper duct sealing to prevent odour leakage.

Ventilation Location

- Position exhaust vents near the primary waste source or at high points where warm, odorous air accumulates.

Additional Considerations

Garbage Compactor Integration

- If a garbage compactor is present, additional localized ventilation (e.g., hoods or direct exhaust) may be required.

Fire Safety

- Ensure ventilation ducts are equipped with fire-rated dampers where ducts penetrate fire-rated partitions.

Energy Efficiency

- Include a timer or occupancy sensor to operate ventilation only during high-usage periods.
- Use variable-speed fans to adjust airflow based on demand.

Noise Control

- Use acoustic linings or silencers to minimize noise from the exhaust system.

Sizing Example

Scenario: Garbage Room Dimensions

- **Room Size**: 4 m × 3 m × 3 m (36 m^3 volume).
- **Required ACH**: 10 ACH.

Exhaust Airflow Calculation

Airflow (m^3/h) = Room Volume (m^3) × ACH

Airflow = 36 m^3 × 10 ACH = 360 m^3/h = 100 L/s

Fan Selection

- Select a fan with a minimum capacity of **360 m^3/h** (100 L/s) and verify against available models.

Negative Pressure Design

- Ensure makeup air supply delivers approximately 90% of the exhaust airflow to maintain negative pressure (e.g., 90 L/s for supply vs. 100 L/s for exhaust).

Rules of Thumb Summary

Use **6–15 ACH** or **10–15 L/s/m^2** as a baseline for ventilation airflow.

Maintain **negative pressure (−5 to −10 Pa)** in the garbage room.

Provide corrosion-resistant exhaust fans and ductwork.

Incorporate odour control (e.g., carbon filters) if required.

Size and position exhaust near the waste source to optimize air capture.

10.9 Stair Pressurization

Stair pressurization systems are critical for life safety in high-rise buildings. They maintain positive pressure in stairwells to prevent smoke infiltration during a fire, enabling safe evacuation. Below are the guidelines and rules of thumb for selecting, sizing, and designing such systems:

System Types

- **Single Injection System**:
  - A single fan delivers air to one point in the stairwell.
  - **Advantages**: Simple and cost-effective.
  - **Disadvantages**: Pressure gradients along the stairwell may occur, especially in tall buildings.

- **Multiple Injection System**:
  - Multiple fans or air inlets deliver air at different levels.
  - **Advantages**: Provides uniform pressure throughout the stairwell.
  - **Disadvantages**: Higher complexity and cost.
- **Variable Air Volume (VAV) System**:
  - Adjusts airflow based on pressure sensors.
  - **Advantages**: Maintains consistent pressure regardless of building conditions (e.g., open/closed doors).
  - **Disadvantages**: Requires advanced controls and regular maintenance.
- **Dedicated vs. Integrated Systems**:
  - **Dedicated**: Pressurization system operates independently of the building's main HVAC system.
  - **Integrated**: Utilizes existing HVAC components to pressurize stairwells.
  - **Rule of Thumb**: Dedicated systems are preferred for reliability in emergencies.

Sizing Guidelines

- **Airflow Rates**:
  - Minimum airflow required: **1.5–2.5 m^3/s** for typical stairwells.
  - Rule of Thumb:
    - **0.75–1.0 m^3/s/10 stories** for single injection systems.
    - Add **10–15% safety margin** for leakage through doors and other pathways.
- **Pressure Differential**:
  - Maintain a pressure differential of **25–50 Pa** between the stairwell and adjacent spaces.
  - **Rule of Thumb**:
    - **25–30 Pa**: Comfortable door operation (opening force ~100 N).
    - **50 Pa**: Maximum pressure for larger buildings; consider door opening challenges.
- **Leakage Allowance**:
  - Account for air leakage through doors (open or closed):
    - **10–15 air changes per hour (ACH)** for open doors.
    - Include leakage from construction gaps and weather-stripping.
- **Fan Selection**:
  - Fan capacity: Size fans for peak flow conditions while ensuring redundancy.
  - Use high-static pressure fans capable of **500–1000 Pa** for taller buildings.

Selection Guidelines

- **Building Height**:
  - For buildings up to **10 stories**: Single injection systems are usually sufficient.
  - For buildings over **10 stories**: Multiple injection or variable systems are preferred to maintain uniformity.
- **Door Usage**:
  - Frequent door opening: Use VAV systems to adapt airflow and maintain pressure.
  - Fire-rated doors: Ensure doors can withstand pressure without excessive deformation.
- **Redundancy**:
  - Provide backup fans for critical life safety systems.
  - Design for **N + 1** redundancy in large buildings or mission-critical facilities.
- **Climate Considerations**:
  - In cold climates, preconditioned air may be required to avoid cold drafts.
  - In hot/humid climates, ensure air supplied is dehumidified to prevent condensation.
- **Integration with Fire Alarm**:
  - The system must be automatically activated by the fire alarm and operate continuously during an emergency.

Design and Installation

- **Duct Sizing**:
  - Size ducts to minimize pressure losses; aim for velocity < **7 m/s**.
  - Include dampers to control flow and pressure at each injection point.
- **Pressure Monitoring**:
  - Install pressure sensors at multiple locations:
    - At the top, middle, and bottom of the stairwell.
    - Adjacent to doors likely to be opened during evacuation.
- **Airflow Balancing**:
  - Use adjustable dampers or VAV boxes to balance airflow across the stairwell.
  - Prevent over-pressurization, which can hinder door operation.
- **Noise Control**:
  - Select low-noise fans and incorporate sound attenuators.
  - Keep operational noise below **55 dB(A)** for occupant comfort.

- **Testing and Commissioning**:
  - Conduct smoke control and door-opening force tests.
  - Measure pressure differentials under various door configurations (e.g., open/closed).
- **Design References**:
  - Refer to ASHRAE, CIBSE or NCC guides for specific calculation methodologies.
- **Door Force Compliance**:
  - Ensure door opening force is within **110 N** to comply with accessibility requirements.

10.10 Acoustic Engineering

Acoustic engineering is critical in HVAC and building services to ensure occupant comfort and productivity. Noise from HVAC systems, equipment, and air distribution must be managed through proper design, equipment selection, and mitigation techniques. Below are the rules of thumb for criteria, sizing, and selection.

Acoustic Criteria for Building Types

Building type	Typical noise criteria (NC)	Comments
Residential (bedrooms)	NC 25–30	Quiet environments; noise levels below 30 dB(A)
Residential (living rooms)	NC 30–35	Slightly higher acceptable noise level
Offices (open plan)	NC 35–40	Balance between comfort and background noise
Offices (private rooms)	NC 30–35	Suitable for focused work; lower background noise
Conference rooms	NC 25–30	Ensure speech intelligibility and comfort
Classrooms	NC 25–30	Minimize distraction for effective learning
Hospitals (patient rooms)	NC 25–30	Quiet environments critical for recovery
Theatres/Auditoriums	NC 20–25	High acoustic sensitivity for speech and music
Industrial spaces	NC 40–50	Background noise tolerance is higher

Notes

- Match the noise criteria (NC levels) to the building's functional requirements.
- Lower NC levels are critical for noise-sensitive spaces like theatres and patient rooms.

Noise Sources in HVAC Systems

Noise source	Frequency characteristics	Mitigation approach
Fans	Low-frequency (63–250 Hz)	Use silencers, vibration isolators, and duct liners
Compressors	Broad-spectrum	Enclose compressors and use acoustic barriers
Duct airflow	Mid-to-high frequency (500–2000 Hz)	Optimize duct design; use attenuators
Terminal devices (diffusers)	High-frequency (1000–4000 Hz)	Select low-noise diffusers; maintain proper airflow
Pumps	Low-frequency	Isolate vibration; use flexible connections
Outdoor equipment (chillers)	Broad-spectrum	Use acoustic enclosures or barriers

Sizing and Selection

Ductwork Design

- **Velocity Limits**:
  - Keep duct velocities below **5–7 m/s** in main ducts and **3–5 m/s** in branch ducts to minimize noise.
  - For noise-sensitive areas, reduce velocities to **3 m/s** or lower.
- **Duct Liners**:
  - Use duct liners for noise attenuation, particularly in ducts serving noise-sensitive areas.
  - Thickness: **12–50 mm**, depending on frequency and required noise reduction.
- **Transitions**:
  - Avoid abrupt transitions to prevent turbulence and noise.
  - Use smooth tapers with angles **<15°**.

Equipment Selection

- **Fans**:
  - Select low-noise fans (e.g., backward-curved or airfoil blades).
  - Fan tip speed: Keep below **30 m/s** for quieter operation.
- **Pumps**:
  - Use variable-speed pumps to reduce noise during partial load conditions.
  - Avoid cavitation by ensuring proper NPSH (Net Positive Suction Head).
- **Compressors and Chillers**:
  - Choose units with noise ratings below **85 dB(A)** for exterior placement and **<70 dB(A)** for interior.

- **Diffusers and Grilles**:
  - Select low-noise diffusers (<NC 25 for sensitive areas).
  - Ensure proper throw distance and avoid direct airflow to occupants.

Silencers and Attenuators

- **Sizing**:
  - Select based on the required insertion loss:
  - **10–15 dB**: Typical for general HVAC noise.
  - **20–30 dB**: For noise-critical environments.
  - Length: **900–1800 mm** depending on space constraints.
- **Placement**:
  - Install close to noise sources, such as fans or equipment.
  - For duct systems, place silencers at transitions or before terminal devices.

Chapter 11
Water Distribution System Design

11.1 Overview

The water distribution system in HVAC and building services is designed to transport chilled water, hot water, or condenser water efficiently between equipment and building spaces. This chapter provides list of equipment and accessories together with sizing and selection notes.

11.2 Primary Equipment

- **Pumps**
  - Circulate water within the system.
  - Types:
    - **Centrifugal Pumps**: Commonly used for HVAC systems.
    - **Vertical Inline Pumps**: Space-saving design for tight mechanical rooms.
    - **Split-Case Pumps**: High-capacity applications.
    - **Variable Speed Pumps**: Optimize energy usage by adjusting flow rates.

V. Vakiloroaya, *HVAC and Building Services Rules of Thumb*, Springer Tracts in Civil Engineering, https://doi.org/10.1007/978-3-032-10285-0_11

An illustration of a Water Pump

- **Chillers**
  - Provide chilled water for cooling.
  - Types: Air-cooled, water-cooled, absorption chillers.
- **Boilers**
  - Generate hot water for heating.
  - Types: Gas-fired, electric, oil-fired, condensing boilers.
- **Cooling Towers**
  - Remove heat from water in condenser loops for water-cooled systems.
- **Heat Exchangers**
  - Transfer heat between water circuits (e.g., plate-and-frame, shell-and-tube).

Piping System Components

- **Pipes**
  - Transport water through the system.
  - Materials: Carbon steel, copper, stainless steel, PEX, CPVC.
- **Pipe Insulation**
  - Prevents heat loss/gain and condensation.
  - Materials: Fiberglass, foam, rubber.
- **Expansion Joints**
  - Absorb thermal expansion and contraction in the piping system.
- **Flexible Connections**
  - Minimize vibrations transmitted from pumps or other equipment.

Control Valves

- **Balancing Valves**
  - Ensure even water distribution in the system.
  - Types: Manual, automatic, and pressure-independent valves.
- **Globe Valves**
  - Control water flow rates.
- **Check Valves**
  - Prevent backflow in the system.
- **Butterfly Valves**
  - Used for isolation or throttling.
- **Ball Valves**
  - Compact isolation valves for on/off control.
- **Pressure-Reducing Valves**
  - Maintain desired pressure levels within the system.
- **Thermostatic Mixing Valves**
  - Blend hot and cold water for temperature regulation.
- **Three-Way Valves**
  - Direct water flow for mixing or diverting.

Tanks and Reservoirs

- **Expansion Tanks**
  - Absorb pressure changes due to water temperature fluctuations.
- **Buffer Tanks**
  - Provide thermal storage and stabilize system operation.
- **Makeup Water Tanks**
  - Store and supply water for system refilling.
- **Condensate Tanks**
  - Collect and return condensate in steam systems.

Water Treatment Equipment

- **Chemical Feed Systems**
  - Add treatment chemicals to prevent corrosion and scaling.
- **Water Softening Units**
  - Remove hardness to reduce scaling.
- **Filtration Systems**
  - Types: Sand filters, cartridge filters, bag filters.
- **Air Separators**
  - Remove air bubbles from water to prevent cavitation.

- **Dirt Separators**
  - Remove solid particles to maintain system cleanliness.

Measurement and Monitoring Devices

- **Flow Meters**
  - Measure water flow rates.
- **Pressure Gauges**
  - Monitor system pressure.
- **Temperature Sensors**
  - Track water temperature at various points.
- **Differential Pressure Sensors**
  - Monitor pressure differences across components like heat exchangers or filters.

Accessories

- **Strainers**
  - Capture debris to protect equipment.
  - Types: Y-strainers, basket strainers.
- **Air Vents**
  - Remove trapped air from the system.
  - Types: Manual, automatic.
- **Drain Valves**
  - Allow for system drainage during maintenance.
- **Isolation Valves**
  - Shut off sections of the system for maintenance or repairs.
- **Gauges and Indicators**
  - Display real-time data for pressure, temperature, and flow.
- **Pipe Supports and Hangers**
  - Secure pipes and allow for thermal movement.
- **Energy Meters**
  - Monitor energy consumption for heating and cooling.

Rules of Thumb for Sizing and Selection

- **Pipe Sizing**:
  - Maintain water velocities between **1.2 and 2.4 m/s** for efficiency and noise control.

- **Pump Sizing**:
  - Select based on system flow rate and head loss, typically using pump curves.
- **Expansion Tanks**:
  - Size to accommodate **3–5% of total water volume** for temperature changes.
- **Valves**:
  - Choose valves that match pipe size and system pressure.
- **Water Treatment**:
  - Ensure total dissolved solids (TDS) levels are under **500 ppm**.

Principles of Piping Design
Piping systems transport water or refrigerant in HVAC systems, serving heating, cooling, or both. Key design principles include:

- **Pressure and Flow Control**:
  - Design piping layouts to minimize pressure drops and maintain adequate flow rates.
  - Include balancing valves and pressure-regulating devices.
- **Material Selection**:
  - Choose materials based on application and fluid type, such as copper for refrigerant lines or steel for hydronic systems.
  - Use corrosion-resistant materials in harsh environments.
- **Thermal Insulation**:
  - Insulate pipes to minimize heat loss or gain and prevent condensation.
  - Ensure compliance with energy codes and standards.
- **Pipe Sizing**:
  - Properly size pipes to prevent velocity-related noise and erosion.
  - Use software tools for detailed hydraulic calculations.

11.3 Typical Pressure Drops for Water Piping System Components

Following table shows typical pressure drops for water piping system components. The values are based on standard assumptions for water velocities commonly used in HVAC systems:

- **Main Piping**: 1.5–2.5 m/s
- **Branch Piping**: 1.0–1.5 m/s
- **Terminal Piping**: 0.5–1.0 m/s

Piping component	Typical pressure drop (kPa)	Water velocity (m/s)	Description
90° Elbow	0.5–2.0 per elbow	1.5–2.5 (Main)	Pressure drop increases with sharpness of turn
45° Elbow	0.2–0.6 per elbow	1.5–2.5 (Main)	Less resistance than a 90° elbow
T-junction (straight flow)	0.2–0.8 per junction	1.0–1.5 (Branch)	For flow through the straight leg
T-junction (branch flow)	1.0–4.0 per junction	1.0–1.5 (Branch)	Higher resistance for flow into the branch leg
Valves (fully open)	2.0–10.0	1.0–1.5 (Branch)	Includes gate, globe, and ball valves (fully open)
Valves (partially closed)	15.0–50.0	1.0–1.5 (Branch)	Increased resistance for throttled valves
Strainers	5.0–20.0	1.0–1.5 (Branch)	Depends on mesh size and fouling
Flexible hoses (per meter)	5.0–15.0	0.5–1.0 (Terminal)	Higher friction due to flexible material
Heat exchangers	20.0–50.0	0.5–1.0 (Terminal)	Includes losses through tubes or plates
Coils (chilled water)	10.0–30.0	0.5–1.0 (Terminal)	Pressure drop depends on coil design
Reducers/ expanders	0.3–1.0	0.5–1.0 (Terminal)	Losses from abrupt changes in pipe diameter
Pipe length (per meter)	0.1–0.6	1.5–2.5 (Main)	Depends on pipe material and diameter

Notes

Friction Factor Dependency: Pipe pressure drop depends on the friction factor, which is a function of Reynolds number and pipe roughness.

Pipe Material: Smooth pipes (e.g., copper, PVC) have lower pressure drops than rough pipes (e.g., steel).

System Complexity: Total pressure drop is the sum of individual components and pipe lengths.

- The pipe size for chilled-water system can be obtained from following table:

Flow rate (l/s)	Velocity (m/s)	Calculated diameter (mm)	Standard pipe size (mm)
0.5	2.0	18	20
1.0	2.0	25	25
5.0	2.0	50	50
10.0	2.0	71	80
20.0	2.0	101	100
50.0	2.0	159	150
100.0	2.0	224	250
200.0	2.0	317	350
300.0	2.0	389	400
500.0	2.0	504	500

Notes

- **Standard Pipe Sizes**:
  - Pipes are selected from nominal pipe sizes in common use.
  - For larger diameters, availability depends on the material (e.g., steel, copper, PVC).
- **Design Assumptions**:
  - Velocity maintained at 2 m/s for efficient water flow.
  - Rounded to the nearest nominal size to ensure compatibility with fittings and ease of procurement.
- **Adjustments for Materials**: If specific materials (e.g., copper, steel, or plastic) are needed, standard diameters may vary slightly.

Below is the **standardized table** for pipe diameters of a condenser water system based on flow rates ranging from **0.5 to 500 l/s**. The design assumes a temperature difference of **5.5 °C** and a velocity of **2.5 m/s**, which is typical for condenser water systems. Nominal pipe sizes are rounded to match standard available sizes.

Flow rate (l/s)	Velocity (m/s)	Calculated diameter (mm)	Standard pipe size (mm)
0.5	2.5	15	20
1.0	2.5	22	25
5.0	2.5	49	50
10.0	2.5	69	80
20.0	2.5	98	100
50.0	2.5	155	150
100.0	2.5	219	250
200.0	2.5	310	350
300.0	2.5	379	400
400.0	2.5	437	450
500.0	2.5	490	500

Assumptions

- **Temperature Difference**:
  - The temperature difference of 5.5 °C leads to higher flow rates compared to chilled water systems.
- **Velocity**:
  - A velocity of 2.5 m/s is chosen for optimal efficiency, ensuring low pressure drops while maintaining pipe size.
- **Standard Pipe Sizes**:
  - Sizes rounded to match nominal diameters as per industry standards for steel and copper pipes.
- Following table shows technical specification for materials used in water piping system:

Material	Common grades/ standards	Size range (mm)	Pressure rating (PN)	Temperature range (°C)	Key properties	Applications	Advantages	Disadvantages
Carbon steel	ASTM A106, A53, API 5 L	15–1200	PN 16–40	−20 to 400	High strength, durable	Chilled and hot water systems, high-pressure systems	High strength, suitable for high temperatures	Requires corrosion protection
Stainless steel	304, 316	15–600	PN 16–40	−100 to 400	Corrosion-resistant, hygienic	Corrosive environments, hospitals, food industries	Excellent corrosion resistance, long lifespan	Expensive, harder to weld
Copper	Type K, L, M (ASTM B88)	6–200	PN 16–25	−20 to 200	Excellent thermal conductivity, antimicrobial	Domestic water, chilled water, medical applications	Corrosion-resistant, hygienic, easy to fabricate	Expensive, prone to theft
PVC (polyvinyl chloride)	ASTM D1785, D2241	15–600	PN 6–16	0 to 60	Lightweight, corrosion-resistant	Condensate lines, low-pressure water systems	Lightweight, inexpensive, corrosion-resistant	Limited temperature and pressure resistance
CPVC (chlorinated PVC)	ASTM D2846, F441	15–150	PN 10–25	0 to 90	High temperature and corrosion resistance	Hot and cold water systems	Inexpensive, easy to install	Brittle at low temperatures
PPR (polypropylene random copolymer)	DIN 8077, 8078	20–160	PN 10–25	0 to 90	Lightweight, heat-resistant, flexible	Domestic hot and cold water, chilled water systems	Corrosion-resistant, long lifespan	Requires specialized welding equipment
HDPE (high-density polyethylene)	ISO 4427, ASTM F714	20–1200	PN 6–25	−40 to 80	Flexible, corrosion-resistant	Underground piping, chilled water systems	Lightweight, flexible, suitable for large diameters	Limited temperature resistance
Ductile iron	ISO 2531, EN 545	80–2000	PN 10–40	−20 to 350	High strength, durable	Cooling tower piping, industrial systems	Suitable for high-pressure and large-diameter pipes	Requires internal/ external coatings
PEX (cross-linked polyethylene)	ASTM F876, F877	15–50	PN 6–16	−40 to 95	Flexible, lightweight	Domestic water systems, chilled water applications	Easy to install, corrosion-resistant	Limited diameter and pressure capabilities
Galvanized steel	ASTM A123, A153	15–300	PN 10–25	−20 to 200	Zinc-coated for corrosion resistance	Fire sprinkler systems, low-pressure water systems	Corrosion-resistant in non-aggressive environments	Heavy, prone to scaling

Additional Notes

- **Pipe Selection**:
  - For chilled water systems: Common materials are carbon steel, stainless steel, and HDPE.
  - For condenser water systems: Carbon steel or ductile iron is often used due to higher temperature and pressure requirements.
- **Insulation**:
  - Pipes for chilled water systems are typically insulated to prevent condensation.
  - Materials such as elastomeric foam or fiberglass are used.
- **Joining Methods**:
  - Carbon steel and stainless steel: Welded or flanged.
  - Plastic pipes: Socket fusion, butt fusion, or threaded fittings.
  - Copper: Soldered or brazed joints.
- **Coatings and Linings**:
  - Carbon steel pipes are often coated with epoxy or lined with cement for corrosion resistance.
  - HDPE pipes do not require additional coatings.

11.4 Piping Configurations

Common piping configurations include:

- **Two-Pipe Systems**:
  - Provide heating or cooling but not both simultaneously.
  - Simple and cost-effective for certain applications.
- **Four-Pipe Systems**:
  - Enable simultaneous heating and cooling by using separate supply and return lines for hot and chilled water.
  - Offer flexibility in multi-zone buildings.
- **Primary-Secondary Loop Systems**:
  - Use a primary loop for generating heating/cooling and a secondary loop for distribution.
  - Reduce pump energy and maintain stable flow rates.

- **Reverse Return Systems**:
  - Ensure balanced flow by making the total pipe length from the supply to return equal for all branches.
  - Ideal for large systems requiring precise flow control.

11.5 Insulation Thickness for Piping: Guidelines and Materials

Proper insulation of HVAC ducts and piping is essential to minimize heat gain or loss, control condensation, and improve energy efficiency. The required insulation thickness depends on factors like the application, operating temperature, ambient conditions, and the type of insulation material. Below is a detailed guide, including rules of thumb, materials, and common practices.

Insulation Thickness for HVAC Piping

Chilled Water Pipes
- **Purpose**: Prevent heat gain and control condensation.
- **Rule of Thumb**: Thickness typically ranges from **25 to 50 mm**, depending on pipe size, ambient temperature, and humidity.

Hot Water and Steam Pipes
- **Purpose**: Minimize heat loss and maintain desired water temperature.
- **Rule of Thumb**: Thickness typically ranges from **25 to 100 mm**, increasing with higher temperatures and larger pipe diameters.

Insulation Thickness Recommendations (Pipe Applications)

Pipe size (diameter, mm)	Chilled water (mm)	Hot water (mm)	Steam (mm)
≤50	25	25–40	50–75
50–150	25–40	40–60	75–100
>150	40–50	60–80	100–150

Common Insulation Materials

Materials for Ducts
- **Fiberglass**:
  - Excellent thermal and acoustic insulation properties.
  - Typically used with a vapor barrier.
- **Mineral Wool (Rock or Slag Wool)**:
  - Fire-resistant and provides good thermal performance.

- **Phenolic Foam**:
  - Lightweight and has high thermal resistance.
- **Polyisocyanurate (PIR) and Polyurethane Foam (PUF)**:
  - High thermal resistance; suitable for ducts in extreme environments.

Materials for Pipes

- **Elastomeric Foam**:
  - Flexible, resistant to moisture, and commonly used for chilled water pipes.
- **Fiberglass**:
  - Common for hot water and steam pipes.
- **Calcium Silicate**:
  - High compressive strength; used for high-temperature applications like steam pipes.
- **Mineral Wool**:
  - Fire-resistant; suitable for both hot and cold applications.
- **Polyethylene Foam**:
  - Lightweight and moisture-resistant; ideal for chilled water systems.

Selecting Insulation Thickness

- **Climate Considerations**:
  - In hot and humid climates, prioritize condensation control for chilled water pipes and air ducts. Use insulation with vapour barriers.
  - In cold climates, focus on minimizing heat loss for hot water and steam systems.
- **Operating Temperature**:
  - Higher operating temperatures require thicker insulation to minimize heat loss (e.g., steam pipes).
  - Low-temperature systems need insulation to control condensation effectively.
- **Ambient Conditions**:
  - For ducts or pipes exposed to outdoor conditions, use thicker insulation and weatherproofing materials.
- **Energy Efficiency Goals**:
  - Align insulation thickness with the desired level of energy conservation. Thicker insulation is more efficient but may increase upfront costs.
- **Space Constraints**:
  - For tight spaces, choose high-performance materials (e.g., phenolic foam or PIR) that achieve the same thermal resistance with less thickness.
- **Cost vs. Performance**:
  - Balance material and installation costs with long-term savings from reduced energy loss.

Chapter 12
Plant Space Allowance

12.1 Overview

When designing HVAC systems and planning building services, it's important to ensure that adequate space allowances are allocated for plant (equipment) rooms and mechanical services. These allowances help ensure that equipment operates efficiently, is accessible for maintenance, and meets safety standards. These rules of thumb for plant and equipment space allowances in HVAC and building services can serve as valuable guidelines during the design and planning phases. However, it's essential to adapt these general recommendations to suit the specific needs of each project, considering factors like equipment type, building occupancy, safety requirements, and local codes and standards. Consulting with design engineers and manufacturers during the planning process will help ensure optimal equipment arrangement and functionality. This chapter discusses some rules of thumb for space allowances for HVAC plant and equipment.

12.2 Spatial Co-ordination

Spatial coordination in HVAC and building services refers to the process of integrating and organizing various mechanical, electrical, and plumbing (MEP) systems within a building's architectural framework. This ensures that all systems work efficiently together and fit well within the constraints of the building's design. Proper spatial coordination is critical to avoid conflicts between systems, optimize operational efficiency, and enhance the overall functionality of the building. Here are some key aspects of spatial coordination in HVAC and building services.

V. Vakiloroaya, *HVAC and Building Services Rules of Thumb*, Springer Tracts in Civil Engineering, https://doi.org/10.1007/978-3-032-10285-0_12

Conflict Resolution

- Identify and resolve conflicts between different systems, such as ductwork, piping, electrical conduits, and structural elements (like beams and columns).
- Utilize tools like **3D modelling** to visualize how different components interact and to identify potential clashes before construction begins.

Clearances and Access

- Ensure that there are adequate **access routes** for maintenance and operations. This includes considering clearances around HVAC units, mechanical rooms, ducts, and other equipment.
- Design service spaces that allow workers easy accessibility for repairs and routine maintenance without interfering with other operational systems.

Integration of Systems

- Coordinate the placement of HVAC systems with other building components such as structural supports, walls, windows, and ceilings.
- Consider how HVAC systems (like ductwork) interact with lighting systems and ceiling layouts to prevent installation problems and ensure aesthetically pleasing arrangements.

Load Distribution

- Ensure that the load from suspended mechanical systems (like ductwork and piping) is appropriately distributed to avoid overloading building structures.
- Utilize structural supports and hangers that do not conflict with other services and comply with building codes.

Collaboration among Disciplines

- Engage in collaboration between architects, engineers (HVAC, plumbing, electrical), and construction teams during the design phase.
- Regular coordination meetings help all stakeholders track progress and address any issues that arise during the design and construction stages.

Building Information Modelling (BIM)

- Use BIM as a powerful tool for spatial coordination. It allows for detailed visualizations, clash detection, and better planning of MEP systems within the architectural space.
- BIM facilitates real-time editing and sharing of designs among various stakeholders, leading to improved accuracy and efficiency.

Compliance with Codes and Standards

- Ensure that all coordinated designs comply with local codes, safety regulations, and standards governing HVAC systems and their installation.
- Consider ergonomics and safety in accessible design layouts to minimize risks for maintenance personnel and future operators.

Environmental Considerations

- Consider the environmental impact of the spatial coordination process. Optimize layouts to improve energy efficiency, facilitate natural ventilation, and enhance indoor air quality.

12.3 Clashes

Clashes between architectural drawings and Mechanical, Electrical, and Plumbing (MEP) or fire services designs are common in building projects. Resolving these clashes requires collaboration, adherence to standards, and clear communication.

Understanding Clashes
Clashes occur when elements from different disciplines occupy the same physical space or are inappropriately coordinated, causing conflicts in installation or functionality. Examples include:

- Ductwork intersecting with structural beams.
- Plumbing pipes conflicting with electrical conduits.
- Fire protection systems obstructed by HVAC equipment.

Solutions to Coordinate and Resolve Clashes

- **Use of BIM (Building Information Modelling):**
  - **Description:** BIM tools like Autodesk Revit allow all disciplines to design within a shared model, identifying clashes early.
  - **Process:** Run clash detection analyses regularly during the design phase to address issues before construction.
  - **Benefit:** Ensures a coordinated and efficient design with reduced rework.
- **Conduct Coordination Meetings:**
  - Regular meetings between architects, mechanical engineers, electrical engineers, and fire safety designers foster collaboration.
  - Use these sessions to discuss and resolve conflicts, ensuring all disciplines align with project goals.
- **Establish Priorities Early:**
  - Develop a hierarchy of priority based on building functionality, safety, and standards. Common priorities:
    - **Architectural Priority:** Structural integrity and aesthetic intent should not be compromised.
    - **Life Safety Systems:** Fire safety and evacuation paths take precedence over other systems.

    - **MEP Coordination:** MEP systems must adapt to fit within the architectural and structural framework unless critical operational needs dictate otherwise.

- **Design Adjustments:**
  - Adjust layouts to resolve overlaps. For example:
    - Reroute ducts or pipes around structural elements.
    - Modify ceiling heights or partition placements.
- **Apply Zoning and Layering Rules:**
  - Divide spaces into zones for specific services, e.g., designate ceiling voids for HVAC or electrical conduits to avoid conflicts.
- **Adopt Modular Solutions:**
  - Use prefabricated modules where MEP systems are integrated and tested off-site, minimizing on-site clashes.
- **On-Site Coordination:**
  - Assign a dedicated coordinator during construction to resolve unforeseen clashes in real-time.

Who Has Priority?

- **Life Safety Systems:**
  - Fire protection systems, emergency lighting, and escape routes must always take precedence to comply with safety regulations.
- **Structural Integrity:**
  - The architectural and structural framework must remain uncompromised, as these are critical to the building's safety and stability.
- **HVAC Systems:**
  - HVAC ductwork and equipment placement are often more challenging to reroute than electrical or plumbing systems due to size and air distribution requirements.
- **Electrical Systems:**
  - Electrical conduits and panels are flexible and can often be adjusted to accommodate other systems.
- **Plumbing Systems:**
  - Plumbing systems are usually more adaptable but require priority in gravity-dependent sections, like drainage.

Best Practices

- **Early Collaboration:** Involve all disciplines in the design phase.
- **Clearly Defined Roles:** Assign responsibilities and define lead roles for resolving clashes.

- **Regular Updates:** Maintain updated models and drawings to reflect design changes.
- **Detailed Documentation:** Keep a record of all clash resolutions for future reference.

12.4 General Space Allowances

- **Mechanical Rooms**:
  - Allocate a minimum of **10–15%** of the area of the space served for mechanical rooms, but this can vary based on the HVAC design and the equipment being used.
  - A typical minimum dimension might be **2.4 m (8 ft)** in height for most mechanical equipment.
- **Chiller Plant Rooms**:
  - Allow at least **7.5 m^2 (80 ft^2)** as a starting point for small to medium-sized systems; larger systems will require significantly more area depending on the units selected.
  - **Clearances around Equipment**
- **Service Clearance**:
  - Always provide clearances around HVAC equipment according to manufacturer recommendations, which typically range from **1.0 to 1.5 m (3 to 5 ft)**.
  - Ensure **minimum clearances of 1.0 m (3 ft)** on all sides of equipment to allow for air circulation and maintenance access.
- **Ductwork and Piping**:
  - Maintain clear access paths of at least **1.0 m (3 ft)** for ductwork and piping to facilitate future maintenance and inspections.
  - **Accessibility**
- Ensure that access doors and panels to equipment rooms are at least **1.2 m (4 ft)** wide to accommodate moving equipment in and out easily.
- If equipment requires regular maintenance or servicing, consider allowing larger spaces or additional access points.
  - **Equipment Layout:**
- **HVAC Units**:
  - For air handling units (AHUs), allocate **0.5–1.0 m^2/1000 L/s** of airflow capacity.

- **Boilers and Chillers**:
  - Allow space for boiler and chiller clearances generally recommended at **1.2–1.5 m (4–5 ft)** around the units to facilitate servicing activities.
  - **Other Consideration**

Noise and Vibration Considerations

- Space should also account for vibration isolation pads or mounting systems that may increase the footprint of equipment.
- Position equipment in locations where noise can be effectively managed without disturbing occupants in adjacent spaces.

Vertical Space Consideration

- Plan for overhead space for ducts, piping, and light fixtures to avoid conflicts that could impede maintenance access. A minimum overhead clearance of **2.4 m (8 ft)** above floor level is often desirable.

Future Expansion Considerations

- If future system expansions are anticipated, allocate additional space (around **15–20% more**) to account for new equipment or system upgrades.

Project-Specific Needs

- Always evaluate specific project requirements, as space allowances can differ for healthcare facilities, laboratories, industrial plants, and residential buildings based on their unique HVAC needs.

12.5 Weight and Space Requirements for Different HVAC Systems

The weight and space requirements of HVAC systems vary widely depending on the type, capacity, and application. These factors are critical for structural design, equipment placement, and ensuring adequate maintenance access. Below is a breakdown of weight and space considerations for various HVAC systems:

Air-Cooled Chillers

- **Application**: Commercial and industrial cooling.
- **Weight**:
  - Small units (10–50 tons): 500–1500 kg.
  - Medium units (50–200 tons): 2000–7000 kg.
  - Large units (200–500 tons): 8000–15,000 kg.

- **Space Requirements**:
  - **Outdoor Installation**: Requires clearances for airflow (usually 1.5–2 m on all sides).
  - **Rooftop Installations**: Ensure the structure can support the weight plus dynamic loads.

Water-Cooled Chillers

- **Application**: Large buildings, data centres, and industrial processes.
- **Weight**:
  - Small units (50–100 tons): 2500–5000 kg.
  - Medium units (100–500 tons): 6000–15,000 kg.
  - Large units (500–1000 tons): 20,000–40,000 kg (with cooling tower).
- **Space Requirements**:
  - **Mechanical Room**: 2.5–3.5 m clearance around the unit for maintenance.
  - **Cooling Tower**: Outdoor space, typically 2–3 m^2/ton of capacity.

Packaged Rooftop Units (RTUs)

- **Application**: Retail, commercial, and industrial spaces.
- **Weight**:
  - Small units (5–20 tons): 300–1500 kg.
  - Medium units (20–50 tons): 2000–5000 kg.
- **Space Requirements**:
  - Requires roof curb or structural support.
  - Allow 1–2 m clearance for air intake and maintenance access.

Split Systems

- **Application**: Residential and light commercial buildings.
- **Weight**:
  - Indoor unit: 20–50 kg.
  - Outdoor unit: 50–300 kg (varies with capacity).
- **Space Requirements**:
  - Indoor unit: Wall- or ceiling-mounted; requires 1 m clearance for airflow.
  - Outdoor unit: 0.5–1.5 m^2, with 1–2 m clearance for ventilation.

Variable Refrigerant Flow (VRF) Systems

- **Application**: Multi-zone applications (commercial, hotels, apartments).
- **Weight**:
  - Outdoor unit: 200–1200 kg.
  - Indoor units: 15–50 kg/unit.

- **Space Requirements**:
  - Outdoor unit: Requires 1–2 m clearance for airflow.
  - Indoor units: Minimal space; mounted on walls, ceilings, or concealed in ducts.

Fan Coil Units (FCUs)

- **Application**: Hotels, offices, and multi-family residential buildings.
- **Weight**:
  - Small units: 10–50 kg.
  - Large units: 50–200 kg.
- **Space Requirements**:
  - Wall-mounted: Minimal clearance required.
  - Ceiling-concealed: Requires 0.5–1 m service access.

Air Handling Units (AHUs)

- **Application**: Large commercial and industrial buildings.
- **Weight**:
  - Small units: 500–1500 kg.
  - Medium units: 2000–5000 kg.
  - Large custom units: 10,000–30,000 kg.
- **Space Requirements**:
  - Requires a mechanical room or dedicated rooftop area.
  - Clearance: 1.5–2 m around the unit for maintenance and airflow.

Airflow rate (L/s)	Required plantroom area (m^2)
1000	8
5000	40
10,000	80
15,000	120
20,000	160
25,000	200
30,000	240
35,000	280
40,000	320
45,000	360
50,000	400

Boilers (Heating Systems)

- **Application**: Commercial and industrial heating.

- **Weight**:
  - Small boilers (100–300 kW): 500–2000 kg.
  - Medium boilers (300–1000 kW): 3000–8000 kg.
  - Large boilers (>1 MW): 10,000–30,000 kg.
- **Space Requirements**:
  - Mechanical room with sufficient headroom for piping and servicing.
  - Clearance: 1–2 m for access and ventilation.

Heat Pumps

- **Application**: Residential, commercial, and industrial heating/cooling.
- **Weight**:
  - Residential units: 50–150 kg.
  - Commercial units: 200–1000 kg.
- **Space Requirements**:
  - Outdoor unit: 1–2 m^2, with 1–2 m clearance for airflow.
  - Indoor unit: Compact design; typically wall- or floor-mounted.

Ductwork

- **Application**: Air distribution in buildings.
- **Weight**:
  - Light-gauge metal ducts: 5–10 kg/m^2.
  - Insulated ducts: 10–20 kg/m^2.
- **Space Requirements**:
  - Plenum or ceiling void with 300–500 mm height for main ducts.
  - Space for access doors and maintenance points.

Exhaust and Ventilation Systems

- **Application**: Kitchens, laboratories, parking garages.
- **Weight**:
  - Fans: 50–1000 kg (depending on capacity).
- **Space Requirements**:
  - Requires roof or wall space for discharge points.
  - Clearance for maintenance access.

General Rules of Thumb

- **Load-Bearing Considerations**:
  - Structural engineers must verify building load capacity for large systems like chillers, boilers, and AHUs.
  - Include dynamic loads for rooftop units.

- **Space Allowance**:
  - Provide sufficient clearance around equipment for maintenance (usually 1.5–2 m).
  - Mechanical rooms typically occupy **5–10% of building floor area** for large projects.
- **Equipment Grouping**:
  - Group similar systems (e.g., AHUs and boilers) to optimize space and serviceability.
- **Noise and Vibration Isolation**:
  - Larger equipment like chillers and RTUs require vibration isolation and dedicated enclosures to minimize sound transmission.
- **Vertical Distribution**:
  - Ducts and piping risers must align with building shafts for space efficiency.

Chapter 13
HVAC Design Specifications, Installation and Commissioning Check List

13.1 Overview

The purpose of an HVAC design and installation checklist is to ensure that every step in the planning, installation, and commissioning of HVAC systems is executed correctly, efficiently, and in compliance with applicable standards. This checklist acts as a comprehensive guide for engineers, contractors, and technicians, reducing the risk of errors and ensuring system performance, safety, and reliability.

13.2 General Design Considerations

- **Project Requirements**:
  - Confirm scope, purpose, and design criteria (e.g., comfort, process cooling, energy efficiency).
  - Verify compliance with local codes, standards (ASHRAE, CIBSE, EN, NCC), and client requirements.
- **Space Assessment**:
  - Analyse building layout, occupancy, and usage patterns.
  - Determine available space for HVAC equipment, ductwork, and utilities.
- **Load Calculations**:
  - Perform accurate heating, cooling, and ventilation load calculations.
  - Account for internal loads (occupants, lighting, and equipment) and external loads (solar, infiltration).

V. Vakiloroaya, *HVAC and Building Services Rules of Thumb*, Springer Tracts in Civil Engineering, https://doi.org/10.1007/978-3-032-10285-0_13

HVAC System Selection

- **System Type**:
  - Evaluate system options (e.g., DX systems, chilled water systems, VRF, heat pumps) based on building type and load.
  - Consider energy efficiency, capital cost, and maintenance.
- **Energy Efficiency**:
  - Select energy-efficient equipment (e.g., high EER/SEER for cooling, high COP for heat pumps).
  - Integrate energy-saving strategies (e.g., VFDs, economizers, heat recovery).
- **Scalability and Flexibility**:
  - Design for future expansion or changes in building usage.
  - Provide zoning for independent control of different areas.

Heating and Cooling Plant

- **Equipment Sizing**:
  - Size chillers, boilers, heat pumps, and cooling towers based on peak load plus safety margin.
  - Consider part-load efficiency (e.g., IPLV or NPLV ratings).
- **Plant Location**:
  - Ensure accessibility for maintenance and proper ventilation.
  - Minimize noise and vibration impact on occupants.
- **Redundancy**:
  - Provide backup capacity for critical systems or buildings with high reliability needs.
- **Piping Systems**:
  - Design hydronic systems with appropriate pipe sizing, insulation, and flow balancing.

Air Distribution

- **Duct Design**:
  - Size ducts to minimize pressure drop and fan power while maintaining adequate airflow.
  - Design to avoid excessive noise and vibration.
- **Air Quality**:
  - Ensure proper filtration (MERV rating per application) and avoid short-circuiting of supply and return air.

- **Zoning**:
  - Divide the system into zones based on usage, occupancy, and thermal requirements.
- **Ventilation**:
  - Meet ventilation requirements as per ASHRAE 62.1 or local standards.
  - Consider demand-controlled ventilation where applicable.

Ventilation and Indoor Air Quality

- **Outdoor Air Requirements**:
  - Calculate required fresh air based on occupancy and building type.
  - Consider energy recovery systems to reduce conditioning loads.
- **Humidity Control**:
  - Maintain relative humidity between **40% and 60%** to ensure comfort and prevent mold growth.
- **Exhaust Systems**:
  - Provide proper exhaust systems for kitchens, bathrooms, and hazardous areas (e.g., labs).
- **Air Monitoring**:
  - Incorporate CO_2 and IAQ sensors for real-time air quality management.

Energy Efficiency and Sustainability

- **Energy Modelling**:
  - Perform energy simulations to optimize system performance and comply with certifications (LEED, BREEAM).
- **Renewable Energy**:
  - Integrate renewable energy sources (e.g., solar thermal, geothermal).
- **Heat Recovery**:
  - Use energy recovery ventilators (ERVs) or heat recovery chillers for energy savings.
- **Controls**:
  - Implement advanced controls for temperature, humidity, and scheduling.
  - Use IoT-enabled systems for real-time monitoring and adaptive control.

Electrical Considerations

- **Power Supply**:
  - Verify electrical load requirements and availability of power.
  - Size electrical feeders, breakers, and transformers for HVAC equipment.

- **Emergency Power**:
  - Provide backup power for critical systems (e.g., hospitals, data centres).
- **Coordination**:
  - Ensure compatibility between HVAC equipment and building power systems.

Plumbing and Drainage

- **Condensate Drainage**:
  - Design proper drainage for cooling coils and dehumidifiers to prevent leaks.
- **Water Supply**:
  - Ensure adequate water supply for chillers, cooling towers, and humidifiers.
- **Expansion Tanks**:
  - Size expansion tanks for both heating and chilled water systems.
- **Pipe Insulation**:
  - Insulate pipes to minimize heat loss/gain and prevent condensation.

Noise and Vibration Control

- **Noise Levels**:
  - Ensure HVAC equipment meets acceptable noise levels for the building type.
  - Use sound attenuators and acoustic insulation where necessary.
- **Vibration Isolation**:
  - Isolate equipment using rubber mounts, spring isolators, or pads to reduce transmission.

Safety and Code Compliance

- **Fire Safety**:
  - Design fire-rated duct systems and provide fire dampers where required.
  - Ensure smoke control systems comply with building codes.
- **Refrigerant Management**:
  - Select refrigerants with low GWP and design leak detection systems.
- **Pressure Relief**:
  - Include safety valves and expansion tanks for hydronic systems.
- **Access and Maintenance**:
  - Provide adequate clearance around equipment for maintenance and inspections.

Building Automation and Controls

- **System Integration**:
  - Use compatible protocols (e.g., BACnet, Modbus) for seamless integration.

- **Monitoring and Diagnostics**:
  - Include remote monitoring and fault diagnostics for energy optimization.
- **Zoning Controls**:
  - Enable independent controls for different areas to enhance comfort and energy savings.

Testing, Commissioning, and Documentation

- **Commissioning Plan**:
  - Develop a detailed commissioning plan to verify system performance.
- **Testing**:
  - Conduct airflow, water flow, and control system tests before handover.
- **O&M Manuals**:
  - Provide comprehensive operation and maintenance documentation.
- **Training**:
  - Train facility managers and operators on system usage and troubleshooting.

Special Applications

- **Data Centres**:
  - Ensure precision cooling, redundancy, and humidity control.
- **Healthcare Facilities**:
  - Design to maintain strict air quality, pressure zoning, and filtration.
- **Industrial Applications**:
  - Account for process-specific requirements (e.g., temperature, humidity, safety).

13.3 Design and Installation Check List

13.3.1 DX Rooftop Package

Design Checklist

Category	Checklist items
System requirements	1. Calculate the total cooling and heating loads using ASHRAE or local standards
	2. Define outdoor air and ventilation requirements based on occupancy and building type
	3. Ensure compliance with energy codes and standards (e.g., ASHRAE 90.1)

Category	Checklist items
Equipment selection	4. Select package capacity (e.g., kW or tons) to meet the design loads with appropriate oversizing factor
	5. Consider energy efficiency ratings such as EER, SEER, or COP
	6. Choose units with heat recovery or economizers for energy savings if applicable
Ductwork design	7. Design supply and return ductwork to minimize pressure losses
	8. Incorporate dampers for balancing and zoning where needed
Electrical requirements	9. Specify electrical power requirements, including voltage and phase
Control system	10. Integrate the rooftop unit with the building management system (BMS) for advanced control

Installation Checklist

Category	Checklist items
Preparation	1. Verify unit placement based on structural load capacity of the rooftop
	2. Confirm that the location minimizes duct length and avoids recirculation zones
Unit placement	3. Use vibration isolators or mounts to prevent structural vibration
	4. Install weatherproof curbs or supports for proper elevation and drainage
Connections	5. Connect ductwork (supply, return, and exhaust) with proper sealing to prevent leakage
	6. Ensure proper electrical connections, including grounding and breaker sizing
	7. Connect refrigerant lines and check for leaks
Condensate drainage	8. Install and slope condensate drain lines for effective water removal
Fresh air intake	9. Ensure proper placement of fresh air intake and install filters if required

Commissioning Checklist

Category	Checklist items
Pre-start-up inspection	1. Verify that all connections (electrical, ductwork, and refrigerant) are secure
	2. Inspect unit components, including coils, compressors, fans, and filters
Functional testing	3. Test the operation of compressors, fans, and dampers
	4. Verify thermostat and control integration
Airflow and balancing	5. Measure and adjust airflow to meet design requirements
	6. Confirm proper operation of economizers and fresh air intakes
Refrigerant system	7. Check refrigerant charge and adjust as per manufacturer specifications
System performance	8. Test cooling and heating modes to verify that they meet load demands
	9. Monitor and log performance metrics like power consumption, temperature, and humidity levels

Troubleshooting and Maintenance Checklist

Issue	Action
Insufficient cooling or heating	Check refrigerant levels and inspect coils for dirt or blockages
High energy consumption	Verify airflow, damper settings, and economizer operation
Noise or vibration	Inspect mounting and isolators; check for loose components
Condensate overflow	Ensure drain lines are clean and sloped correctly

Safety Tips

- **Refrigerant Handling:** Always follow safety protocols for refrigerant storage and handling.
- **Electrical Safety:** Ensure power connections are made by licensed electricians and comply with local codes.
- **Weather Protection:** Use weatherproofing materials to protect units from outdoor elements.

Technical Specification for DX Rooftop Package

Parameter	Unit	Specification
Manufacturer		
Model number		
Cooling capacity	kW or TR	
Heating capacity	kW or BTU/h	
Type		(e.g., Cooling Only, Heat Pump, Gas Heating)
Refrigerant type		(e.g., R410A, R134a)
Airflow rate	L/s or CFM	
External static pressure (ESP)	Pa or in. wg	
Cooling coil type		(e.g., DX Coil)
Heating coil type		(e.g., Electric, Gas-Fired)
Supply air temperature range	°C	
Return air temperature range	°C	
EER (energy efficiency ratio)		
SEER (seasonal energy efficiency ratio)		
Power supply	V/Ph/Hz	
Electrical consumption	kW	
Compressor type		(e.g., Scroll, Rotary)
Number of compressors		
Fan type		(e.g., Centrifugal, Axial)
Fan motor power	kW	

Parameter	Unit	Specification
Noise level	dB(A)	
Condensate drain type		(e.g., Pump, Gravity)
Dimensions (L × W × H)	mm or inches	
Weight	kg or lbs	
Material of construction		(e.g., Galvanized Steel)
Filter type		(e.g., MERV 8, HEPA)
Control type		(e.g., Thermostat, BMS Integration)
Installation type		(e.g., Roof-Mounted, Outdoor)
Duct connection size	mm or inches	
Certification		(e.g., CE, UL, AHRI)
Accessories included		(e.g., Economizer, Dampers)
Special features		(e.g., Variable Speed, High Efficiency)

Technical Specification of Split Air Conditioning System

Parameter	Unit	Specification
Manufacturer		
Model number		
System type		(e.g., Cooling Only, Heat Pump)
Cooling capacity	kW or BTU/h	
Heating capacity (if applicable)	kW or BTU/h	
Compressor type		(e.g., Rotary, Scroll, Inverter)
Refrigerant type		(e.g., R410A, R32)
Outdoor unit dimensions	mm (H × W × D)	
Outdoor unit weight	kg	
Indoor unit dimensions	mm (H × W × D)	
Indoor unit weight	kg	
System EER (energy efficiency ratio)		
System COP (coefficient of performance)		
Power supply voltage	V	(e.g., 220–240 V, 50 Hz)
Power consumption (cooling mode)	kW	
Power consumption (heating mode)	kW	
Airflow rate (indoor unit)	L/s or m^3/h	
Noise level (indoor unit)	dB(A)	
Noise level (outdoor unit)	dB(A)	
Temperature range (cooling)	°C	

Parameter	Unit	Specification
Temperature range (heating)	°C	
Control type		(e.g., Wired, Wireless, Smart Control)
Air filter type		(e.g., Washable, HEPA)
Operating range (outdoor unit)	°C	(e.g., −10 °C to +46 °C)
Refrigerant charge	kg	
Refrigerant pipe diameter	mm	
Maximum piping length	m	
Maximum height difference	m	
Maintenance interval	Months	
Warranty	Years	
Special features		(e.g., Sleep Mode, Energy Saver, Smart App Control)

13.3.2 *Variable Refrigerant Flow (VRF) Systems*

Design Checklist

Category	Checklist items
System requirements	1. Calculate heating and cooling loads for each zone using standard methods (e.g., ASHRAE)
	2. Define the number of indoor zones and their respective capacities
	3. Specify the type of VRF system: heat pump or heat recovery
Equipment selection	4. Select outdoor units with appropriate capacity and efficiency ratings (e.g., COP, EER, or SEER)
	5. Choose indoor units (ducted or ductless) to match zone requirements
Piping design	6. Design refrigerant piping layout ensuring total pipe length and height differences are within limits
	7. Use appropriate pipe sizes to minimize pressure losses
Control system	8. Integrate with a central control system or BMS for monitoring and energy management
	9. Plan for user-friendly local controls in each zone
Ventilation requirements	10. Ensure adequate fresh air supply, either by integrating with a dedicated ventilation system or using DOAS
Electrical design	11. Verify power requirements and plan for independent circuit breakers for outdoor and indoor units

Installation Checklist

Category	Checklist items
Preparation	1. Inspect and prepare installation areas for outdoor and indoor units
	2. Verify that the building structure supports the weight and vibration of outdoor units
Piping installation	3. Install refrigerant piping, ensuring proper insulation and leak testing
	4. Maintain recommended bending radius for pipes to prevent kinks
	5. Install branch selector units (if heat recovery system) correctly
Unit placement	6. Install outdoor units in well-ventilated locations away from direct sunlight or obstructions
	7. Place indoor units securely and verify alignment with drain connections
Electrical connections	8. Ensure all electrical connections comply with local codes and manufacturer guidelines
	9. Connect communication wiring for control signals between units
Drainage System	10. Install condensate drains for indoor units with proper slope for gravity flow

Commissioning Checklist

Category	Checklist Items
Pre-startup inspection	1. Verify refrigerant piping connections and insulation
	2. Check electrical connections, including voltage and grounding
Refrigerant system testing	3. Conduct pressure testing and evacuation to remove moisture and air
	4. Charge refrigerant as per manufacturer specifications
System configuration	5. Program the control system for zone settings and operational modes
	6. Set operational parameters like temperature setpoints and schedules
Performance testing	7. Test cooling and heating modes for proper operation
	8. Monitor system performance metrics (e.g., refrigerant pressure, temperatures, power consumption)

Troubleshooting and Maintenance Checklist

Issue	Action
Insufficient cooling/heating	Check refrigerant levels, piping insulation, and air filters for blockages
Communication errors	Verify wiring connections between indoor and outdoor units
Water leakage from indoor unit	Inspect and clean the condensate drain line and tray
High energy consumption	Check for improperly set controls or malfunctioning components like compressors

Troubleshooting a VRF

Following is a structured approach for troubleshooting a VRF system. The process is presented in the form of questions a technician might ask, along with potential answers and solutions.

General Preliminary Checks

1. **Is the system powered on?**
   - **Answer**: No.
     - Check the power supply, circuit breakers, and ensure the system is receiving voltage.
   - **Answer**: Yes.
     - Proceed to the next question.
2. **Are there any error codes on the indoor or outdoor units?**
   - **Answer**: Yes (e.g., E1, P3, etc.).
     - Refer to the manufacturer's error code manual to diagnose the fault.
   - **Answer**: No.
     - Move to operational checks.

Common Operational Issues

1. System Does Not Cool or Heat Properly

3. **Is the setpoint temperature correct?**
   - **Answer**: No.
     - Adjust the temperature to the desired level.
   - **Answer**: Yes.
     - Check other factors.
4. **Are the indoor unit filters clean?**
   - **Answer**: No.
     - Clean or replace the filters and retest the system.
   - **Answer**: Yes.
     - Inspect further.
5. **Are the outdoor unit fans operational?**
   - **Answer**: No.
     - Check for obstructions, fan motor failure, or control board issues.
   - **Answer**: Yes.
     - Verify refrigerant flow.

6. **Is the refrigerant charge correct?**
    - **Answer**: No (undercharged or overcharged).
        - Recover or add refrigerant as per system specifications.
    - **Answer**: Yes.
        - Investigate system piping or valve issues.
7. **Are there any refrigerant leaks?**
    - **Answer**: Yes.
        - Locate the leak using an electronic leak detector or soap solution, repair the leak, and recharge the system.
    - **Answer**: No.
        - Check the expansion valves.

2. Some Indoor Units Work, Others Don't

8. **Are the non-operational units receiving power?**
    - **Answer**: No.
        - Inspect the wiring and connectors for loose or damaged connections.
    - **Answer**: Yes.
        - Investigate communication issues.
9. **Is the communication between indoor and outdoor units stable?**
    - **Answer**: No.
        - Check communication cables for continuity, correct wiring, and proper connection.
    - **Answer**: Yes.
        - Test individual units.
10. **Are the non-working units showing any error codes?**
    - **Answer**: Yes.
        - Diagnose using the error code and resolve the specific issue.
    - **Answer**: No.
        - Check indoor unit sensors and actuators.

3. System Shuts Down Randomly

11. **Is the power supply stable?**
    - **Answer**: No.
        - Inspect for voltage fluctuations and stabilize the power source.

- **Answer**: Yes.
  - Check for overheating or overcurrent protection activation.

12. **Are the outdoor unit coils clean?**
    - **Answer**: No.
      - Clean the coils to improve heat transfer efficiency.
    - **Answer**: Yes.
      - Test system load conditions.
13. **Is the system overloaded?**
    - **Answer**: Yes.
      - Reduce the number of operating units or resize the system for load requirements.
    - **Answer**: No.
      - Inspect compressor functionality.

Compressor Issues

14. **Does the compressor start?**
    - **Answer**: No.
      - Check the contactor, start capacitor, and control board.
    - **Answer**: Yes.
      - Ensure it is running at the correct capacity.
15. **Is the compressor overheating?**
    - **Answer**: Yes.
      - Inspect oil levels, refrigerant charge, and cooling fans for proper operation.
    - **Answer**: No.
      - Test suction and discharge pressures.
16. **Are suction/discharge pressures within the design range?**
    - **Answer**: No.
      - Investigate blockages, refrigerant flow issues, or valve malfunctions.
    - **Answer**: Yes.
      - Evaluate compressor efficiency.

Sensor and Control Issues

17. **Are all temperature and pressure sensors operational?**
    - **Answer**: No.
      - Replace or recalibrate faulty sensors.

- **Answer**: Yes.
  - Check control algorithms.

18. **Are the electronic expansion valves (EEVs) operating correctly?**
    - **Answer**: No.
      - Clean or replace faulty valves and check for proper control signals.
    - **Answer**: Yes.
      - Ensure refrigerant flow paths are not obstructed.

Airflow Issues

19. **Are the supply and return air paths unobstructed?**
    - **Answer**: No.
      - Remove obstructions and ensure dampers are positioned correctly.
    - **Answer**: Yes.
      - Verify fan performance.
20. **Are the indoor unit fans running at the correct speed?**
    - **Answer**: No.
      - Inspect fan motors, capacitors, or control boards.
    - **Answer**: Yes.
      - Check duct static pressure.

Example Scenario
Symptom: The VRF system is not cooling properly.

Questions and Actions

1. Is the system powered on? → Yes.
2. Are there error codes? → No.
3. Is the refrigerant charge correct? → No, it's low.
   - Action: Locate and repair a refrigerant leak, then recharge the system.
4. Are the indoor filters clean? → Yes.

After recharging refrigerant, the system operates normally, and cooling is restored.

Refrigerant to Be Charged to a VRF
The procedure to calculate the amount of refrigerant to charge a VRF system involves several steps and depends on key factors such as the type of refrigerant, total pipe length, and the thermal capacity of the system. Here's a detailed step-by-step guide:

Step 1: Gather Necessary Data

1. **Manufacturer's Specifications**:

Obtain the VRF system's installation manual or technical datasheet. It typically includes:

- Base refrigerant charge for the outdoor unit.
- Additional refrigerant charge per meter of piping.
- Maximum and minimum allowable refrigerant charge.

2. **Type of Refrigerant**:

Determine the refrigerant type (e.g., R410A, R32). Each refrigerant has specific properties affecting charge calculations.

3. **Total Piping Length**:

Measure the total equivalent pipe length, including vertical and horizontal segments. Include fittings in the calculation using equivalent lengths for elbows, tees, etc.

4. **Thermal Capacity of the System**:

Identify the cooling/heating capacity (e.g., in kW or BTU/hr) from the system's specifications.

Step 2: Calculate Total Equivalent Pipe Length

1. Add the actual length of all horizontal and vertical pipes.
2. Account for equivalent lengths of fittings:
   - Elbows, tees, and valves add to the total length.
3. Use the manufacturer's guidelines for equivalent lengths.

Step 3: Base Refrigerant Charge

The outdoor unit is pre-charged with a base amount of refrigerant. This charge is sufficient for a specific pipe length, typically 30–50 ft (9–15 m), depending on the manufacturer.

Example

- Base charge: 5 kg for the first 15 m of piping.

Step 4: Additional Refrigerant for Pipe Length

For piping beyond the base length, additional refrigerant must be added based on the refrigerant type and system specifications.

Formula:

$$Q_{add} = \text{Additional Charge per Meter} \times (\text{Total Pipe Length} - \text{Base Length}).$$

Where:

- Q_{add}: Additional refrigerant charge (kg).
- Additional Charge per Meter: Given in the manufacturer's manual (e.g., 0.02 kg/m for R410A).

Step 5: Consider Indoor Units

Some VRF systems require extra refrigerant for each indoor unit based on their capacity. Check the manufacturer's recommendations.

Step 6: Total Refrigerant Charge

The total refrigerant charge is the sum of the base charge, the additional charge for piping, and any extra charge for indoor units:

$$Q_{total} = Q_{base} + Q_{add} + Q_{indoor}$$

Where:

- Q_{total}: Total refrigerant charge (kg).
- Q_{indoor}: Additional charge for indoor units (kg).

Step 7: Example Calculation

Scenario

- Refrigerant type: R410A.
- Base charge: 5 kg for the first 15 m.
- Additional charge: 0.02 kg/m.
- Total pipe length: 50 m.
- Indoor units: No additional charge required.

Calculation

1. Pipe length exceeding base:
$50 - 15 = 35$ m
2. Additional charge:
$35 \times 0.02 = 0.7$ kg
3. Total refrigerant charge:
$5 + 0.7 = 5.7$ kg

Step 8: Charging the System

1. Use a digital weighing scale to measure the refrigerant accurately.
2. Charge refrigerant in liquid form to avoid compressor damage.
3. Monitor system pressure and temperature to ensure optimal performance during charging.

Notes

1. Always refer to the manufacturer's manual for precise values and recommendations.
2. Overcharging or undercharging refrigerant can lead to system inefficiencies and potential damage.

Technical Specification of VRF System

Parameter	Unit	Specification
Manufacturer		
Model number		
System type		(e.g., Heat Pump, Heat Recovery)
Cooling capacity	kW or BTU/h	
Heating capacity	kW or BTU/h	
Compressor type		(e.g., Inverter, Scroll)
Refrigerant type		(e.g., R410A, R32)
Outdoor unit dimensions	mm (H × W × D)	
Outdoor unit weight	kg	
Indoor unit types		(e.g., Wall-Mounted, Ducted, Cassette)
Number of indoor units		
Indoor unit capacity range	kW or BTU/h	
Total piping length	m	
Maximum piping length per unit	m	
Maximum height difference	m	
System EER (energy efficiency ratio)		
System COP (coefficient of performance)		
Power supply voltage	V	(e.g., 220–240 V, 50 Hz)
Power consumption (cooling mode)	kW	
Power consumption (heating mode)	kW	
Noise level (outdoor unit)	dB(A)	
Noise level (indoor unit)	dB(A)	
Airflow rate (indoor unit)	L/s or m^3/h	
Control type		(e.g., Wired, Wireless, Centralized)
Temperature range (cooling)	°C	
Temperature range (heating)	°C	
Integration with BMS		(Yes/No)
Defrosting method		(e.g., Automatic, Manual)
Operating range (outdoor unit)	°C	(e.g., −15 °C to +46 °C)
Refrigerant charge	kg	
Refrigerant pipe diameter	mm	
Maintenance interval	Months	
Warranty	Years	
Special features		(e.g., Zoning Control, Energy Monitoring)

13.3.3 Air-Cooled and Water-Cooled Chillers

Design Checklist

Category	Checklist items
System requirements	1. Calculate the total cooling load using standard methods (e.g., ASHRAE guidelines)
	2. Specify chiller type (air-cooled or water-cooled) based on application, climate, and site constraints
Capacity selection	3. Select chiller capacity considering peak and part-load conditions
	4. Factor in redundancy if required (e.g., N + 1 configuration)
Site considerations	5. For air-cooled chillers: Ensure adequate space for air circulation and access for maintenance
	6. For water-cooled chillers: Plan mechanical room space and cooling tower placement
Piping design	7. Design chilled water and condenser water loops with proper pipe sizes and insulation
	8. Include provisions for expansion tanks and air vents in the piping layout
Electrical design	9. Confirm electrical power supply, voltage, and backup power requirements
	10. Plan for separate circuits for each chiller
Control integration	11. Incorporate chiller controls into the Building Management System (BMS) for monitoring and optimization

Installation Checklist

Category	Checklist items
Preparation	1. Inspect the installation site for structural and accessibility requirements
	2. Verify clearances for airflow (air-cooled) or cooling tower placement (water-cooled)
Equipment placement	3. Install air-cooled chillers on a flat, vibration-isolated surface with adequate airflow on all sides
	4. Place water-cooled chillers inside a mechanical room with easy access for maintenance
Piping connections	5. Connect chilled water and condenser water piping with appropriate valves, strainers, and insulation
	6. Install flexible connections to account for vibration
Electrical connections	7. Connect power supply to the chiller, ensuring compliance with local electrical codes
	8. Wire controls for integration with the BMS
Cooling tower setup	9. For water-cooled systems, install cooling towers with proper piping for makeup, overflow, and drainage
Drainage and ventilation	10. Ensure provision for condensate drainage and adequate ventilation for mechanical rooms

Commissioning Checklist

Category	Checklist items
Pre-startup inspection	1. Inspect piping for leaks and ensure proper insulation
	2. Check all electrical connections and verify grounding
Refrigerant and water systems	3. Confirm refrigerant charge and water loop parameters
	4. Test cooling tower water flow and chemical treatment system (if water-cooled)
Control system configuration	5. Program operational parameters like supply water temperature and scheduling
Performance testing	6. Verify chiller performance under different load conditions
	7. Measure and document key parameters like power consumption, water flow rate, and outlet temperatures

Troubleshooting and Maintenance Checklist

Issue	Action
Low cooling performance	Check refrigerant levels, water flow rate, and fouling in heat exchangers
High energy consumption	Verify condenser operation, piping insulation, and chiller efficiency settings
Water leakage	Inspect chilled water piping and cooling tower connections (if water-cooled)
Vibration or noise	Check for misaligned components or insufficient vibration isolation

Safety Tips

- **For Air-Cooled Systems:** Ensure unobstructed airflow around the chiller.
- **For Water-Cooled Systems:** Regularly treat cooling tower water to prevent scaling and biological growth.
- **Electrical Safety:** Always use lockout-tagout procedures during maintenance.

Technical Specification of a Water-Cooled Chiller

Parameter	Unit	Specification
Manufacturer		
Model number		
Cooling capacity	kW or Tons	
Compressor type		
Number of compressors		
Condenser type		
Chilled water flow rate	L/s or GPM	
Chilled water inlet temp	°C or °F	
Chilled water outlet temp	°C or °F	

Parameter	Unit	Specification
Condenser water flow rate	L/s or GPM	
Condenser water inlet temp	°C or °F	
Condenser water outlet temp	°C or °F	
Power input	kW	
Energy efficiency ratio (EER)	kW/kW or BTU/W	
Coefficient of performance (COP)		
Voltage	V	
Frequency	Hz	
Refrigerant type		
Refrigerant charge	kg	
Sound level	dB(A)	
Weight (operating)	kg	
Dimensions (L × W × H)	mm or inches	
Piping connection size	mm or inches	
Operating conditions		
Control type		
Special features		

Technical Specification of Air-Cooled Chiller

Parameter	Unit	Specification
Manufacturer		
Model number		
Cooling capacity	kW or Tons	
Compressor type		
Number of compressors		
Air-cooled condenser type		
Chilled water flow rate	L/s or GPM	
Chilled water inlet temp	°C or °F	
Chilled water outlet temp	°C or °F	
Power input	kW	
Energy efficiency ratio (EER)	kW/kW or BTU/W	
Coefficient of performance (COP)		
Voltage	V	
Frequency	Hz	
Refrigerant type		
Refrigerant charge	kg	
Sound level	dB(A)	
Weight (operating)	kg	
Dimensions (L × W × H)	mm or inches	

Parameter	Unit	Specification
Piping connection size	mm or inches	
Fan type		
Number of fans		
Airflow rate (condenser)	L/s or CFM	
Operating conditions		
Control type		
Special features		

13.3.4 *Heat Pumps*

Design Checklist

Category	Checklist items
System requirements	1. Calculate heating and cooling loads using standard methods (e.g., ASHRAE, CIBSE)
	2. Select the appropriate type of heat pump (air-source, water-source, or ground-source) based on application and climate
Capacity selection	3. Match heat pump capacity to the building load requirements with consideration for efficiency at partial load conditions
	4. Ensure sufficient capacity for defrost cycles (air-source) or peak demand
Site considerations	5. Determine outdoor unit location for proper airflow and noise compliance (air-source)
	6. Plan for ground loops or water source connections if using ground-source or water-source systems
Piping design	7. Design refrigerant and/or water loop piping for minimal pressure drop and proper insulation
	8. Include expansion tanks, air vents, and isolation valves as needed
Electrical design	9. Confirm power supply requirements, voltage compatibility, and provisions for backup power
	10. Ensure electrical circuits are designed to handle startup loads
Control integration	11. Integrate controls with the Building Management System (BMS) for temperature regulation and efficiency monitoring

Installation Checklist

Category	Checklist items
Preparation	1. Inspect the site for accessibility, structural integrity, and drainage for outdoor units
	2. Verify compliance with local codes for refrigerant handling and electrical connections

Category	Checklist items
Equipment placement	3. Install outdoor units with sufficient clearance for airflow and maintenance (air-source)
	4. For water-source or ground-source, install indoor units and associated piping in an accessible mechanical room
Piping connections	5. Connect refrigerant or water piping with proper fittings, insulation, and leak tests
	6. Install flexible connections to minimize vibration transmission
Electrical connections	7. Connect power supply and control wiring, ensuring proper grounding
	8. Follow manufacturer specifications for wiring configurations
Ventilation and drainage	9. Provide adequate ventilation for outdoor units and drainage for condensate water

Commissioning Checklist

Category	Checklist items
Pre-startup inspection	1. Inspect piping and connections for leaks and insulation integrity
	2. Check refrigerant charge and water loop parameters
System configuration	3. Program control parameters for heating and cooling modes
	4. Verify defrost cycle settings (air-source)
Performance testing	5. Measure system performance at various loads and compare with design parameters
	6. Test thermostat and BMS integration for accurate control

Troubleshooting and Maintenance Checklist

Issue	Action
Poor heating/cooling performance	Check refrigerant charge, water flow, and heat exchanger cleanliness
Excessive noise or vibration	Inspect for loose connections, misalignment, or obstructions around the outdoor unit
Short cycling	Verify thermostat placement, system sizing, and refrigerant levels

Advantages of Heat Pumps

Type	Advantages
Air-source	1. Easier installation and lower upfront cost
	2. Suitable for moderate climates
Ground-source	1. Higher efficiency and stable performance across seasons
	2. Lower long-term operational costs
Water-source	1. Suitable for large-scale applications with water bodies or closed-loop systems

Safety Tips

- **Refrigerant Safety:** Ensure proper handling of refrigerants to avoid leaks and environmental impact.
- **Electrical Safety:** Use lockout-tagout procedures during maintenance or installation.
- **Drainage:** Avoid water pooling by ensuring proper slope and drainage connections.

13.3.5 Heat Recovery Systems

Design Checklist

Category	Checklist items
Load calculation	1. Assess ventilation airflows for both supply and exhaust
	2. Calculate heating and cooling loads based on the outdoor and indoor air temperature difference
System selection	3. Choose the appropriate heat recovery system type (e.g., plate heat exchanger, rotary wheel, or heat pipe)
	4. Verify compatibility with existing HVAC equipment and system design
Efficiency goals	5. Determine the heat recovery efficiency target based on energy savings goals
	6. Specify materials to ensure thermal performance and corrosion resistance
Airflow design	7. Ensure proper balancing of supply and exhaust airflows
	8. Consider pressure drops across the heat recovery system
Frost control	9. Evaluate frost risk in cold climates and design frost prevention strategies if needed
Integration	10. Ensure compatibility with Building Management Systems (BMS) for control and monitoring

Installation Checklist

Category	Checklist items
Site preparation	1. Verify the location for mounting the heat recovery system, ensuring structural support and accessibility
	2. Ensure proper duct routing for supply and exhaust air streams
Duct connections	3. Connect supply and exhaust ducts as per design, ensuring no cross-contamination
	4. Install dampers to control airflow and prevent backdrafts
Filters	5. Install air filters upstream to protect the heat recovery core from dust and debris
	6. Verify filter type and pressure drop according to system requirements
Insulation	7. Insulate ducts and heat recovery units to prevent energy losses and condensation

Category	Checklist items
Electrical and controls	8. Provide power connections as per equipment specifications
	9. Connect sensors and actuators for monitoring temperatures and airflows

Commissioning Checklist

Category	Checklist items
Pre-start checks	1. Verify all duct connections are sealed and leak-free
	2. Confirm proper installation of filters, dampers, and insulation
System balancing	3. Balance supply and exhaust airflows to achieve design specifications
	4. Check for proper operation of fans and control dampers
Performance testing	5. Measure air temperatures at the supply and exhaust to evaluate heat recovery efficiency
	6. Verify pressure drops across the heat recovery core match design conditions
Frost control	7. Test frost prevention mechanisms, such as bypass dampers or preheaters, in cold climate conditions
Control system	8. Confirm the functionality of sensors, actuators, and control algorithms

Troubleshooting and Maintenance Checklist

Issue	Action
Low heat recovery	Check for leaks, dirty filters, or imbalanced airflows
Cross-contamination	Inspect gaskets and seals in the heat recovery core
High pressure drop	Clean or replace filters and inspect the heat recovery core for blockages
Control malfunction	Recalibrate sensors and verify control settings

Safety and Efficiency Tips

Maintenance Access: Ensure that the heat recovery system is installed with sufficient access for cleaning and maintenance.

Bypass Option: Include bypass dampers for economizer mode during favourable outdoor conditions.

Monitoring: Install real-time sensors for airflow, temperature, and humidity to optimize performance.

Advantages and Disadvantages

Aspect	Advantages	Disadvantages
Energy savings	Reduces heating and cooling loads by recovering thermal energy	May require additional ductwork and space
Indoor air quality	Ensures continuous fresh air supply without significant energy penalties	Cross-contamination risk if not properly installed and maintained
Operational costs	Low operational costs after installation	Initial costs can be high depending on the system type

Technical Specification

Parameter	Unit	Specification
Manufacturer		
Model number		
Type of heat recovery		(e.g., Rotary, Plate, Run-around)
Heat recovery efficiency	%	
Airflow capacity	L/s or CFM	
Pressure drop (Supply)	Pa or in. wg	
Pressure drop (Exhaust)	Pa or in. wg	
Material of heat exchanger		(e.g., Aluminium, Plastic, Stainless Steel)
Energy recovery type		(e.g., Sensible, Latent, Both)
Operating temperature range	°C	
Operating relative humidity	%	
Maximum air leakage rate	%	
Bypass damper included	Yes/No	
Motor type		(e.g., EC, AC)
Motor power	kW or hp	
Sound level	dB(A)	
Control options		(e.g., Manual, Automatic, Remote)
Dimensions (L × W × H)	mm or inches	
Weight	kg or lbs	
Mounting type		(e.g., Wall, Ceiling, Floor)
Certification		(e.g., AHRI, Eurovent)
Maintenance interval	Months	
Warranty	Years	
Special features		(e.g., Anti-corrosion Coating, Integrated Filters)

13.3.6 Cooling Towers

Design Checklist

Category	Checklist items
System requirements	1. Calculate heat rejection load based on the cooling system's total capacity and condenser water flow rate
	2. Select cooling tower type (induced draft, forced draft, crossflow, or counter flow) based on application and site constraints
Capacity selection	3. Ensure the cooling tower is sized for peak summer conditions, including design wet-bulb temperature
	4. Verify capacity for part-load operation and ensure redundancy for critical applications
Site considerations	5. Choose a location with adequate clearance for airflow and access for maintenance
	6. Verify structural support for rooftop or elevated installations

Category	Checklist items
Water quality and treatment	7. Incorporate a water treatment system to prevent scaling, corrosion, and biological growth
	8. Plan for a blowdown system to control water concentration cycles
Piping design	9. Design condenser water piping for minimal pressure drop and correct sizing to prevent cavitation
	10. Include isolation, balancing, and control valves in the piping layout
Electrical design	11. Confirm power requirements for fan motors and control systems
	12. Include Variable Frequency Drives (VFDs) for fan speed control where applicable

Installation Checklist

Category	Checklist items
Preparation	1. Verify site readiness, including structural supports and access for crane or lifting equipment
	2. Inspect cooling tower components for shipping damage
Placement	3. Install the tower with proper orientation to prevailing winds and unobstructed airflow
	4. Ensure sufficient clearance from nearby structures to prevent recirculation of exhaust air
Piping connections	5. Connect supply and return pipes with proper alignment and support
	6. Install flexible connections to accommodate thermal expansion and vibration
Electrical connections	7. Connect power supply to fan motors and control panels, ensuring proper grounding
	8. Install control wiring for temperature sensors and fan speed control
Ventilation and drainage	9. Provide adequate drainage for overflow, blowdown, and condensate
	10. Ensure proper installation of louvers to prevent water carryover and splash

Commissioning Checklist

Category	Checklist items
Pre-start up inspection	1. Inspect piping, valves, and connections for leaks
	2. Verify fan motor alignment and blade clearance
System configuration	3. Set fan speed and temperature controls to match design parameters
	4. Test water treatment systems for proper operation
Performance testing	5. Measure condenser water temperatures and flow rates to ensure compliance with design specifications
	6. Test fan operation and monitor for vibrations or unusual noise

Troubleshooting and Maintenance Checklist

Issue	Action
High approach temperature	Check for fouled fill media, inadequate airflow, or insufficient water flow
Excessive drift	Inspect drift eliminators and adjust water flow distribution
Fan noise or vibration	Verify fan alignment, motor bearings, and blade balance
Scaling or fouling	Test water treatment system and perform chemical cleaning as needed

Advantages of Cooling Towers

Type	Advantages
Induced draft	Efficient heat transfer and low energy consumption
Forced draft	Compact design and flexibility in installation
Crossflow	Lower fan power requirements and easy access for maintenance
Counter flow	Higher thermal efficiency and smaller footprint

Safety Tips

- **Fall Protection:** Use guardrails and safety harnesses during installation and maintenance.
- **Electrical Safety:** Follow lockout-tagout procedures for fan motors and control panels.
- **Chemical Handling:** Use appropriate PPE when working with water treatment chemicals.

Technical Specification

Parameter	Unit	Specification
Manufacturer		
Model number		
Cooling capacity	kW or Tons	
Design water flow rate	L/s or GPM	
Water inlet temperature	°C or °F	
Water outlet temperature	°C or °F	
Ambient wet-bulb temperature	°C or °F	
Cooling range	°C or °F	
Approach temperature	°C or °F	
Fan type		
Number of fans		
Fan motor power	kW or HP	
Airflow rate	L/s or CFM	
Drift loss	%	

Parameter	Unit	Specification
Make-up water requirement	L/s or GPM	
Blowdown rate	% or L/s	
Noise level	dB(A)	
Voltage	V	
Frequency	Hz	
Dimensions (L × W × H)	mm or inches	
Weight (Operating)	kg	
Piping connection size	mm or inches	
Construction material		
Fill type		
Basin capacity	Liters or Gallons	
Water distribution type		
Control type		
Special features		

Technical Specification of Cooling Tower Chemical Treatment

Parameter	Unit	Specification
Manufacturer		
Model number		
Type of chemical treatment		(e.g., Scale Inhibitor, Biocide, Corrosion Inhibitor)
Chemical name/composition		(e.g., Sodium Hypochlorite, Polyphosphates)
Purpose		(e.g., Algae Control, Scale Prevention, Corrosion Control)
Dosage rate	ppm	
Water flow rate	L/s or m^3/h	
Circulation volume	m^3	
pH range		(e.g., 7.0–8.5)
Operating temperature range	°C	
Corrosion protection efficiency	%	
Scale inhibition efficiency	%	
Biological control efficiency	%	
Injection system type		(e.g., Continuous, Batch)
Injection pump flow rate	L/h or gph	
Injection point location		(e.g., Basin, Inlet Pipe)
Chemical storage tank capacity	L or m^3	
Material of storage tank		(e.g., HDPE, Stainless Steel)
Control system type		(e.g., pH Controller, ORP Controller)
Monitoring system		(e.g., Real-Time Sensors, Manual Testing)
Chemical compatibility		(e.g., Compatible with Chiller, Heat Exchanger)
Environmental compliance		(e.g., EPA Standards, Local Regulations)

Parameter	Unit	Specification
Safety measures		(e.g., Spill Containment, Ventilation)
Maintenance interval	Months	
Warranty	Years	
Special features		(e.g., Automated Dosing, Remote Monitoring)

13.3.7 Evaporative Coolers

Design Checklist

Category	Checklist items
System requirements	1. Calculate the required cooling capacity based on ambient dry-bulb and wet-bulb temperatures
	2. Assess the type of evaporative cooler needed (direct or indirect)
Capacity selection	3. Select a unit that meets the airflow and cooling capacity requirements of the space
	4. Consider cooling effectiveness based on design conditions and equipment efficiency (typically 75–90%)
Water supply	5. Ensure a continuous and reliable water supply with adequate pressure for proper operation
	6. Plan for water usage, including evaporation and bleed-off rates, to size the water line appropriately
Location and placement	7. Install in a location with access to fresh air and ensure unobstructed airflow around the unit
	8. Avoid areas with contamination sources, such as exhaust outlets or dusty environments
Drainage requirements	9. Provide drainage to handle overflow, bleed-off, and condensate effectively
	10. Verify drain sizing and route to an appropriate disposal location
Electrical design	11. Confirm power requirements for the fan motor and pump system
	12. Include control systems for fan speed and water flow regulation

Installation Checklist

Category	Checklist items
Preparation	1. Inspect all components for damage during transportation
	2. Verify the mounting structure for proper support and vibration isolation
Placement and mounting	3. Install the cooler at an elevation that ensures uniform airflow distribution
	4. Securely mount the unit with appropriate fasteners and ensure level installation

Category	Checklist items
Water connections	5. Connect the water supply line to the float valve with proper fittings
	6. Install a backflow prevention device to protect the water supply
Electrical connections	7. Wire the motor, pump, and controls according to the manufacturer's guidelines
	8. Verify grounding and circuit protection to avoid electrical hazards
Air distribution	9. Connect ductwork or air diffusers, ensuring airtight seals to avoid leakage
	10. Install dampers or grilles for airflow control and distribution

Commissioning Checklist

Category	Checklist items
Pre-startup inspection	1. Inspect water connections for leaks and test the water supply system
	2. Verify fan and pump alignment and confirm they are operational
System configuration	3. Adjust the water flow to maintain proper wetting of the media
	4. Test fan speed and airflow settings to match design specifications
Performance testing	5. Measure outlet air temperature and humidity to ensure performance meets expectations
	6. Verify even distribution of water over the cooling pads

Troubleshooting and Maintenance Checklist

Issue	Action
Uneven cooling	Check for clogged pads or improper water distribution
Reduced airflow	Inspect the fan motor, bearings, and ductwork for obstructions or damage
Excessive water usage	Adjust the bleed-off rate and inspect the float valve for proper operation
Foul odours	Clean and disinfect the water reservoir and replace water regularly to prevent microbial growth

Advantages of Evaporative Coolers

Type	Advantages
Direct evaporative coolers	Low initial and operating costs, environmentally friendly (no refrigerants), and simple maintenance
Indirect evaporative coolers	Provides cooler air without adding humidity, suitable for integration with existing HVAC systems

Safety Tips

- **Electrical Safety:** Ensure all electrical connections are properly grounded and protected from water.

- **Water Quality:** Use filtered water to prevent scaling and fouling of the cooling media.
- **Structural Safety:** Verify that the mounting structure can support the weight of the unit, including water.

Technical Specification

Parameter	Unit	Specification
Manufacturer		
Model number		
Airflow rate	L/s or CFM	
Cooling capacity	kW or BTU/h	
Water consumption rate	L/h	
Fan type		(e.g., Axial, Centrifugal)
Fan motor power	kW or HP	
Efficiency	%	
Static pressure	Pa or in. WG	
Cooling media type		(e.g., Cellulose Pads)
Cooling efficiency	%	
Pump type		
Pump power	kW or HP	
Operating voltage	V	
Frequency	Hz	
Dimensions (L × W × H)	mm or inches	
Weight (Dry)	kg or lbs	
Weight (Operating)	kg or lbs	
Water tank capacity	L	
Noise level	dB(A)	
Control type		(e.g., Manual, Remote)
Mounting type		(e.g., Roof, Wall, Window)
Application		(e.g., Industrial, Residential)
Air discharge type		(e.g., Down, Side, Up)
Operating temperature range	°C or °F	
Humidity increase	%	
Special features		(e.g., Energy Saving, UV Protection)

13.3.8 Absorption Chillers

Design Checklist

Category	Checklist items
System requirements	1. Determine the cooling load requirements and select the appropriate capacity for the chiller
	2. Identify the type of absorption chiller needed (single-effect, double-effect, or direct-fired)
Energy source	3. Ensure availability of the heat source (e.g., steam, hot water, natural gas)
	4. Calculate the required heat input and assess energy efficiency
Water quality	5. Evaluate the water quality for both chilled water and condenser water to prevent scaling
	6. Plan for water treatment systems to maintain system efficiency
Condenser type	7. Select between air-cooled or water-cooled condensers based on site conditions and efficiency goals
Space and layout	8. Verify the available space for the chiller, including clearances for maintenance and piping
	9. Ensure the mechanical room can handle the weight and dimensions of the absorption chiller
Cooling tower selection	10. Size the cooling tower based on condenser heat rejection requirements
	11. Account for water make-up and treatment needs for the cooling tower
Integration with HVAC	12. Plan the integration of the absorption chiller with the chilled water system, including pumps and piping

Installation Checklist

Category	Checklist items
Preparation	1. Inspect the chiller components for shipping damage before installation
	2. Verify the foundation and mounting supports are adequate for the chiller's weight
Placement and alignment	3. Install the chiller in a level position to ensure proper operation
	4. Provide vibration isolation pads or mounts to minimize noise and vibration
Piping connections	5. Connect chilled water and condenser water pipes using flexible joints to allow for thermal expansion
	6. Include strainers in the piping to prevent debris from entering the system
Electrical connections	7. Ensure proper wiring of the control panel and any auxiliary equipment
	8. Provide backup power or emergency systems if required by design
Control systems	9. Install and configure control systems for monitoring and automation
	10. Set up sensors for temperature, pressure, and flow rate monitoring

Commissioning Checklist

Category	Checklist items
Pre-startup inspection	1. Verify all piping and electrical connections are complete and secure
	2. Check the water flow rates and pressure drops through the chiller
System configuration	3. Calibrate control systems and ensure proper operation of temperature and pressure sensors
	4. Adjust the heat source to provide the required energy input
Performance testing	5. Measure cooling capacity and efficiency to confirm compliance with design specifications
	6. Test condenser and evaporator performance under various load conditions

Troubleshooting and Maintenance Checklist

Issue	Action
Low cooling capacity	Check for insufficient heat input, low refrigerant level, or fouled tubes
Excessive energy consumption	Inspect for heat exchanger fouling, improper flow rates, or incorrect control settings
Water leakage	Verify piping connections and inspect for damaged gaskets or seals
Noise or vibration	Ensure proper alignment of components and check for loose connections

Advantages and Disadvantages

Aspect	Advantages	Disadvantages
Energy efficiency	Uses waste heat or renewable energy, reducing electricity usage	Lower Coefficient of Performance (COP) compared to compression chillers
Environmental impact	Does not use synthetic refrigerants, reducing ozone depletion potential	Requires water and heat sources, limiting application in some settings
Maintenance	Fewer moving parts, leading to lower maintenance costs	Requires strict water quality management to prevent scaling and corrosion

Safety Tips

- **Heat Source Safety:** Ensure that the heat source does not exceed the chiller's design temperature to prevent damage.
- **Water Quality:** Implement a water treatment plan to minimize scaling and corrosion in the system.
- **Ventilation:** Ensure adequate ventilation in the mechanical room to dissipate any heat generated by the chiller.

Technical Specification

Parameter	Unit	Specification
Manufacturer		
Model number		
Cooling capacity	kW or TR	
Type		(e.g., Single Effect, Double Effect)
Working principle		(e.g., Lithium Bromide-Water, Ammonia-Water)
Input heat source		(e.g., Steam, Hot Water, Natural Gas)
Input heat energy	kW	
COP (coefficient of performance)		
Chilled water flow rate	L/s	
Chilled water inlet/outlet temp	°C	
Cooling water flow rate	L/s	
Cooling water inlet/outlet temp	°C	
Hot water/steam flow rate	L/s or kg/h	
Hot water/steam inlet temp	°C	
Nominal operating pressure	kPa	
Refrigerant type		(e.g., Water, Ammonia)
Absorbent type		(e.g., Lithium Bromide)
Noise level	dB(A)	
Material of construction		(e.g., Carbon Steel, Stainless Steel)
Dimensions (L × W × H)	mm or inches	
Weight	kg or lbs	
Energy source consumption	MJ/h or kW	
Startup time	Minutes	
Maintenance interval	Hours or Months	
Control interface		(e.g., Digital, Analog)
Installation type		(e.g., Indoor, Outdoor)
Certification		(e.g., CE, UL, ISO)
Accessories included		(e.g., Pumps, Valves, Controls)
Special features		(e.g., High Efficiency, Low Noise)

13.3.9 *Adsorption Chillers*

Design Checklist

Category	Checklist items
Load calculation	1. Calculate cooling loads based on building type, occupancy, and usage patterns
	2. Determine chilled water supply and return temperatures to meet cooling demand
Chiller selection	3. Select an adsorption chiller with suitable capacity and COP for the application
	4. Confirm the compatibility of the heat source (e.g., hot water, steam, or waste heat)
Heat source design	5. Verify that the heat source has sufficient capacity and stable supply temperature
	6. Design the heat source loop for optimal heat recovery and energy efficiency
Cooling water system	7. Specify the required cooling water flow rate and temperature range
	8. Ensure the cooling tower or dry cooler is appropriately sized to reject heat
Chilled water system	9. Calculate chilled water flow rate and pressure drop
	10. Select piping, valves, and pumps to ensure efficient and reliable operation
System integration	11.Confirm compatibility with the building management system (BMS) for control and monitoring
	12. Incorporate redundancy in the design, if critical cooling is required

Installation Checklist

Category	Checklist items
Site preparation	1. Verify structural support for the chiller's weight and dimensions
	2. Ensure proper ventilation and access for maintenance in the installation area
Piping connections	3.Install chilled water, heat source, and cooling water piping according to manufacturer specifications
	4. Include flexible connectors to mitigate vibration and thermal expansion
Insulation	5. Insulate chilled water piping to prevent heat gain and condensation
	6. Ensure hot water and cooling water pipes are insulated where necessary to avoid energy loss
Electrical connections	7. Provide power supply as per chiller's electrical requirements
	8. Ensure proper grounding and electrical safety measures
Control integration	9. Connect the chiller to the control system for automation and remote monitoring
	10. Configure sensors and actuators for temperature, pressure, and flow rate monitoring

Commissioning Checklist

Category	Checklist items
Pre-start checks	1. Verify that all piping connections are leak-free and pressure-tested
	2. Confirm proper alignment and mounting of all components
System balancing	3. Adjust flow rates in chilled water, cooling water, and heat source loops to meet design specifications
	4. Verify proper operation of control valves and pumps
Performance testing	5. Measure leaving chilled water and cooling water temperatures to confirm performance
	6. Verify the chiller's COP and capacity against design conditions
Heat source optimization	7. Ensure the heat source delivers sufficient energy at the required temperature
Safety tests	8. Check for alarms, pressure relief devices, and emergency shutdown functionality

Troubleshooting and Maintenance Checklist

Issue	Action
Low cooling output	Check heat source temperature and flow rate; inspect chilled water and cooling water flow rates
High energy consumption	Clean heat exchangers, cooling towers, and water strainers to improve efficiency
Leaks	Inspect connections and seals for damage and repair as necessary
Control issues	Recalibrate sensors and verify control logic in the BMS

Safety and Efficiency Tips

Energy Recovery: Use waste heat sources, such as exhaust gases or industrial processes, to maximize system efficiency.

Water Treatment: Implement a water treatment system to minimize scaling and corrosion in cooling water circuits.

Monitoring: Install advanced monitoring tools for real-time performance tracking and predictive maintenance.

Advantages and Disadvantages

Aspect	Advantages	Disadvantages
Energy efficiency	Utilizes low-grade heat sources, reducing reliance on electricity	Lower COP compared to conventional vapour compression systems
Environmental impact	Uses environmentally friendly refrigerants like water or silica gel	Requires significant cooling water infrastructure
Reliability	Simple operation with fewer moving parts, reducing maintenance needs	Larger size and weight compared to other chiller types

Technical Specification

Parameter	Unit	Specification
Manufacturer		
Model number		
Cooling capacity	kW or TR	
Type		(e.g., Silica Gel-Water, Zeolite-Water)
Input heat source		(e.g., Hot Water, Solar Thermal, Waste Heat)
Input heat energy	kW	
COP (Coefficient of Performance)		
Chilled water flow rate	L/s	
Chilled Water Inlet/outlet temp	°C	
Cooling water flow rate	L/s	
Cooling water inlet/outlet temp	°C	
Hot water flow rate	L/s	
Hot water inlet/outlet temp	°C	
Nominal operating pressure	kPa	
Adsorbent material		(e.g., Silica Gel, Zeolite)
Refrigerant type		(e.g., Water)
Noise level	dB(A)	
Material of construction		(e.g., Stainless Steel, Aluminium)
Dimensions (L × W × H)	mm or inches	
Weight	kg or lbs	
Energy source consumption	MJ/h or kW	
Startup time	Minutes	
Maintenance interval	Hours or Months	
Control interface		(e.g., Digital, Analog)
Installation type		(e.g., Indoor, Outdoor)
Certification		(e.g., CE, UL, ISO)
Accessories included		(e.g., Pumps, Valves, Controls)
Special features		(e.g., High Efficiency, Eco-Friendly)

13.3.10 Cooling and Heating Coils

Design Checklist

Category	Checklist items
Load calculation	1. Perform accurate cooling and heating load calculations for the space (consider peak conditions)
	2. Account for fresh air requirements and latent heat loads for cooling coil design
Coil selection	3. Select cooling coils with appropriate rows and fin density (e.g., based on required leaving air temperature)
	4. Choose coil material (copper/aluminium) based on corrosion resistance and application
	5. For heating, consider medium (e.g., hot water, steam, electric) and required outlet temperature
Airflow requirements	6. Verify coil face velocity is within acceptable limits (typically 2.5–3.0 m/s to prevent water carryover)
	7. Ensure sufficient coil depth and fin spacing to handle required airflows without excessive pressure drops
Temperature requirements	8. Design cooling coil to achieve required dew point temperature for humidity control
	9. Ensure heating coil can provide sufficient temperature rise to meet the design set-point
Drainage and condensate	10. Provide proper slope for cooling coil drain pans to prevent standing water
	11. Design for appropriate condensate disposal or recovery systems
Pipe connections	12. Specify proper pipe sizing for chilled/hot water flow to maintain coil performance
	13. Include pressure drop and flow balancing considerations in piping design
Energy efficiency	14. Consider energy recovery options (e.g., preheating/cooling with return air or waste heat)
	15. Use variable flow systems to optimize energy efficiency during part-load conditions
Documentation	16. Include cooling and heating coil details in HVAC drawings and specifications
	17. Provide clear schedules and data sheets for selected coils

Installation Checklist

Category	Checklist items
Preparation	1. Verify coil type, size, and capacity match the approved design and submittals
	2. Inspect the coil for damage or defects before installation

Category	Checklist items
Installation process	3. Install the coil in the correct orientation (e.g., airflow direction and pitch for drainage)
	4. Ensure proper alignment with ductwork or air-handling units
	5. Connect piping using proper techniques (e.g., brazing, threading) and avoid excessive stress on coil headers
Drainage and condensate	6. Install a drain pan with proper slope and connect to a condensate disposal system
	7. Check for potential condensate leaks during coil operation
Insulation	8. Insulate piping and connections to prevent condensation and heat loss/gain
	9. Verify insulation is sealed properly to prevent moisture ingress
Testing and balancing	10. Pressure-test coil connections to ensure no leaks before operation
	11. Perform water or airflow balancing as per design specifications
	12. Verify the flow rates and temperature differences match the design parameters
Final checks	13. Confirm fin cleanliness and ensure no blockage of airflow
	14. Label coil connections (e.g., supply and return lines) for future maintenance
	15. Complete and document final inspections and testing

Practical Notes and Tips

- **Cooling Coils:** Use extended drain pans for systems with high latent loads to ensure all condensate is captured.
- **Heating Coils:** Verify that the heating medium temperature (e.g., hot water or steam) matches the design specifications.
- **Access:** Ensure sufficient space around coils for cleaning and maintenance.
- **Energy Recovery:** For cooling coils, consider pairing with energy recovery wheels or plates for enhanced efficiency.

Technical Specification

Parameter	Heating coil	Cooling coil
Material	Copper/Aluminium	Copper/Aluminium
Tube diameter	12.7 or 15.9 mm	12.7 or 15.9 mm
Fin material	Aluminium/Copper	Aluminium/Copper
Fin thickness	0.2–0.3 mm	0.2–0.3 mm
Number of fins per mm	3.15–5.51 fins/mm	3.15–5.51 fins/mm
Operating pressure	Up to 1.72 MPa	Up to 1.72 MPa
Maximum temperature	93–149 °C	4–15 °C
Fluid type	Hot Water/Steam	Chilled Water/Refrigerant
Face area	As per design requirements	As per design requirements
Airflow rate	0.24–2.36 m^3/s	0.24–2.36 m^3/s
Heat transfer rate	Based on system design	Based on system design
Pressure drop	<249 Pa	<249 Pa

Parameter	Heating coil	Cooling coil
Coil circuiting	Single/Multiple Pass	Single/Multiple Pass
Connection type	Threaded/Flanged	Threaded/Flanged
Coating (optional)	Epoxy/Anti corrosion	Epoxy/Anti-corrosion

13.3.11 Fans and Jet Fans

Design Checklist

Category	Checklist items
Load calculation	1. Perform airflow calculations for the space based on ventilation and cooling/heating requirements
	2. For jet fans, calculate airflow based on car park volume and clearance requirements
Fan selection	3. Select fan type (e.g., axial, centrifugal, mixed flow) based on application and static pressure requirements
	4. Choose a jet fan type (e.g., unidirectional, reversible) and thrust based on car park layout
	5. Verify fan efficiency and motor power align with energy codes
Airflow requirements	6. Design for required air changes per hour (ACH) based on the occupancy and usage of the space
	7. For jet fans, ensure proper airflow velocity and coverage for effective smoke and fume extraction
Noise and vibration	8. Include acoustic analysis and specify silencers if necessary to control noise levels
	9. Account for vibration isolation with proper mounts and flexible connections
Ductwork integration	10. Ensure adequate space for duct connections, transitions, and fittings near the fan
	11. Verify duct size and pressure drop calculations for compatibility with fan performance
Safety and controls	12. Specify fire-rated fans where required (e.g., smoke extraction or pressurization fans)
	13. Design control systems with speed control (e.g., VFDs) and fire-mode overrides for jet fans
	14. Include redundancy in critical systems such as stairwell pressurization fans
Jet fan layout	15. Ensure proper positioning of jet fans for effective airflow coverage and minimal dead zones
	16. Design for thrust overlap between jet fans for optimal flow distribution
Documentation	17. Provide detailed fan schedules, layouts, and specifications for installation and maintenance

Installation Checklist

Category	Checklist items
Preparation	1. Verify fan model, size, and capacity match approved design documents
	2. Inspect fans and components for damage before installation
Installation process	3. Ensure proper orientation and alignment of fans with ductwork or open spaces
	4. Use vibration isolators and flexible duct connections to minimize vibration transmission
	5. Install jet fans at the correct height and angle for optimal airflow direction
Electrical connections	6. Verify power supply matches fan motor voltage and phase requirements
	7. Install control panels and wiring for speed control and fire-mode operation
	8. Test wiring connections for continuity and proper operation
Ductwork integration	9. Seal all duct joints and connections to prevent air leakage
	10. Ensure adequate clearance for fan inlet and outlet for proper airflow
Testing and balancing	11. Perform airflow testing and balancing to meet design specifications
	12. Measure static pressure, airflow velocity, and motor amperage for each fan
	13. For jet fans, test thrust and airflow coverage to ensure effective operation
Safety checks	14. Test fire-mode operation and override controls for fans in critical areas
	15. Confirm that all safety labels and access points are clearly marked
Final checks	16. Verify noise levels and ensure compliance with design specifications
	17. Document installation details and performance test results for future maintenance

Practical Notes and Tips

- **HVAC Fans:**
  - Use high-efficiency motors and variable speed drives (VSDs) to optimize energy consumption.
  - Consider dual fans for redundancy in critical systems such as ventilation shafts.
- **Jet Fans in Car Parks:**
  - Position fans to eliminate stagnant zones and provide adequate thrust to direct fumes to exhaust points.
  - Include CO and NOx sensors to modulate fan operation based on air quality.
- **Noise Mitigation:**
  - Install sound attenuators or silencers for fans in noise-sensitive areas like offices or residential spaces.

Technical Specification Fans

Parameter	Unit	Specification
Manufacturer		
Model number		
Fan type		(e.g., Centrifugal, Axial, Inline)
Airflow rate	L/s or CFM	
Static pressure	Pa or in. WG	
Fan efficiency	%	
Motor power	kW or HP	
Speed	RPM	
Impeller material		
Casing material		
Connection size (Inlet)	mm or inches	
Connection size (Outlet)	mm or inches	
Operating temperature range	°C or °F	
Operating pressure	kPa or psi	
Voltage	V	
Frequency	Hz	
Noise level	dB(A)	
Fan dimensions (L × W × H)	mm or inches	
Weight (operating)	kg or lbs	
Blade type		(e.g., Forward Curved, Backward Curved)
Drive type		(e.g., Direct, Belt Drive)
Control type		(e.g., VFD, Manual)
Mounting type		(e.g., Floor, Wall, Ceiling)
Application		(e.g., Supply, Exhaust, Return)
Special features		

Technical Specification of Jet-Fans

Parameter	Unit	Specification
Manufacturer		
Model number		
Fan type		(e.g., Axial, Mixed Flow)
Airflow rate	L/s or CFM	
Thrust force	N	
Static pressure	Pa or in. WG	
Fan efficiency	%	
Motor power	kW or HP	
Speed	RPM	
Impeller material		
Casing material		
Diameter	mm or inches	
Noise level	dB(A)	

Parameter	Unit	Specification
Control type		(e.g., VFD, On/Off, Manual)
Mounting type		(e.g., Ceiling, Wall)
Operating temperature range	°C or °F	
Voltage	V	
Frequency	Hz	
Weight (operating)	kg or lbs	
Dimensions (L × W × H)	mm or inches	
Drive type		(e.g., Direct, Belt Drive)
Application		(e.g., Car Park, Tunnel)
Jet velocity	m/s	
Throw distance	m	
Special features		

13.3.12 Attenuators

Design Checklist

Category	Checklist items
Noise level assessment	1. Identify spaces with critical noise requirements (e.g., offices, meeting rooms, hospitals)
	2. Measure or estimate sound power levels of HVAC equipment (e.g., fans, compressors)
Attenuator selection	3. Select attenuator type based on noise reduction requirements (e.g., straight, elbow, circular)
	4. Choose attenuators with appropriate insertion loss values for required frequency ranges
	5. Ensure pressure drop caused by the attenuator does not exceed system limits
Duct integration	6. Design placement of attenuators within ducts to minimize turbulence and maintain airflow
	7. Verify sufficient duct length before and after the attenuator for optimal performance
Sizing	8. Calculate attenuator dimensions based on duct size, airflow rate, and acoustic requirements
	9. Consider selecting attenuators with acoustic splitters for large airflow rates
Environmental factors	10. For outdoor systems, select weather-resistant attenuators with proper casing material
	11. Evaluate thermal insulation needs to prevent condensation in cold climates
Maintenance access	12. Ensure access points near attenuators for cleaning and maintenance

Category	Checklist items
Documentation	13. Provide detailed noise reduction calculations and specifications in project documents

Installation Checklist for HVAC Attenuators

Category	Checklist items
Preparation	1. Inspect attenuators for any physical damage before installation
	2. Verify attenuator model, size, and specifications against approved design documents
Installation process	3. Install attenuators in the correct orientation as specified by the manufacturer
	4. Ensure secure attachment to ductwork with proper sealing to prevent air leakage
	5. For flexible ducts, verify that connections to the attenuator are tight and leak-free
Alignment	6. Align the attenuator with the duct centerline to minimize turbulence and resistance
	7. Avoid abrupt changes in duct size or shape near the attenuator
Acoustic integrity	8. Seal gaps around the attenuator with acoustic sealants or gaskets to maintain noise attenuation
Pressure drop testing	9.Measure system pressure drop after installation to confirm compliance with design values
Noise level testing	10. Test noise levels at sensitive areas to ensure attenuator performance meets acoustic requirements
Safety checks	11. Verify that installation does not obstruct airflow or introduce excessive resistance
Final checks	12. Conduct a visual inspection to ensure proper alignment, sealing, and secure connections
	13. Document installation details, including noise level test results and pressure drop measurements

Practical Notes and Tips

- **Straight Attenuators:**
  Best for long duct runs with minimal bends and where space is sufficient.
- **Elbow Attenuators:**
  Useful in tight spaces or areas where the duct makes sharp turns.
- **Circular Attenuators:**
  Ideal for round duct systems, especially in industrial or specialized HVAC applications.
- **Maintenance Considerations:**
  - Ensure that attenuators are accessible for cleaning and inspections.
  - Regularly check for dirt and dust build-up that can affect performance.
- **Energy Efficiency:**
  - Balance noise reduction needs with acceptable pressure drop levels to avoid excessive fan energy consumption.

13.3.13 Filters

Design Checklist

Category	Checklist items
Application requirements	1. Identify the purpose of filtration (e.g., indoor air quality, equipment protection, or process air)
	2. Determine required filter efficiency (e.g., MERV, HEPA, ULPA ratings) based on application needs
Airflow design	3. Calculate system airflow to size filters properly
	4. Ensure filter face velocity does not exceed recommended values (typically ≤ 2.5 m/s)
Pressure drop	5. Select filters with acceptable initial and final pressure drops
	6. Consider energy costs related to pressure drop and fan power
Filter type	7. Choose filter type (e.g., panel, bag, and cartridge) based on application and maintenance frequency
Environment	8. Assess environmental factors (e.g., humidity, temperature, particulate load)
Standards compliance	9. Verify compliance with local and international standards (e.g., ASHRAE 52.2, EN 779)
	10. Plan for redundancy in critical applications

Installation Checklist for HVAC Filters

Category	Checklist items
Site preparation	1. Verify filter housing dimensions match filter specifications
	2. Clean the filter housing to remove dust or debris before installation
Filter installation	3. Ensure filters are installed in the correct orientation (airflow direction marked on filters)
	4. Seal filters properly to prevent bypass airflow
	5. Verify gaskets and clamping mechanisms for proper sealing
Pressure monitoring	6. Install pressure drop gauges (e.g., manometers, differential pressure sensors) across the filters
	7. Check for uniform airflow distribution across the filter bank
Safety	8. Follow proper handling procedures to avoid damage to high-efficiency filters (e.g., HEPA filters)

Commissioning Checklist

Category	Checklist items
Performance testing	1. Measure pressure drop across the filters under design airflow
	2. Inspect for air leaks or bypass around filters
	3. Verify filtration efficiency meets design specifications

Category	Checklist items
System balancing	4. Adjust airflow to ensure uniform distribution through filters
	5. Confirm fan performance with added resistance from filters
Documentation	6. Record filter specifications, initial pressure drop, and airflow

Troubleshooting and Maintenance Checklist

Issue	Action
High pressure drop	Inspect filters for clogging or damage and replace as needed
Bypass airflow	Check gaskets and clamping mechanisms for proper sealing
Low filtration efficiency	Verify correct filter installation and check for damaged filter media
Uneven airflow	Adjust system balancing to ensure even distribution

Safety and Efficiency Tips

Energy Efficiency: Use low-pressure-drop filters to reduce fan energy consumption.

Redundancy: Install pre-filters to extend the life of final filters in high-dust environments.

Monitoring: Regularly monitor pressure drop to identify maintenance needs.

Sealing: Ensure proper sealing to prevent unfiltered air bypass.

Advantages and Disadvantages

Aspect	Advantages	Disadvantages
Indoor air quality	Filters improve air quality by removing particulates and contaminants	High-efficiency filters may increase pressure drop and energy use
Equipment protection	Protects HVAC components from dust and debris	Regular maintenance is required for effective performance
Energy savings	Proper selection and maintenance reduce energy costs	Improperly selected filters can lead to higher operational costs

Technical Specification

Parameter	Unit	Specification
Manufacturer		
Model number		
Filter type		(e.g., HEPA, MERV, Carbon, Pleated)
Efficiency rating	% or MERV Rating	
Filter material		(e.g., Fiberglass, Synthetic Fiber, Carbon)
Dimensions (L × W × H)	mm or inches	
Thickness	mm or inches	

Parameter	Unit	Specification
Airflow capacity	L/s or CFM	
Initial pressure drop	Pa or in. wg	
Final pressure drop	Pa or in. wg	
Dust holding capacity	g or g/m^2	
Frame material		(e.g., Cardboard, Aluminium, Plastic)
Temperature resistance	°C	
Relative humidity resistance	%	
Service life	Months or Hours	
Certification		(e.g., ISO 16890, ASHRAE 52.2)
Weight	kg or lbs	
Installation type		(e.g., Panel, Bag, Cartridge)
Application		(e.g., Residential, Commercial, Industrial)
Replacement Interval	Months or Hours	
Pressure drop at rated airflow	Pa or in. wg	
Special features		(e.g., Anti-Microbial Coating, Washable)

13.3.14 Fire Dampers

Design Checklist

Category	Checklist items
Code compliance	1. Ensure compliance with local fire and building codes
	2. Verify if fire dampers are required for specific ducts penetrating fire-rated walls or floors
Type selection	3. Choose appropriate fire damper type (curtain blade, multi-blade, or combination fire/smoke damper)
	4. Consider static or dynamic fire damper based on airflow condition during fire scenarios
Location and placement	5. Identify locations where ducts penetrate fire-rated barriers (walls, ceilings, floors)
	6. Ensure dampers are accessible for inspection, testing, and maintenance
	7. Place dampers as close as possible to the fire-rated barrier
Sizing	8. Size dampers based on duct dimensions and airflow requirements
	9. Verify pressure drop and airflow resistance to match system performance
Integration	10. Ensure coordination with other systems, such as fire alarms and smoke control systems
	11. Specify appropriate fusible links, actuators, or sensors as required by the design
	12. Ensure proper sealing between ducts and fire-rated barriers

Category	Checklist items
Documentation	13. Include fire damper details in the HVAC design drawings and specifications
	14. Provide detailed installation instructions and testing requirements

Installation Checklist

Category	Checklist items
Preparation	1. Verify fire damper type, size, and location match the approved design and submittals
	2. Inspect fire-rated barriers for compliance with design requirements
Installation process	3. Install dampers in accordance with manufacturer's instructions
	4. Use approved sleeves, retaining angles, and sealants for fire-rated assembly compliance
	5. Avoid any field modifications to the fire damper without approval
	6. Secure dampers to ensure they are properly aligned with ductwork
Accessibility	7. Ensure access doors are installed for damper inspection and maintenance
Inspection	8. Inspect fusible links or actuators for proper placement and functionality
	9. Ensure no obstructions impede the closing of the fire damper blades
	10. Verify fire caulking and sealant applications maintain the fire rating of the barrier
Testing	11. Conduct operational testing to ensure the damper closes fully under simulated fire conditions
	12. Perform airflow tests to confirm minimal pressure drop across the damper when open
Final checks	13. Label each damper for identification and maintenance purposes
	14. Complete and document a final inspection checklist for handover

Practical Notes and Tips

- **Maintenance:** Include maintenance schedules in the project handover documentation, ensuring dampers are inspected regularly as per NFPA or local codes.
- **Coordination:** Coordinate damper locations with structural, electrical, and architectural teams to avoid clashes.
- **Record Keeping:** Maintain accurate as-built drawings showing the exact location of each fire damper for future reference.

Technical Specification

Parameter	Specification
Material	Galvanized Steel/Stainless Steel
Blade type	Single-blade or Multi-blade
Blade thickness	1.5–3 mm
Frame thickness	1.5–3 mm
Maximum operating temperature	Up to 1000 °C

Parameter	Specification
Fire rating	1–4 h
Pressure class	1000–2500 Pa
Velocity class	Up to 15 m/s
Damper type	Curtain type/Multi-blade type
Actuator type	Manual, Electric, or Pneumatic
Mounting orientation	Vertical or Horizontal
Leakage class	Class I, II, or III as per EN 1751 or similar standards
Standard compliance	BS EN 1366-2, UL 555, NFPA 80, or equivalent
Connection type	Flanged/Slip joint
Finish (optional)	Powder Coating/Anti-corrosion Coating
Dimensions	Customizable as per duct design requirements

13.3.15 Volume Control Dampers

Design Checklist

Category	Checklist items
System requirements	1. Determine the purpose of the damper (e.g., balancing airflow, isolating zones, or regulating pressure)
	2. Identify locations requiring dampers based on airflow and system balancing requirements
Damper type selection	3. Choose damper type based on application (e.g., opposed blade, parallel blade, or single-blade)
	4. Consider low-leakage dampers for critical zones or high-efficiency systems
Sizing	5. Size dampers to match duct dimensions and maintain design airflow rates
	6. Ensure damper capacity matches the maximum airflow of the connected duct section
Pressure and velocity	7. Confirm that the damper can handle the system's pressure and air velocity
Noise and vibration	8. Select dampers designed to minimize noise and vibration, particularly in sensitive spaces
Control requirements	9. Specify manual or motorized operation based on system requirements
	10. For automated systems, define control integration with building management systems (BMS)
Access and maintenance	11. Plan damper locations for ease of access for future adjustments and maintenance
	12. Avoid placing dampers in hard-to-reach or confined spaces
Environmental factors	13. For outdoor systems, select dampers with weather-resistant materials and seals
	14. In corrosive environments, use dampers with corrosion-resistant coatings or stainless steel

Category	Checklist items
Documentation	15. Provide damper specifications, airflow calculations, and pressure drop data in design documents

Installation Checklist for HVAC Volume Control Dampers

Category	Checklist items
Preparation	1. Inspect dampers for damage and verify specifications before installation
	2. Confirm that duct sections are properly aligned and sealed before damper installation
Installation process	3. Install dampers in the correct orientation as per design documents or manufacturer's instructions
	4. Ensure proper alignment of the damper blades with the duct centreline to avoid leakage
	5. Use appropriate mounting hardware and gaskets to secure dampers and prevent air leakage
Sealing	6. Seal damper perimeters with approved materials to avoid bypass airflow or leakage
Operation check	7. Test manual or motorized operation of the damper to ensure smooth opening and closing
	8. Verify that motorized dampers are correctly wired and integrated with the BMS
Airflow balancing	9. Adjust damper blades to achieve design airflow rates during system balancing
Pressure drop testing	10. Measure pressure drop across the damper to confirm compliance with design values
Noise testing	11. Check for noise generation from damper operation, especially in sensitive areas
Safety checks	12. Ensure dampers are not obstructing airflow or causing excessive turbulence
Final checks	13. Conduct a visual inspection to confirm proper installation and sealing
	14. Record damper positions, airflow rates, and test results for system documentation

Practical Notes and Tips

- **Opposed Blade Dampers:**
  - Best for airflow modulation due to smoother flow and reduced turbulence.
- **Parallel Blade Dampers:**
  - Suitable for applications requiring quick changes in airflow (e.g., isolation zones).
- **Material Selection:**
  - Use galvanized steel for standard applications.
  - Use aluminium or stainless steel for lightweight or corrosive environments.

- **Automation:**
  - Motorized dampers with BMS integration enhance precision and ease of control in large systems.
- **Noise Control:**
  - Ensure dampers are not installed near bends or transitions to minimize turbulence and noise.

Technical Specification

Parameter	Specification
Material	Galvanized Steel/Stainless Steel/Aluminium
Blade type	Single-blade or Multi-blade
Blade thickness	1.2–3 mm
Frame thickness	1.2–3 mm
Blade operation	Parallel or Opposed
Maximum operating temperature	−10 °C to 100 °C (varies based on material)
Pressure class	Up to 1500 Pa
Velocity class	Up to 15 m/s
Actuator type	Manual, Electric, or Pneumatic
Mounting orientation	Vertical or Horizontal
Leakage class	Class I, II, or III as per EN 1751 or similar standards
Connection type	Flanged/Slip joint
Finish (optional)	Powder Coating/Anti-corrosion Coating
Dimensions	Customizable as per duct design requirements
Airflow regulation	Fully open to fully closed (infinite adjustment positions)
Standard compliance	EN 1751, AMCA 500-D, or equivalent

13.3.16 Waterproof Louvers

Design Checklist

Category	Checklist items
System requirements	1. Determine the purpose of the louver (e.g., intake, exhaust, or ventilation protection)
	2. Ensure the louver is designed to block water infiltration under expected weather conditions
Type selection	3. Choose waterproof louvers with appropriate blade profiles (e.g., Z-shaped or J-shaped blades)
	4. Specify wind-driven rain-resistant louvers for high-exposure areas
Sizing	5. Size the louver to match airflow requirements without excessive pressure drop
	6. Consider a 20–25% free area ratio to account for water deflection and airflow

Category	Checklist items
Pressure and velocity	7. Confirm the louver can handle the system's design air velocity (typically 2.5–3.5 m/s through free area)
Drainage and waterproofing	8. Ensure louvers include drainage systems for water runoff
Material selection	9. Specify materials suitable for environmental conditions (e.g., aluminium for corrosion resistance)
	10. Use protective coatings in highly corrosive or coastal environments
Integration	11. Plan louver integration with adjacent walls, ducts, and flashing for weatherproof sealing
	12. Account for bird or insect screens if required
Noise control	13. Evaluate potential noise from airflow through louvers and use silencers or acoustic treatments if necessary
Documentation	14. Include specifications, performance data (airflow, pressure drop, water penetration), and installation guidelines in design documents

Installation Checklist

Category	Checklist items
Preparation	1. Verify louver dimensions, orientation, and specifications before installation
	2. Inspect louvers for damage or defects during delivery
Installation process	3. Install louvers at the correct orientation and slope to facilitate water drainage
	4. Secure louvers using appropriate fasteners as per manufacturer guidelines
	5. Seal gaps between the louver frame and surrounding structure to prevent water leakage
Drainage check	6. Ensure the drainage system is unobstructed and slopes away from the louver for proper water runoff
Pressure drop testing	7. Measure pressure drop across the louver to confirm compliance with design values
Performance testing	8. Simulate rain or water spray to test the louver's water resistance and drainage functionality
Alignment and fit	9. Ensure the louver blades and frame are properly aligned and free from deformation
Screens and accessories	10. Install bird or insect screens (if required) ensuring they are securely fastened and easy to maintain
Final checks	11. Conduct a final inspection to confirm proper installation, alignment, and sealing
	12. Record airflow, pressure drop, and water penetration test results for documentation

Practical Notes and Tips

- **Blade Orientation:**
  Blade orientation is critical for water deflection. Ensure Z- or J-blades are installed in the correct orientation.

- **Pressure Drop:**
  Minimize pressure drop by selecting louvers with a high free area ratio while maintaining waterproofing performance.
- **Weather Resistance:**
  In areas prone to extreme weather, select louvers certified to AMCA 550 (water penetration) and AMCA 500-L (performance).
- **Integration with Walls:**
  Use flashing and sealant around the louver frame to prevent leaks at the interface with walls or ducts.
- **Maintenance:**
  Ensure accessibility for periodic cleaning of blades and drainage channels to maintain performance.

Technical Specification

Parameter	Specification
Material	Aluminium/Stainless Steel/Galvanized Steel
Blade type	Fixed, Horizontal, or Vertical
Blade thickness	1.5–3 mm
Frame thickness	1.5–3 mm
Water resistance	Designed to prevent water penetration at specified airflow rates
Free area	50–80% of the total louver area
Pressure drop	≤75 Pa at design airflow rate
Air velocity	Up to 3.5 m/s (typical operating velocity)
Maximum operating temperature	−20 °C to 80 °C
Mounting orientation	Vertical or Horizontal
Coating/finish	Powder Coating/Anodized/Anti-corrosion Coating
Drainage system	Built-in channels to divert water away
Standard compliance	AMCA 500-L, EN 13030, or equivalent
Connection type	Flanged/Slip joint
Weather resistance	Resistant to wind-driven rain and harsh environmental conditions
Optional features	Bird Screen/Insect Mesh
Dimensions	Customizable as per project requirements

13.3.17 *Grilles and Diffusers*

Design Checklist

Category	Checklist items
System requirements	1. Determine the type of grille or diffuser required (e.g., supply, return, exhaust)
	2. Identify the airflow type (laminar, turbulent) and distribution pattern
	3. Consider noise levels and pressure drops in the design phase
Sizing	4. Size grilles and diffusers based on required airflow (CFM or L/s) and coverage area
	5. Maintain recommended face velocity (2–2.5 m/s for diffusers, 2.5–3 m/s for grilles)
	6. Account for throw and spread distances to ensure even airflow distribution
Placement	7. Position supply diffusers to avoid drafts and ensure optimal mixing of supply air with room air
	8. Locate return grilles to facilitate efficient air extraction
Material selection	9. Use corrosion-resistant materials (e.g., aluminium or steel) for areas prone to moisture or chemicals
	10. Select aesthetically pleasing designs for exposed applications
Integration	11. Ensure compatibility with the ductwork and mounting surfaces
	12. Plan for balancing dampers behind grilles and diffusers for airflow adjustments
Performance standards	13. Design to meet ASHRAE standards for indoor air quality and thermal comfort
	14. Consider noise level requirements (NC levels $\leq$ 35 for offices, $\leq$25 for hospitals, etc.)

Installation Checklist

Category	Checklist items
Preparation	1. Verify grille and diffuser dimensions and specifications before installation
	2. Inspect for damage or defects during delivery
Installation process	3. Install supply diffusers and return grilles at the designated locations as per the design drawings
	4. Ensure proper alignment and secure mounting to the ceiling, wall, or floor
	5. Seal connections to ductwork to prevent air leakage
Balancing and adjustments	6. Use balancing dampers to regulate airflow and achieve uniform distribution
	7. Test throw, spread, and pressure drop to confirm compliance with design specifications
Noise control	8. Ensure installation minimizes vibration and operational noise
	9. Add acoustic insulation around diffusers and grilles in noise-sensitive areas if necessary

Category	Checklist items
Final adjustments	10. Confirm that all adjustable blades (if applicable) are set to achieve the desired airflow pattern
	11. Check the aesthetic appearance and alignment with architectural finishes
Documentation	12. Record airflow measurements and balancing data for future reference

Practical Notes and Tips

- **Face Velocity Recommendations:**
  - For supply diffusers: Maintain a face velocity of 2–2.5 m/s.
  - For return grilles: Allow slightly higher face velocities of 2.5–3 m/s.
- **Selection by Type:**
  - **Supply Diffusers:** Use ceiling diffusers for uniform air distribution in offices and other enclosed spaces.
  - **Return Grilles:** Use wall-mounted grilles for efficient air extraction.
- **Avoid Drafts:**
  Avoid placing supply diffusers directly above workstations or seating areas to prevent discomfort.
- **Airflow Balancing:**
  Use opposed-blade dampers or volume control dampers behind grilles and diffusers to fine-tune airflow.
- **Ease of Maintenance:**
  Select models with removable cores for easier cleaning and maintenance.
- **Special Environments:**
  For areas like hospitals or cleanrooms, use HEPA-filter-compatible diffusers and antimicrobial finishes.

Technical Specification

Parameter	Unit	Specification
Manufacturer		
Model number		
Type		(e.g., Return Grille, Supply Diffuser, Linear Slot)
Material		(e.g., Aluminium, Steel, Plastic)
Finish/coating		(e.g., Powder Coated, Anodized)
Mounting type		(e.g., Ceiling, Wall, Floor)
Dimensions (L × W)	mm or inches	
Neck size	mm or inches	
Face size	mm or inches	
Core type		(e.g., Fixed, Removable, Adjustable)

Parameter	Unit	Specification
Air flow pattern		(e.g., 4-Way, Linear, Vortex)
Airflow capacity	L/s or CFM	
Pressure drop	Pa or inches w.g.	
Throw distance	m or ft	
Noise Criteria (NC)		
Damper included		(Yes/No)
Damper type		(e.g., Opposed Blade, Sliding)
Deflection angles	Degrees	
Volume control		(Yes/No)
Insulation included		(Yes/No)
Fire rating		(e.g., UL Rated)
Weight	kg or lbs	
Mounting accessories included		(e.g., Screws, Clamps)
Certification		(e.g., CE, UL)
Special features		(e.g., Anti-Corrosion, Adjustable Louvers)

13.3.18 Humidifiers and Dehumidifiers

Design Checklist

Category	Checklist items
System requirements	1. Determine the humidity control requirements (target relative humidity range)
	2. Assess building usage and occupancy levels to estimate moisture generation or removal needs
	3. Evaluate ambient conditions and seasonal humidity variations
Sizing	4. Size humidifiers based on required moisture addition rate (kg/h)
	5. Size dehumidifiers based on moisture removal rate (L/day or kg/h) and airflow (CFM or L/s)
	6. Consider latent and sensible heat load impacts on cooling and heating systems
Placement	7. Locate humidifiers upstream of critical zones for even moisture distribution
	8. Place dehumidifiers in high-humidity zones or integrate with return ductwork
Integration	9. Ensure compatibility with existing ductwork, piping, or standalone configurations
	10. Plan for electrical connections, drainage systems, and water supply lines
Type selection	11. Select the humidifier type based on application (e.g., steam, ultrasonic, evaporative)
	12. Choose refrigerant-based or desiccant dehumidifiers based on capacity and application requirements

Category	Checklist items
Performance standards	13. Design to meet ASHRAE 55 for thermal comfort and ASHRAE 62.1 for indoor air quality
	14. Ensure humidity control aligns with equipment-specific moisture tolerance (e.g., IT rooms, archives)

Installation Checklist for HVAC Humidifiers

Category	Checklist items
Preparation	1. Verify humidifier size, capacity, and specifications before installation
	2. Ensure water quality meets manufacturer recommendations to prevent scaling and deposits
Installation process	3. Install humidifiers in the supply air duct or plenum for optimal dispersion
	4. Position the unit to avoid water pooling or condensation within ducts
	5. Connect water supply lines and drainage systems securely
Electrical connections	6. Verify electrical connections and circuit protection
	7. Test humidifier control systems and integration with building automation systems
Safety measures	8. Install overflow protection and safety switches to prevent over-humidification
	9. Ensure access for routine cleaning and maintenance

Installation Checklist for HVAC Dehumidifiers

Category	Checklist items
Preparation	1. Verify dehumidifier capacity and specifications before installation
	2. Inspect for sufficient airflow and clearance in the installation area
Installation process	3. Install standalone units or ducted systems in appropriate locations to optimize moisture removal
	4. Connect condensate drainage systems to a suitable outlet
Electrical connections	5. Verify electrical connections and circuit protection
	6. Test the dehumidifier control system and ensure compatibility with existing HVAC systems
Safety measures	7. Ensure filters and coils are accessible for routine cleaning
	8. Check for proper airflow across the coils to avoid freezing or inefficiency

Technical Specification of Humidifiers

Parameter	Unit	Specification
Manufacturer		
Model number		
Type		(e.g., Steam, Ultrasonic, Evaporative)
Humidification capacity	kg/h or lbs./h	

Parameter	Unit	Specification
Airflow compatibility	m^3/h or cfm	
Operating voltage	V	
Power consumption	kW	
Water supply requirement	L/min or GPM	
Water quality requirement	ppm (TDS), pH	
Operating pressure range	kPa or psi	
Control method		(e.g., On/Off, Modulating, Sensor-Based)
Relative humidity range	%	
Operating temperature range	°C or °F	
Drainage requirement	L/hr. or GPM	
Noise level	dB(A)	
Material of construction		(e.g., Stainless Steel, Plastic)
Installation type		(e.g., Duct-Mounted, Standalone)
Maintenance interval	Hours or Months	
Weight	kg or lbs	
Dimensions (L × W × H)	mm or inches	
Control interface		(e.g., Digital, Analog)
Certification		(e.g., CE, UL, ISO)
Corrosion resistance		(Yes/No)
Accessories included		(e.g., Sensors, Drain Kit)
Special features		(e.g., Energy Saving, Auto Clean)

Technical Specification of Dehumidifiers

Parameter	Unit	Specification
Manufacturer		
Model number		
Type		(e.g., Refrigerant, Desiccant)
Dehumidification capacity	L/day or pints/day	
Airflow rate	m^3/h or cfm	
Operating voltage	V	
Power consumption	kW	
Water removal efficiency	L/kWh	
Operating temperature range	°C or °F	
Relative humidity range	%	
Water drainage method		(e.g., Gravity, Pump)
Drainage connection size	mm or inches	
Refrigerant type		(if applicable)
Desiccant material		(if applicable)
Noise level	dB(A)	
Material of construction		(e.g., Stainless Steel, Plastic)
Dimensions (L × W × H)	mm or inches	

Parameter	Unit	Specification
Weight	kg or lbs	
Installation type		(e.g., Wall-Mounted, Portable)
Control interface		(e.g., Digital, Analog)
Control method		(e.g., On/Off, Humidity Sensor)
Filter type		(e.g., Washable, HEPA)
Maintenance interval	Hours or Months	
Certification		(e.g., CE, UL, ISO)
Accessories included		(e.g., Humidity Sensor, Drain Kit)
Special features		(e.g., Energy Efficiency, Auto Restart)

13.3.19 Expansion Tanks

Design Checklist

Category	Checklist items
System requirements	1. Identify the system type (chilled water, hot water, or condenser water) to determine the expansion tank need
	2. Calculate the system's total water volume, including piping and equipment
	3. Determine the temperature range and maximum pressure of the system
Sizing	4. Use the expansion tank formula
	5. Consider safety factors for maximum operating pressures
Type selection	6.Select the type of expansion tank: diaphragm, bladder, or plain (non-diaphragm)
	7. Choose a pre-pressurized tank for closed-loop systems to maintain consistent pressure
Placement	8. Locate the tank near the pump suction side for optimal performance
	9. Ensure the placement allows easy access for inspection and maintenance
Performance standards	10. Ensure compatibility with ASHRAE guidelines and local building codes
	11. Confirm the tank material can withstand system water quality (e.g., treated water or glycol mixtures)

Installation Checklist

Category	Checklist items
Preparation	1. Verify tank capacity, type, and specifications against the system design
	2. Inspect the tank for damage and ensure pressure settings align with system requirements

Category	Checklist items
Installation process	3. Install the tank in a vertical position to ensure proper operation
	4. Connect to the system at the appropriate location (near pump suction)
	5. Use properly sized piping and fittings to handle the expected pressure
Pressure setting	6. Pre-charge the tank to the required static system pressure using nitrogen or air
	7. Ensure the tank pressure matches the system's cold fill pressure
Safety measures	8. Install pressure relief valves downstream to protect the system from over-pressurization
	9. Provide a clear and accessible location for periodic pressure checks and maintenance
Testing and inspection	10. Conduct a hydrostatic pressure test to ensure system integrity
	11. Verify no leaks in connections and ensure the tank operates within specified pressure limits

Commissioning Checklist

Category	Checklist items
Initial operation	1. Check for proper tank operation during system startup
	2. Monitor pressure levels during heating/cooling cycles
Final inspection	3. Verify the tank absorbs thermal expansion as expected
	4. Ensure no abnormal noises or vibrations occur during operation

Troubleshooting and Maintenance

Issue	Action
Tank pressure too high/low	Check the pre-charge pressure and adjust as necessary
Waterlogging in non-bladder tank	Drain and refill the tank; inspect for internal corrosion or leaks
Corrosion or damage	Replace the tank if structural integrity is compromised

Tips for Optimal Performance

- **Diaphragm Tanks:** Require less maintenance due to separation of air and water.
- **Plain Tanks:** Need regular monitoring to prevent waterlogging.
- **All Tanks:** Periodic inspection and pressure adjustment are necessary to maintain system efficiency.

Technical Specification of Expansion Tanks

Parameter	Unit	Specification
Manufacturer		
Model number		
Tank type		(e.g., Diaphragm, Bladder)
Volume (capacity)	L or gallons	

Parameter	Unit	Specification
Maximum working pressure	kPa or psi	
Pre-charge pressure	kPa or psi	
Temperature range	°C or °F	
Material (tank)		(e.g., Steel, Stainless Steel)
Material (bladder/diaphragm)		(e.g., EPDM, Butyl)
Connection size	mm or inches	
Mounting type		(e.g., Floor, Wall)
Dimensions (Diameter × Height)	mm or inches	
Weight (empty)	kg or lbs	
Weight (full)	kg or lbs	
Coating/finish		(e.g., Epoxy, Powder Coated)
Certification		(e.g., ASME, CE)
Operating medium		(e.g., Water, Glycol Mixture)
Application		(e.g., Chilled Water, Heating)
Inlet/outlet type		(e.g., Threaded, Flanged)
Burst pressure	kPa or psi	
Accessories included		(e.g., Mounting Bracket, Gauge)
Expansion ratio		
Service life expectancy	Years	
Special features		(e.g., Maintenance-Free)

13.3.20 *Buffer Tanks*

Design Checklist

Category	Checklist items
System requirements	1. Identify the system type (chilled water, hot water, or heat pump application)
	2. Determine the purpose: increase water volume, stabilize temperature fluctuations, or improve system cycling
	3. Calculate the required tank volume based on system water volume and design flow rate
Sizing	4. Use the buffer tank sizing formula
	5. Ensure sufficient capacity to minimize short-cycling of the chiller or boiler
Material selection	6. Choose tank material based on system water quality (e.g., stainless steel for corrosive environments)
	7. Ensure the tank is insulated to reduce heat losses in hot water systems or heat gains in chilled water systems
Connections and ports	8. Verify the number of inlet/outlet ports and their sizes for system compatibility
	9. Include provisions for air vents and drain ports

Category	Checklist items
Standards compliance	10. Ensure compliance with local building codes and industry standards (e.g., ASHRAE)

Installation Checklist

Category	Checklist items
Preparation	1. Verify tank specifications against system design requirements
	2. Inspect the tank for damage and confirm the presence of all required connections and fittings
Placement	3. Locate the tank as close as possible to the primary pump or chiller/boiler for optimal performance
	4. Ensure sufficient clearance for insulation, maintenance, and inspection
Piping connections	5. Use properly sized piping and ensure ports align with the system layout
	6. Install isolation valves on tank connections to allow for maintenance
Air management	7. Install an automatic air vent or manual air vent at the highest point of the tank
Safety measures	8. Include a pressure relief valve and temperature sensor to monitor tank conditions

Commissioning Checklist

Category	Checklist items
Startup	1. Verify that the tank is filled and free of air pockets
	2. Check for leaks at all connections and fittings
	3. Monitor system temperatures and flow rates to ensure proper operation
Final inspection	4. Verify that the buffer tank stabilizes system temperature fluctuations
	5. Ensure the tank's insulation is intact and effective

Troubleshooting and Maintenance

Issue	Action
Insufficient volume impact	Recalculate required volume and consider adding or replacing the buffer tank
Leaks or corrosion	Inspect connections and replace damaged components or tank
Temperature fluctuations	Check tank placement, insulation, and flow rates for optimization

Tips for Optimal Performance

- **Insulation:** Use high-quality insulation to minimize energy losses.
- **Multiple Ports:** Select tanks with sufficient ports for multi-point connection to the system.
- **Air Vents:** Ensure proper air elimination to maintain efficiency and avoid water hammer.

Technical Specification of Buffer Tanks

Parameter	Unit	Specification
Manufacturer		
Model number		
Tank type		(e.g., Vertical, Horizontal)
Volume (capacity)	L or gallons	
Maximum working pressure	kPa or psi	
Temperature range	°C or °F	
Material (tank)		(e.g., Steel, Stainless Steel)
Insulation material		(e.g., Polyurethane, Fiberglass)
Insulation thickness	mm or inches	
Connection size	mm or inches	
Number of ports		
Port configuration		(e.g., Top/Bottom, Side Ports)
Mounting type		(e.g., Floor, Stand-Mounted)
Dimensions (Diameter × Height)	mm or inches	
Weight (empty)	kg or lbs	
Weight (full)	kg or lbs	
Coating/finish		(e.g., Epoxy, Powder Coated)
Certification		(e.g., ASME, CE)
Operating medium		(e.g., Water, Glycol Mixture)
Application		(e.g., Chilled Water, Heating)
Inlet/outlet type		(e.g., Threaded, Flanged)
Heat loss (insulated)	W or BTU/h	
Thermal conductivity of insulation	W/m·K or BTU·in/(hr·ft.^{2}·°F)	
Accessories included		(e.g., Mounting Bracket, Gauge)
Service life expectancy	Years	
Special features		(e.g., Maintenance-Free)

13.3.21 Valves

Design Checklist

Category	Checklist items
System requirements	1. Determine the valve's purpose (isolation, control, balancing, or pressure relief)
	2. Identify the type of fluid (e.g., water, glycol mixtures, or steam) and its properties (temperature, pressure)

Category	Checklist items
Valve type selection	3. Select the appropriate valve type (e.g., globe, ball, butterfly, gate, pressure-independent, etc.)
	4. Match the valve function with system needs, e.g., modulating valves for precise control or ball valves for quick isolation
Sizing	5. Use the Cv (flow coefficient) to size control valves accurately
	6. Ensure the valve size matches or is slightly smaller than the pipe diameter for control applications
Pressure and temperature	7. Ensure valves can handle system design pressure and temperature
	8. Consider pressure drop across the valve when selecting
Actuation	9. Determine if valves need manual, electric, pneumatic, or hydraulic actuation
Material selection	10. Choose valve material suitable for the fluid and operating environment (e.g., stainless steel for corrosive conditions)
Standards compliance	11. Ensure compliance with local codes and standards (e.g., ASHRAE, ISO, ANSI)

Installation Checklist

Category	Checklist items
Preparation	1. Verify valve specifications against design requirements
	2. Inspect valves for damage and ensure all accessories (actuators, seals) are included
Placement and orientation	3. Confirm valve orientation aligns with the flow direction indicated on the valve body
	4. Ensure easy access for operation and maintenance
Connections	5. Use compatible fittings (e.g., threaded, flanged, or welded) based on the valve type
	6. Include isolation valves upstream and downstream for maintenance
Installation details	7. Support valves properly to avoid stress on pipes
	8. Ensure pipe cleanliness before valve installation to prevent debris damage
Testing and sealing	9. Perform pressure testing to check for leaks after installation
	10. Use appropriate gaskets or sealants to prevent leaks

Commissioning Checklist

Category	Checklist items
Startup	1. Operate valves through their full range to confirm proper movement and operation
	2. Verify actuator functionality and calibration (if applicable)
Flow and pressure checks	3. Measure flow rates and pressure drops to ensure compliance with design
	4. Verify valve shutoff tightness
System integration	5. Confirm compatibility with building automation systems (BAS) if applicable

Category	Checklist items
Final inspection	6. Ensure all valves are properly labelled for identification and functionality

Troubleshooting and Maintenance

Issue	Action
Leaks	Tighten connections or replace seals/gaskets
Valve sticking	Check for debris or inspect actuator for malfunction
Inconsistent flow	Verify valve sizing and inspect for obstructions
Excessive noise	Check for improper sizing or excessive pressure drop

Tips for Optimal Performance

- **Proper Sizing:** Oversized valves can cause control instability; undersized valves lead to excessive pressure drops.
- **Material Compatibility:** Select materials based on fluid composition to avoid corrosion or scaling.
- **Routine Checks:** Regularly inspect actuators and seals to ensure reliable operation.

Technical Specification of HVAC Valves

Parameter	Unit	Specification
Manufacturer		
Model number		
Valve type		(e.g., Ball, Butterfly, Globe)
Application		(e.g., Chilled Water, Heating)
Material (body)		(e.g., Brass, Cast Iron, Steel)
Material (seal)		(e.g., EPDM, PTFE)
Nominal diameter (DN)	mm or inches	
Connection type		(e.g., Threaded, Flanged)
Maximum working pressure	kPa or psi	
Temperature range	°C or °F	
Flow coefficient (Cv/Kv)		
Valve action		(e.g., Normally Open/Closed)
Actuator type		(e.g., Electric, Pneumatic)
Actuator voltage	V	
Control signal		(e.g., 0–10 V, 4–20 mA)
Positioning		(e.g., Modulating, On/Off)
Leakage rate	% or L/hr	
Flow direction		(e.g., Unidirectional)
End connection standard		(e.g., ANSI, ISO)
Pressure drop	kPa or psi	
Operating medium		(e.g., Water, Glycol Mixture)

Parameter	Unit	Specification
Dimensions (L × H × W)	mm or inches	
Weight	kg or lbs	
Insulation included		(Yes/No)
Certification		(e.g., CE, UL, ASME)
Accessories included		(e.g., Stem Extension)
Service life expectancy	Years	
Special features		(e.g., Anti-Corrosion Coating)

13.3.22 Electrical Heaters

Design Checklist

Category	Checklist items
System requirements	1. Determine the purpose (e.g., space heating, duct heating, or specific zone heating)
	2. Calculate the heating load based on space size, insulation, and ventilation rates
Heater type selection	3. Choose the appropriate type of electrical heater (e.g., finned tubular, open coil, or infrared)
	4. Select based on the application (e.g., duct-mounted, standalone, or unit heaters)
Sizing	5. Use heating load calculations (W/m^2 or BTU/h) to size the heater
	6. Ensure heaters have a suitable wattage to meet the required temperature rise
Voltage and phase	7. Select heaters compatible with the building's electrical supply (e.g., single-phase or three-phase)
Control requirements	8. Determine thermostat or controller type (e.g., on/off control, proportional control)
Safety features	9. Specify overheat protection (e.g., thermal cutoffs or limit switches)
	10. Include airflow interlock for duct heaters to prevent operation without adequate airflow
Energy efficiency	11. Evaluate options for staged or modulated heating to reduce energy consumption

Installation Checklist

Category	Checklist items
Preparation	1. Verify heater specifications against design requirements
	2. Inspect heaters for damage or defects before installation

Category	Checklist items
Placement and orientation	3. Install duct heaters downstream of filters and upstream of coils
	4. Maintain minimum clearance around heaters as per manufacturer recommendations
Connections	5. Use properly sized electrical wiring and circuit breakers
	6. Ensure grounding is completed as per electrical codes
Installation details	7. Ensure proper insulation in duct areas where heaters are installed to prevent heat loss
	8. Verify that airflow direction aligns with the heater's design
Testing and sealing	9. Test all electrical connections for continuity and tightness
	10. Seal duct penetrations to prevent air leakage

Commissioning Checklist

Category	Checklist items
Startup	1. Verify that heaters turn on and off according to control settings
	2. Check the functionality of airflow interlock systems
Temperature checks	3. Measure the temperature rise across the heater to ensure it matches design specifications
	4. Verify thermostat accuracy and response
System integration	5. Confirm heaters integrate properly with building automation systems (BAS) if applicable
Safety testing	6. Test safety features such as thermal cutoffs and airflow switches

Troubleshooting and Maintenance

Issue	Action
Heater not turning on	Check electrical connections, thermostat settings, and airflow interlock
Uneven heating	Inspect duct layout and airflow distribution
Overheating	Verify that airflow is sufficient and safety features are functioning
Excessive power consumption	Check heater wattage and usage patterns; consider staged heating

Safety Tips

- **Overheat Protection:** Always include thermal cutoffs or high-limit switches to prevent fire hazards.
- **Airflow Interlock:** Ensure heaters operate only when adequate airflow is detected.
- **Regular Inspection:** Inspect heaters and wiring periodically to avoid overheating or electrical failures.

Technical Specification of Electrical Heaters

Parameter	Specification
Material	Stainless Steel (SS304/SS316) for heating elements and housing
Heating element type	Finned Tubular/Open Coil
Heating capacity	1–100 kW (or as required)
Voltage rating	230 V/400 V/Customizable
Phase	Single-phase/Three-phase
Control method	Step Control/Thyristor Control (SCR)
Temperature range	−10 °C to 60 °C (ambient), output up to 120 °C or higher
Air velocity range	2–10 m/s
Pressure drop	≤75 Pa
Protection class	IP20 to IP65, depending on application
Insulation	High-temperature insulation material
Overheat protection	Thermal Cutouts, Limit Switches, or Built-in Thermostats
Standard compliance	IEC 60335, EN 60204-1, or equivalent
Mounting type	Flanged/Slip-in/Custom
Control accessories	Integrated temperature sensors (thermocouples, RTDs), relays, and controllers
Optional features	Humidity control, integrated BMS compatibility
Dimensions	Customizable based on duct or AHU requirements
Coating/finish	Anti-corrosion coating or powder coating for housing

13.3.23 HVAC Isolators

Design Checklist

Category	Checklist items
System requirements	1. Determine the type of equipment to be isolated (e.g., fans, pumps, chillers)
	2. Assess the vibration and noise levels produced by the equipment
Isolator type selection	3. Choose the isolator type based on application:
	– **Spring Isolators**: For high vibration equipment like chillers or large AHUs
	– **Rubber Pads/Isolators**: For lower vibration applications like small pumps
	– **Hangers**: For suspended equipment like duct fans or piping
	– **Inertia Bases**: For large, dynamic loads like centrifugal pumps
Load analysis	4. Calculate static and dynamic loads for proper sizing of isolators
	5. Factor in equipment weight, operating conditions, and load distribution

Category	Checklist items
Deflection requirements	6. Specify required deflection (e.g., 25 mm for chillers, 10 mm for fans) based on vibration criteria
Environmental factors	7. Ensure isolators are compatible with environmental conditions (e.g., corrosion-resistant materials)

Installation Checklist

Category	Checklist items
Preparation	1. Verify isolator specifications against design requirements
	2. Inspect isolators for defects or damage prior to installation
Placement and alignment	3. Position isolators at recommended mounting points to ensure even load distribution
	4. Align isolators properly with the equipment base
Fixing and securing	5. Secure isolators to the floor or structure using bolts or anchors as per manufacturer instructions
	6. For inertia bases, ensure proper concrete curing before placing equipment
Adjustment and leveling	7. Adjust isolators to ensure the equipment is level
	8. Verify deflection of spring isolators matches design specifications under load
Connections	9. Ensure flexible connections for piping and ductwork to avoid transmitting vibrations

Commissioning Checklist

Category	Checklist items
Visual inspection	1. Inspect isolators for proper alignment and secure fixing
	2. Check for visible wear or deformation after the equipment is operational
Load verification	3. Measure deflection under load to confirm compliance with design criteria
	4. Ensure even load distribution across all isolators
Vibration testing	5. Perform vibration analysis to ensure isolation meets specified levels
Noise testing	6. Measure noise levels to verify proper damping

Troubleshooting and Maintenance

Issue	Action
Excessive vibration or noise	Check for improper alignment, loose fixings, or insufficient deflection
Uneven deflection	Recalculate load distribution or inspect isolators for damage
Wear or corrosion	Replace damaged or degraded isolators; use corrosion-resistant materials if necessary

Safety Tips

- **Flexible Connections:** Always use flexible connections for ducts and pipes attached to isolated equipment to prevent vibration bridging.
- **Periodic Maintenance:** Inspect isolators regularly for wear, loosening, or corrosion.
- **Alignment:** Ensure proper alignment to avoid unnecessary stress on equipment and isolators.

Technical Specification of HVAC Isolators

Parameter	Unit	Specification
Manufacturer		
Model number		
Type		(e.g., Spring Isolator, Rubber Mount, Vibration Pad)
Application		(e.g., Fan, Chiller, Pump, AHU)
Load capacity	kg or lbs	
Deflection rating	mm or inches	
Static deflection	mm or inches	
Dynamic deflection	mm or inches	
Material		(e.g., Steel, Rubber, Neoprene)
Finish/coating		(e.g., Powder Coated, Galvanized)
Operating temperature range	°C or °F	
Isolation efficiency	%	
Natural frequency	Hz	
Spring constant	N/mm or lb./in	
Height	mm or inches	
Base size	mm or inches	
Fixing method		(e.g., Bolt-Down, Adhesive)
Anchoring accessories included		(Yes/No)
Corrosion resistance		(Yes/No)
Fire resistance		(e.g., UL Rated, None)
Noise isolation rating	dB	
Weight	kg or lbs	
Certification		(e.g., CE, UL, ISO)
Special features		(e.g., Anti-Vibration, Adjustable Height)

13.3.24 *Water Pumps*

Design Checklist

Category	Checklist items
System requirements	1. Determine the total water flow rate (in cubic meters per hour or L/s) based on HVAC load calculations
	2. Calculate the total dynamic head (TDH), including pipe friction losses, equipment head, and fittings
Pump selection	3. Select pump type (e.g., end-suction, vertical inline, centrifugal, or split-case) based on application
	4. Verify pump capacity matches system requirements for flow rate and TDH
Motor and drive	5. Select a motor with sufficient power to meet pump requirements, accounting for efficiency
	6. Consider variable frequency drives (VFDs) for improved energy efficiency and variable load operation
Piping design	7. Design suction and discharge piping to minimize pressure losses and avoid cavitation
	8. Include isolation valves, check valves, and pressure gauges at appropriate locations
Energy efficiency	9. Select pumps with high-efficiency ratings and consider low life-cycle cost options
	10. Ensure the system meets energy standards (e.g., MEPS or ASHRAE 90.1)

Installation Checklist

Category	Checklist items
Preparation	1. Verify pump specifications and inspect for shipping damage
	2. Confirm foundation and baseplates are level and properly aligned
Alignment	3. Check and correct the alignment of the pump and motor shafts after installation
	4. Use proper coupling and alignment tools for precise installation
Piping connections	5. Install flexible connectors to absorb thermal expansion and vibration
	6. Ensure suction and discharge piping has adequate straight lengths to minimize turbulence
Electrical connections	7. Ensure proper wiring for motor starters, variable frequency drives (VFDs), and control systems
	8. Provide proper earthing/grounding of the pump motor
Vibration isolation	9. Install vibration isolators or rubber pads to reduce noise and vibrations
	10. Check tightness of bolts and connections after initial operation

Commissioning Checklist

Category	Checklist items
Pre startup inspection	1. Verify that all piping connections are secure and free of leaks
	2. Check that all isolation valves are in the correct position for start-up
System configuration	3. Prime the pump to remove air from the suction line before starting
	4. Set and calibrate variable frequency drives (if installed)
Performance testing	5. Measure pump flow rate, head, and power consumption to ensure compliance with design specifications
	6. Monitor for excessive vibration, noise, or overheating during operation

Troubleshooting and Maintenance Checklist

Issue	Action
Low flow rate	Check for clogged filters, airlocks, or improper pump sizing
Excessive noise or vibration	Inspect for misalignment, cavitation, or worn bearings
Overheating	Ensure adequate cooling for the motor and verify proper lubrication
Leaks in piping	Tighten connections and replace damaged gaskets or seals

Advantages and Disadvantages

Aspect	Advantages	Disadvantages
Flexibility	Wide range of types and sizes for various applications	Requires regular maintenance to ensure performance
Energy efficiency	High-efficiency models and VFDs can significantly reduce energy costs	Initial cost may be high for advanced pumps with VFDs
Durability	Robust construction for long service life in demanding environments	Susceptible to wear and tear in systems with poor water quality

Safety Tips

- **Pressure Monitoring:** Install pressure relief valves to prevent over-pressurization in the system.
- **Motor Safety:** Use overload protection to safeguard the pump motor from electrical faults.
- **Water Quality:** Regularly inspect and maintain water quality to minimize corrosion and scaling.

Technical Specification for Pumps

Parameter	Unit	Specification
Manufacturer		

Parameter	Unit	Specification
Model number		
Pump type		(e.g., Centrifugal, Inline)
Flow rate	L/s or GPM	
Head (Pressure)	kPa or ft	
Pump efficiency	%	
Motor power	kW or HP	
Speed	RPM	
Impeller material		
Pump casing material		
Connection size (inlet)	mm or inches	
Connection size (outlet)	mm or inches	
Working temperature range	°C or °F	
Operating pressure	kPa or psi	
Voltage	V	
Frequency	Hz	
Seal type		
Noise level	dB(A)	
Pump dimensions (L × W × H)	mm or inches	
Weight (operating)	kg or lbs	
NPSH required	m or ft	
Drive type		(e.g., Direct, Belt Drive)
Control type		(e.g., VFD, On/Off)
Application		(e.g., Chilled Water, Condenser Water)
Special features		

13.3.25 Underground Hydronic Heating Systems

Design Checklist

Category	Checklist items
System requirements	1. Calculate heating load based on building heat loss and local climate conditions
	2. Determine required supply and return water temperatures for the system (e.g., 35–55 °C)
Pipe selection	3. Choose piping material (e.g., PEX, PEX-AL-PEX, or copper) suitable for underground use
	4. Determine pipe diameter and length based on flow rate and pressure drop calculations
Insulation	5. Select pipe insulation material to minimize heat loss, especially in high water table areas
	6. Ensure insulation complies with local energy efficiency standards

Category	Checklist items
Heat source	7. Choose appropriate heat source (e.g., gas boiler, heat pump, or solar thermal system)
	8. Ensure the heat source capacity meets system load requirements
Control systems	9. Design zoning and controls to provide flexibility and efficiency (e.g., thermostats, actuators)
	10. Include weather compensation controls to adjust supply temperature based on outdoor conditions

Installation Checklist

Category	Checklist items
Preparation	1. Prepare the subgrade and ensure proper grading for drainage
	2. Lay out the system according to the design, marking loop locations and manifold placement
Pipe installation	3. Install pipes at the specified depth (typically 200–600 mm below the surface, depending on use)
	4. Maintain a consistent pipe spacing to ensure even heat distribution (e.g., 150–300 mm)
	5. Secure pipes with staples, clips, or mats to prevent movement during concrete pouring
Manifold installation	6. Position the manifold in an accessible location for maintenance and balancing
	7. Connect pipes to the manifold with proper fittings and ensure watertight connections
Pressure testing	8. Pressure test the system (e.g., 6 bar for 24 h) to detect leaks before covering
	9. Monitor pressure gauge to ensure system integrity
Final covering	10. Cover pipes with appropriate screed or sand layer before installing flooring or landscaping

Commissioning Checklist

Category	Checklist items
System start-up	1. Fill the system with water or a glycol mix (if required for frost protection)
	2. Bleed air from the system using automatic or manual air vents
Temperature control	3. Verify supply and return temperatures match design specifications
	4. Calibrate thermostats and zone controls for optimal performance
Performance testing	5. Check system balance to ensure even heat distribution across all loops
	6. Monitor pump operation and adjust flow rates as necessary

Troubleshooting and Maintenance Checklist

Issue	Action
Uneven heating	Check for airlocks in pipes and ensure proper system balancing

Issue	Action
Leaks or pressure drops	Inspect connections at the manifold and piping joints for damage or poor sealing
Inadequate heating	Verify water temperature settings and inspect the heat source for proper operation
Reduced flow rate	Clean filters or strainers and check for debris or scaling in the piping system

Advantages and Disadvantages

Aspect	Advantages	Disadvantages
Efficiency	Provides high energy efficiency with low operating costs when paired with heat pumps or condensing boilers	Installation cost can be high, especially for retrofitting existing buildings
Comfort	Delivers uniform heating without drafts or noise	Response time can be slower compared to forced-air systems
Aesthetic	Invisible system eliminates the need for radiators or visible ductwork	Requires careful design to prevent cold spots or overheating in some zones

Safety Tips

- **Frost Protection:** Use glycol antifreeze in cold climates to prevent freezing in pipes.
- **Expansion Management:** Include an expansion tank to manage pressure fluctuations.
- **Electrical Safety:** Install ground fault protection for pumps and controls.

Technical Specification of Underground Hydronic Heating Systems

Parameter	Specification
Pipe material	Cross-linked Polyethylene (PEX), Polypropylene (PP-R), or Multilayer Composite Pipes (MLCP)
Pipe diameter	16–32 mm (commonly used)
Pipe insulation	High-density polyethylene foam or equivalent, 20–50 mm thick
System type	Closed-loop (most common)/Open-loop
Heat source	Boiler, Heat Pump, or Solar Thermal System
Operating temperature	30–60 °C
Operating pressure	Up to 600 kPa
Flow rate	0.1–5 m^3/h (depending on system size)
Thermal output	30–100 W/m^2 (depending on insulation and ground conditions)
Heat transfer fluid	Water or Water-Glycol Mixture (antifreeze protection)
Control system	Thermostats, Zone Valves, Flow Meters, and Control Panels
Energy efficiency	Up to 95% (depending on heat source and system design)
Installation depth	Typically 20–50 cm below ground surface
Manifold type	Stainless Steel or Plastic, with individual flow control per loop
Standard compliance	ISO 11855, EN 1264, or equivalent

Parameter	Specification
Pipe connections	Compression fittings, Press fittings, or Push-fit connectors
Optional features	Leak Detection, Smart Controls, and BMS Integration
Coating/protection	UV-resistant pipe coating for exposed sections
Warranty	Typically 10–25 years for piping components

13.3.26 Switchboards (MSSB)

Technical Specification of Main Switchboards (MSSB)

Category	Specification	Details/options
General	Type	Indoor, Outdoor, Weatherproof, Custom
	Application	HVAC Equipment Power Distribution, Chiller, AHU, Pumps, Fans
	Voltage Rating	Low Voltage (400 V AC), Medium Voltage (up to 11 kV)
	Current Rating	100–5000 A
	Frequency	50 Hz, 60 Hz
Construction	Enclosure Material	Mild Steel, Galvanized Steel, Stainless Steel, Aluminium
	Enclosure Protection	IP44, IP55, IP65
	Paint Finish	Powder Coated, Epoxy Coated
	Cable Entry	Top Entry, Bottom Entry, Custom
	Form of Separation	Form 1, Form 2, Form 3, Form 4
Electrical components	Circuit Breakers	Molded Case Circuit Breaker (MCCB), Air Circuit Breaker (ACB)
	Switches	Isolators, Changeover Switches
	Busbars	Copper, Aluminium, Tinned Copper
	Control Relays	Electromechanical, Solid-State
	Monitoring Devices	Voltage Meters, Current Meters, Power Factor Meters, Energy Meters
Performance	Power Factor Correction	Capacitor Bank Integration, Automatic PF Controller
	Short Circuit Rating	25 kA, 50 kA, 75 kA
	Surge Protection	Surge Protection Device (SPD)
Controls and automation	Control Method	Manual, Semi-Automatic, Fully Automatic
	PLC/SCADA Integration	Optional
	BMS Connectivity	BACnet, Modbus, Proprietary Protocols
	Indicators	LED/Lamp Indicators for Status and Faults
Safety	Certifications	CE, UL, IEC, ISO, AS/NZS
	Safety Devices	Earth Fault Relay, Overload Protection, Emergency Stop
	Arc Flash Protection	Optional

Category	Specification	Details/options
Installation	Mounting	Floor-Mounted, Wall-Mounted, Skid-Mounted
	Accessibility	Front Access, Rear Access
Standards compliance	Standards	IEC 61439, IEC 60947, AS/NZS 3000
	Testing	Factory Acceptance Test (FAT), Site Acceptance Test (SAT)
Additional features	Maintenance Requirements	Low Maintenance, Modular Design for Easy Replacement
	Accessories	Metering Devices, Alarm Systems, Communication Modules
	Customization Options	Custom Enclosure Dimensions, Colour Coding for Wiring

Chapter 14
HVAC Standards

14.1 Overview

This collection of standards demonstrates the global diversity in HVAC practices, tailored to regional needs and climate conditions. For international projects, understanding and complying with local standards is critical for successful system design and implementation. These standards are essential for verifying the performance, efficiency, and reliability of HVAC equipment. They provide a consistent basis for product evaluation, ensuring that equipment performs as expected under real-world conditions. These standards are also particularly valuable for equipment selection, energy efficiency compliance, and ensuring occupant comfort. Always consult the latest standards and coordinate with other guidelines for a comprehensive approach to HVAC design and system integration. This chapter lists the standards and codes used globally in HVAC design and building services, categorized by regions and countries. These standards cover various aspects of HVAC systems, energy efficiency, indoor air quality (IAQ), fire safety, and building performance.

14.2 HVAC Standard in Different Countries

United States

- **ASHRAE Standards**
  - Focus: HVAC design, energy efficiency, IAQ, thermal comfort.
  - Example: ASHRAE 62.1 (Ventilation), ASHRAE 90.1 (Energy Efficiency).
- **AHRI Standards**
  - Focus: Equipment performance ratings and specifications.

V. Vakiloroaya, *HVAC and Building Services Rules of Thumb*, Springer Tracts in Civil Engineering, https://doi.org/10.1007/978-3-032-10285-0_14

  - Example: AHRI 550/590 (Chillers), AHRI 700 (Refrigerants).
- **NFPA (National Fire Protection Association)**
  - Focus: Fire safety in HVAC systems.
  - Example: NFPA 90A (Air-Conditioning Systems), NFPA 101 (Life Safety Code).
- **SMACNA (Sheet Metal and Air Conditioning Contractors' National Association)**
  - Focus: Duct design, fabrication, and installation.
  - Example: HVAC Duct Construction Standards.
- **IBC (International Building Code)**
  - Focus: General building construction, including HVAC integration.

United Kingdom

- **CIBSE Guides (Chartered Institution of Building Services Engineers)**
  - Focus: HVAC design, thermal comfort, IAQ, energy efficiency.
  - Example: CIBSE Guide A (Environmental Design), CIBSE Guide B (HVAC Systems).
- **BS EN Standards (British Standards/European Norms)**
  - Focus: Product specifications and testing.
  - Example: BS EN 15251 (Indoor Environmental Input), BS EN 13779 (Ventilation).
- **Building Regulations Part F and L**
  - Focus: Ventilation and energy efficiency in buildings.

European Union

- **EN Standards (European Norms)**
  - Focus: Unified standards across EU nations for HVAC and energy efficiency.
  - Example: EN 16798 (Energy Performance of Buildings), EN 779 (Air Filters).
- **EPBD (Energy Performance of Buildings Directive)**
  - Focus: Energy efficiency and building certification.

Canada

- **CSA (Canadian Standards Association)**
  - Focus: HVAC equipment and system design.
  - Example: CSA B52 (Mechanical Refrigeration Code).

- **NBC (National Building Code of Canada)**
  - Focus: Building design and safety, including HVAC.
- **ASHRAE Standards (Adopted in Canada)**
  - Widely referenced in HVAC practices.

Australia

- **NCC (National Construction Code)**
  - Focus: Building standards, including HVAC and energy efficiency.
  - Example: NCC Volume 1 Section J (Energy Efficiency).
- **Australian Standards (AS)**
  - Focus: Product testing, safety, and performance.
  - Example: AS 1668 (Ventilation), AS 4254 (Ductwork).

India

- **National Building Code of India (NBC)**
  - Focus: Building design, HVAC, fire safety.
- **IS Standards (Indian Standards)**
  - Example: IS 7873 (Duct Design), IS 659 (Air Conditioning Systems).
- **ECBC (Energy Conservation Building Code)**
  - Focus: Energy efficiency in commercial buildings.

China

- **GB Standards (Guobiao Standards)**
  - Focus: HVAC, energy efficiency, and IAQ.
  - Example: GB 50736 (Design Code for HVAC), GB 50189 (Energy Code for Buildings).

Japan

- **JIS (Japanese Industrial Standards)**
  - Focus: Product and system standards.
  - Example: JIS B8616 (Air-Conditioning Systems).
- **SHASE (Society of Heating, Air-Conditioning and Sanitary Engineers of Japan)**
  - Focus: HVAC design practices and standards.

Middle East

- **GCC Code (Gulf Cooperation Council)**
  - Focus: Building standards for Gulf countries, including HVAC.
- **ASHRAE Standards (Widely Adopted)**

Singapore

- **SS (Singapore Standards)**
  - Focus: Ventilation, IAQ, and energy efficiency.
  - Example: SS 553 (Code of Practice for Ventilation).
- **Green Mark Certification**
  - Focus: Sustainable building practices.

Germany

- **DIN Standards (Deutsches Institut für Normung)**
  - Focus: Product and system standards.
  - Example: DIN 1946 (Ventilation and Air Conditioning).
- **VDI Guidelines (Verein Deutscher Ingenieure)**
  - Focus: HVAC design and IAQ.

South Africa

- **SANS (South African National Standards)**
  - Focus: Building performance and HVAC.
  - Example: SANS 10400 (Building Code).

Global Standards

- **ISO (International Organization for Standardization)**
  - Focus: International HVAC and building service practices.
  - Example: ISO 16890 (Air Filters), ISO 17772 (Indoor Environmental Quality).
- **IEC (International Electrotechnical Commission)**
  - Focus: Electrical components in HVAC systems.

14.3 ASHRAE Standards

Followings are the **list of key ASHRAE standards** related to HVAC systems and a brief description of what they address. These standards collectively guide HVAC engineers in designing safe, efficient, and compliant systems. Always consult the latest version of the standards for project-specific applications.

ASHRAE Standard 15: "Safety Standard for Refrigeration Systems"

- **Purpose**: Provides safety requirements for the design, construction, installation, and operation of refrigeration systems.

- **Key Points:**
  - Specifies refrigerant classifications and allowable quantities.
  - Details ventilation, pressure relief, and leak detection requirements.
  - Addresses safety measures for different occupancy classifications.

ASHRAE Standard 34: "Designation and Safety Classification of Refrigerants"

- **Purpose**: Establishes a uniform system for refrigerant identification and safety classification.
- **Key Points:**
  - Classifies refrigerants based on toxicity (A or B) and flammability (1, 2 L, 2, or 3).
  - Defines refrigerant properties and acceptable use cases.

ASHRAE Standard 55: "Thermal Environmental Conditions for Human Occupancy"

- **Purpose**: Specifies acceptable indoor thermal comfort ranges for various environments.
- **Key Points:**
  - Defines acceptable temperature, humidity, and air velocity ranges.
  - Provides comfort models and guidelines for naturally and mechanically ventilated spaces.

ASHRAE Standard 62.1: "Ventilation for Acceptable Indoor Air Quality"

- **Purpose**: Defines minimum ventilation rates for indoor air quality in non-residential buildings.
- **Key Points:**
  - Establishes outdoor air requirements for different occupancy types.
  - Addresses air contaminants and ventilation system design criteria.
  - Includes natural and mechanical ventilation guidelines.

ASHRAE Standard 62.2: "Ventilation and Acceptable Indoor Air Quality in Residential Buildings"

- **Purpose**: Focuses on ventilation and indoor air quality for low-rise residential buildings.
- **Key Points:**
  - Sets ventilation rates for healthy indoor environments.
  - Considers pollutant sources, exhaust systems, and air filtration.

ASHRAE Standard 90.1: "Energy Standard for Buildings Except Low-Rise Residential Buildings"

- **Purpose**: Provides minimum energy efficiency requirements for buildings.

- **Key Points:**
  - Covers HVAC systems, building envelope, lighting, and service water heating.
  - Includes prescriptive and performance compliance paths for HVAC design.

ASHRAE Standard 90.2: "Energy Efficient Design of Low-Rise Residential Buildings"

- **Purpose**: Establishes energy efficiency guidelines for low-rise residential structures.
- **Key Points:**
  - Focuses on HVAC system efficiency, insulation, and airtightness.
  - Promotes renewable energy integration.

ASHRAE Standard 100: "Energy Efficiency in Existing Buildings"

- **Purpose**: Targets energy use reduction in existing buildings.
- **Key Points:**
  - Provides operational and retrofit strategies for improving HVAC performance.
  - Includes benchmarking and monitoring energy usage.

ASHRAE Standard 170: "Ventilation of Health Care Facilities"

- **Purpose**: Specifies ventilation requirements for healthcare environments.
- **Key Points:**
  - Addresses airflow, filtration, and pressure control for critical spaces.
  - Covers operating rooms, patient rooms, and isolation rooms.

ASHRAE Standard 189.1: "Standard for the Design of High-Performance Green Buildings"

- **Purpose**: Provides sustainable building design requirements.
- **Key Points:**
  - Focuses on energy-efficient HVAC systems and IAQ.
  - Encourages renewable energy integration and water conservation.

ASHRAE Standard 202: "Commissioning Process for Buildings and Systems"

- **Purpose**: Guides the commissioning process for building systems, including HVAC.
- **Key Points:**
  - Details procedures for verifying HVAC system performance.
  - Ensures systems meet design intent and operational requirements.

ASHRAE Guideline 0: "The Commissioning Process"

- **Purpose**: Provides a framework for commissioning building systems.

- **Key Points:**
  - Covers documentation, functional testing, and performance evaluation of HVAC systems.

ASHRAE Guideline 1.1: "HVAC&R Technical Requirements for the Commissioning Process"

- **Purpose**: Expands on HVAC-specific commissioning requirements.
- **Key Points:**
  - Details testing procedures for HVAC systems.
  - Ensures compliance with energy efficiency and performance standards.

ASHRAE Standard 188: "Legionellosis: Risk Management for Building Water Systems"

- **Purpose**: Establishes measures to prevent Legionella in building water systems.
- **Key Points:**
  - Addresses HVAC cooling towers and humidifiers as potential risk sources.
  - Recommends water management plans and monitoring.

ASHRAE Standard 15.2P

- **Purpose**: Focused on refrigerant safety and application in residential systems.
- **Key Points:**
  - Tailored for smaller systems and residential buildings.

Summary Table for HVAC Engineers

Standard	Focus area	Key HVAC application
ASHRAE 15	Refrigeration safety	System design, refrigerant management, and safety compliance
ASHRAE 34	Refrigerant classification	Selection of refrigerants based on toxicity and flammability
ASHRAE 55	Thermal comfort	Ensuring occupant comfort in indoor environments
ASHRAE 62.1	Non-residential ventilation	Outdoor air requirements for commercial spaces
ASHRAE 62.2	Residential ventilation	Ventilation design for homes and apartments
ASHRAE 90.1	Energy efficiency for buildings	HVAC system efficiency, performance compliance
ASHRAE 90.2	Residential energy efficiency	Energy-efficient HVAC for homes
ASHRAE 100	Existing building energy performance	Retrofit strategies for HVAC systems
ASHRAE 170	Healthcare facility ventilation	Filtration, airflow, and pressure for critical spaces
ASHRAE 189.1	Sustainable building design	High-performance HVAC design and operation
ASHRAE 202	Commissioning for buildings	Ensures HVAC systems are functional and efficient
ASHRAE 188	Legionella risk management	Cooling towers and water system hygiene

14.4 CIBSE Standard

Followings are the **list of key CIBSE (Chartered Institution of Building Services Engineers) guides and standards** relevant to HVAC systems and building services, along with a brief description of their focus. These CIBSE guides provide a comprehensive foundation for designing efficient and sustainable HVAC systems. Always consult the latest versions for specific project requirements.

CIBSE Guide A: "Environmental Design"

- **Purpose**: Provides fundamental design data for environmental conditions in buildings.
- **Key Points:**
  - Covers thermal comfort, ventilation rates, and cooling/heating loads.
  - Includes design parameters for internal and external climates.
  - Details U-values, solar gains, and thermal mass considerations.

CIBSE Guide B: "Heating, Ventilating, Air Conditioning and Refrigeration"

- **Purpose**: Comprehensive guide to HVAC system design and operation.
- **Key Points:**
  - Divided into sections covering heating, ventilation, air conditioning, and refrigeration.
  - Includes design guidance for system efficiency, control, and integration.
  - Discusses ductwork and pipework design, plant sizing, and equipment selection.

CIBSE Guide C: "Reference Data"

- **Purpose**: Provides essential reference data for building services engineers.
- **Key Points:**
  - Includes psychrometric properties, heat transfer coefficients, and material properties.
  - Offers tables and charts for design calculations.

CIBSE Guide D: "Transportation Systems in Buildings"

- **Purpose**: Focuses on the design of transportation systems like lifts, escalators, and moving walkways.
- **Key Points:**
  - While not HVAC-specific, it informs integration with building services.
  - Provides data on heat gains and ventilation requirements for lift systems.

CIBSE Guide E: "Fire Safety Engineering"

- **Purpose**: Addresses fire safety in building services engineering.

- **Key Points:**
  - Covers smoke control, stair pressurization, and fire compartmentation.
  - Provides guidance on HVAC integration with fire safety systems.

CIBSE Guide F: "Energy Efficiency in Buildings"

- **Purpose**: Focuses on strategies for energy-efficient building services.
- **Key Points:**
  - Highlights HVAC energy-saving measures, system optimization, and control strategies
  - Discusses renewable energy integration and lifecycle energy analysis.

CIBSE Guide G: "Public Health Engineering"

- **Purpose**: Provides design principles for water supply, drainage, and sanitary systems.
- **Key Points:**
  - Relevant to HVAC for integration of chilled water and hot water systems.
  - Includes considerations for legionella prevention.

CIBSE Guide H: "Building Control Systems"

- **Purpose**: Explains the principles and design of building control systems.
- **Key Points:**
  - Covers control strategies for HVAC systems.
  - Provides guidance on Building Management Systems (BMS) and sensors.

CIBSE Guide J: "Weather, Solar and Illuminance Data"

- **Purpose**: Offers detailed climatic data for building design.
- **Key Points:**
  - Provides external temperature, humidity, and solar radiation data.
  - Supports accurate HVAC load and system sizing.

CIBSE TM31: "Building Logbook Toolkit"

- **Purpose**: Aids in maintaining a comprehensive log of building operations.
- **Key Points:**
  - Ensures accurate documentation of HVAC and building systems.
  - Assists in energy monitoring and system optimization.

CIBSE TM39: "Building Energy Metering"

- **Purpose**: Guides metering of energy consumption in buildings.

- **Key Points:**
  - Encourages detailed energy monitoring for HVAC systems.
  - Recommends strategies for sub-metering and data analysis.

CIBSE AM10: "Natural Ventilation in Non-Domestic Buildings"

- **Purpose**: Offers guidance on designing natural ventilation systems.
- **Key Points:**
  - Provides principles for airflow paths, stack effects, and wind-driven ventilation.
  - Discusses hybrid ventilation systems integrating mechanical and natural methods.

CIBSE AM11: "Building Performance Modelling"

- **Purpose**: Provides guidelines for using simulation tools in building design.
- **Key Points:**
  - Supports HVAC system modelling for thermal comfort and energy efficiency.
  - Discusses the application of computational fluid dynamics (CFD).

CIBSE AM13: "Mixed Mode Ventilation"

- **Purpose**: Combines natural and mechanical ventilation strategies.
- **Key Points:**
  - Provides design principles for energy-efficient ventilation systems.
  - Includes case studies of successful implementations.

CIBSE TM52: "Limits of Thermal Comfort: Avoiding Overheating in European Buildings"

- **Purpose**: Defines criteria for avoiding overheating in buildings.
- **Key Points:**
  - Identifies key metrics for thermal comfort.
  - Suggests HVAC solutions for mitigating overheating risks.

CIBSE TM54: "Evaluating Operational Energy Performance of Buildings at the Design Stage"

- **Purpose**: Ensures buildings meet energy performance expectations post-construction.
- **Key Points:**
  - Discusses HVAC system impact on operational energy use.
  - Offers a framework for performance modelling during design.

Summary Table for HVAC Engineers

CIBSE guide	Focus area	Key HVAC application
Guide A	Environmental Design	Thermal comfort, load calculations, and U-values for envelope design
Guide B	HVAC and Refrigeration	Detailed design and operation of HVAC systems
Guide C	Reference Data	Psychrometrics, heat transfer, and material properties for calculations
Guide E	Fire Safety Engineering	Smoke control, pressurization, and fire-rated ductwork
Guide F	Energy Efficiency	HVAC system optimization and energy-saving measures
Guide H	Building Control Systems	Integration of HVAC with BMS and control strategies
TM31	Building Logbook Toolkit	Documentation for HVAC systems and operational monitoring
AM10	Natural Ventilation	Design of natural and hybrid ventilation systems
TM52	Avoiding Overheating	Ensures HVAC solutions mitigate overheating risks

14.5 Australian Standard

Followings are the **list of key Australian Standards (AS) relevant to HVAC and building services** and a brief description of what each standard covers. These Australian Standards ensure that HVAC systems meet safety, energy efficiency, and performance requirements specific to the region. Always consult the latest versions for compliance and accuracy.

AS/NZS 1668.1: "The Use of Ventilation and Air Conditioning in Buildings—Fire and Smoke Control in Buildings"

- **Purpose**: Focuses on fire and smoke control through ventilation systems.
- **Key Points:**
  - Provides design criteria for smoke spill systems and fire compartmentation.
  - Ensures safe operation of HVAC systems during fire events.

AS/NZS 1668.2: "The Use of Ventilation and Air Conditioning in Buildings—Mechanical Ventilation for Acceptable Indoor Air Quality"

- **Purpose**: Establishes ventilation rates and methods for maintaining indoor air quality.
- **Key Points:**

- Specifies minimum outdoor air requirements for various building types.
- Provides guidelines for duct and equipment design to ensure proper air circulation.

AS/NZS 3666 Series: "Air-Handling and Water Systems of Buildings—Microbial Control"

- **Purpose**: Addresses microbial control in HVAC systems.
- **Key Points:**
  - Ensures the design, installation, and maintenance of systems to minimize microbial contamination.
  - Divided into parts focusing on system cleanliness, cooling towers, and operational practices.

AS 4254 Series: "Ductwork for Air Handling Systems in Buildings"

- **Purpose**: Specifies requirements for ductwork used in HVAC systems.
- **Key Points:**
  - Covers materials, construction, installation, and testing of ductwork.
  - Ensures compliance with air leakage and fire safety standards.

AS/NZS 1894: "The Measurement of Airflow in Air-Conditioning Ducts"

- **Purpose**: Provides methods for measuring airflow rates in ducted systems.
- **Key Points:**
  - Ensures accurate airflow measurements for system commissioning and testing.
  - Includes guidelines for equipment calibration and placement.

AS/NZS 5149 Series: "Refrigerating Systems and Heat Pumps—Safety and Environmental Requirements"

- **Purpose**: Covers safety and environmental requirements for refrigerating systems.
- **Key Points:**
  - Ensures proper design, installation, and operation of refrigeration and heat pump systems.
  - Includes requirements for refrigerant handling, system pressures, and leak detection.

AS/NZS 4324: "Energy Efficiency of Air-Conditioners and Heat Pumps"

- **Purpose**: Specifies performance testing and energy efficiency standards for air-conditioning equipment.
- **Key Points:**

  - Provides a framework for calculating and labelling equipment efficiency (EER, COP).
  - Encourages energy-efficient designs and product selection.

AS/NZS 3000: "Electrical Installations (Wiring Rules)"

- **Purpose**: Details requirements for electrical installations, including HVAC systems.
- **Key Points:**
  - Ensures safe wiring practices and compliance with electrical codes.
  - Includes requirements for motor controls, switchgear, and electrical protection.

AS 1170.4: "Structural Design Actions—Earthquake Actions in Australia"

- **Purpose**: Provides guidelines for HVAC equipment and systems in earthquake-prone areas.
- **Key Points:**
  - Ensures secure mounting and bracing of HVAC equipment.
  - Mitigates risks associated with seismic activity.

AS 1682 Series: "Fire Dampers in Buildings"

- **Purpose**: Specifies requirements for the design and installation of fire dampers.
- **Key Points:**
  - Ensures fire dampers comply with fire-resistance ratings.
  - Details testing methods and installation criteria.

AS/NZS 4755: "Demand Response Capabilities and Equipment for HVAC Systems"

- **Purpose**: Focuses on demand-side management in HVAC systems.
- **Key Points:**
  - Supports integration of HVAC systems into smart grid technologies.
  - Provides guidelines for reducing peak energy demand.

AS/NZS 4859.1: "Thermal Insulation Materials for Buildings"

- **Purpose**: Specifies requirements for thermal insulation in HVAC systems.
- **Key Points:**
  - Includes material properties, performance, and installation methods.
  - Covers duct and pipe insulation for energy efficiency.

AS/NZS ISO 9001: "Quality Management Systems—Requirements"

- **Purpose**: Ensures quality assurance in the design and construction of HVAC systems.
- **Key Points:**

– Encourages systematic design processes and high construction standards.
– Focuses on customer satisfaction and regulatory compliance.

Summary Table for HVAC Engineers

Australian standard	Focus area	Key HVAC application
AS/NZS 1668.1	Fire and Smoke Control	Smoke spill systems and fire compartmentation in HVAC design
AS/NZS 1668.2	Indoor Air Quality	Ventilation rates and acceptable IAQ levels
AS/NZS 3666 Series	Microbial Control	Legionella prevention in cooling towers and air-handling systems
AS 4254	Ductwork	Design, construction, and testing of HVAC duct systems
AS/NZS 1894	Airflow Measurement	Accurate airflow measurement for commissioning
AS/NZS 5149 Series	Refrigeration Safety	Refrigerant handling and safety requirements for refrigeration systems
AS/NZS 4324	Energy Efficiency	Performance testing of air conditioners and heat pumps
AS/NZS 3000	Electrical Installations	Safe wiring and electrical practices for HVAC systems
AS 1170.4	Earthquake Actions	Seismic resilience of HVAC systems
AS 1682	Fire Dampers	Fire safety in ductwork with proper damper installation
AS/NZS 4755	Demand Response	Energy management for HVAC systems with smart grid integration
AS/NZS 4859.1	Thermal Insulation	Duct and pipe insulation to improve energy efficiency
AS/NZS ISO 9001	Quality Management	Systematic design and regulatory compliance for HVAC systems

14.6 Australian National Construction Code (NCC)

The **National Construction Code (NCC)** is Australia's overarching building regulation framework, containing mandatory performance standards and guidelines for safe, efficient, and sustainable construction. The NCC integrates requirements The NCC provides essential, climate-specific guidance for designing HVAC systems that comply with Australian standards for energy efficiency, safety, and performance. It serves as a legal framework to ensure sustainable and resilient designs tailored to Australian building codes. Always refer to the latest NCC version and verify compliance with local regulations for various building systems, including HVAC and building services, across three volumes.

NCC Structure Overview

1. **Volume 1**: Commercial and industrial buildings (Class 2–9 buildings).
2. **Volume 2**: Residential buildings (Class 1 and 10 buildings).
3. **Volume 3**: Plumbing and drainage provisions.

Following is a summary of key sections and provisions within the NCC that pertain to HVAC and building services, along with their relevance to system design and construction.

NCC Provisions Related to HVAC

Energy Efficiency (Section J, Volume 1)

- **Purpose**: Sets minimum energy performance standards for HVAC systems and other building services.
- **Key Points:**
  - Thermal performance of building envelopes, including insulation, glazing, and sealing.
  - Energy efficiency criteria for HVAC equipment, such as chillers, boilers, and fans.
  - Mandates the use of energy-efficient components, including variable speed drives (VSDs) and energy recovery systems.
  - Specifies system zoning to avoid over-conditioning.

Indoor Air Quality and Ventilation (F4, Volume 1)

- **Purpose**: Ensures acceptable indoor air quality (IAQ) for occupant health and comfort.
- **Key Points:**
  - Specifies minimum outdoor air ventilation rates based on building type and occupancy.
  - Sets requirements for natural and mechanical ventilation systems, including fresh air intakes and exhausts.
  - Ensures that air recirculation and filtration comply with IAQ standards.

Fire Safety (Part E, Volume 1)

- **Purpose**: Outlines requirements for fire resistance, detection, and suppression in building systems.
- **Key Points:**
  - Guidelines for the integration of smoke control systems with HVAC systems.
  - Specifies the use of fire and smoke dampers in ductwork to prevent fire spread.
  - Includes provisions for pressurization systems in fire stairwells and escape routes.

Sound Insulation and Acoustics (F5, Volume 1)

- **Purpose**: Reduces noise transmission from HVAC systems.
- **Key Points:**
  - Sets maximum noise levels for air-conditioning systems and mechanical equipment in different building spaces.

– Requires acoustic treatment for ductwork, fan rooms, and mechanical plantrooms.

Plumbing and Drainage (Volume 3)

- **Purpose**: Regulates water systems for HVAC equipment, including cooling towers, boilers, and heat exchangers.
- **Key Points:**
  - Guidelines for water quality control and legionella prevention in cooling towers.
  - Pipe sizing and layout for water-based HVAC systems, including chilled water and condenser water loops.
  - Backflow prevention and proper drainage for HVAC equipment.

Condensation Management (F6, Volume 1)

- **Purpose**: Addresses risks related to moisture and condensation in building envelopes.
- **Key Points:**
  - Ensures HVAC systems manage humidity levels to prevent mold and structural damage.
  - Guidelines for proper insulation and vapour barrier placement in ductwork and pipes.

Seismic and Structural Safety (Part B, Volume 1)

- **Purpose**: Ensures HVAC systems can withstand seismic and structural loads.
- **Key Points:**
  - Mandates secure mounting and bracing of mechanical equipment.
  - Sets criteria for rooftop units and external HVAC installations.

NCC Requirements for Specific Building Types

Building type	Relevant NCC provisions	Key HVAC considerations
High-rise buildings	Section J (Energy), F4 (Ventilation), E2 (Fire Safety)	Energy zoning, stairwell pressurization, smoke control systems
Hospitals	Section J (Energy), F4 (Ventilation), Part E (Fire Safety)	High outdoor air ventilation rates, air filtration, critical cooling requirements
Schools	Section J (Energy), F4 (Ventilation), F6 (Condensation)	IAQ management, natural ventilation integration, acoustic control in classrooms
Warehouses	Section J (Energy), F4 (Ventilation)	Natural ventilation, minimal mechanical cooling, insulation for temperature regulation
Industrial facilities	Section J (Energy), Volume 3 (Plumbing)	Exhaust systems for hazardous fumes, cooling systems for equipment, and water management for HVAC systems
Residential buildings	Volume 2 (Thermal performance), F4 (Ventilation)	Natural ventilation, energy-efficient HVAC units, and reduced noise levels

Key HVAC Design Rules of Thumb from NCC Guidance

- **Energy Efficiency:**
  - Use insulated ductwork with thermal resistance (R-value) meeting Section J requirements.
  - Incorporate energy recovery ventilators (ERVs) to improve overall system efficiency.
- **Ventilation:**
  - Outdoor air ventilation rates: 10–15 L/s/person for offices; higher for health-care and laboratories.
  - Use high-efficiency filters (MERV 13 or above) for improved IAQ.
- **Fire Safety:**
  - Locate smoke dampers at fire-rated barriers in duct systems.
  - Design stairwell pressurization systems to maintain a pressure difference of at least 50 Pa.
- **Acoustics:**
  - Limit HVAC noise to below 45 dB(A) in offices and below 35 dB(A) in bedrooms.
  - Apply silencers or acoustic lagging for noisy fans or equipment.
- **Plumbing:**
  - Cooling towers should have proper drift eliminators and water treatment systems to comply with microbial control guidelines.
  - HVAC drainage must connect to sanitary drainage per plumbing codes.

Comparison to NCC, ASHRAE and CIBSE

Aspect	NCC	ASHRAE	CIBSE
Climate-specific focus	Tailored to Australian conditions	Global focus, mainly for North America	Focused on the UK and European climates
Energy efficiency	Section J emphasizes energy compliance	Emphasizes building energy performance	Provides similar guidance for the UK
Fire safety integration	Includes robust smoke and fire controls	Covers fire but less regionally specific	Covers fire but not always mandatory

14.7 AHRI Standards

The **Air-Conditioning, Heating, and Refrigeration Institute (AHRI)**, formerly known as the **American Refrigeration Institute (ARI)**, publishes performance standards and guidelines for HVAC and refrigeration equipment. These standards are recognized internationally for evaluating the performance, efficiency, and reliability of HVAC systems and components. Below is a summary of key AHRI standards relevant to HVAC and building services, including their titles and brief descriptions of their content.

Key AHRI Standards and Their Applicability

AHRI 210/240: Performance Rating of Unitary Air-Conditioning and Air-Source Heat Pump Equipment

- **Purpose**: Establishes performance ratings for central air conditioners and heat pumps up to 65,000 BTU/h (19 kW).
- **Key Points:**
  - Provides criteria for measuring energy efficiency (SEER, EER, and HSPF).
  - Defines testing conditions and requirements for rating equipment under real-world scenarios.

AHRI 310/380: Packaged Terminal Air Conditioners and Heat Pumps

- **Purpose**: Covers packaged terminal equipment commonly used in hotels and motels.
- **Key Points:**
  - Specifies performance ratings and energy efficiency standards.
  - Addresses sound levels and capacity under standard conditions.

AHRI 340/360: Commercial Unitary Air-Conditioning and Heat Pump Equipment

- **Purpose**: Applies to commercial air-conditioning and heat pump systems from 65,000 to 250,000 BTU/h (19–73 kW).
- **Key Points:**
  - Includes provisions for rating cooling and heating performance.
  - Focuses on energy efficiency (IEER) for part-load conditions.

AHRI 550/590: Performance Rating of Water-Chilling and Heat Pump Water-Heating Packages Using the Vapour Compression Cycle

- **Purpose**: Evaluates the performance of chillers and water-cooled heat pumps.
- **Key Points:**
  - Defines methods to determine the cooling capacity and efficiency (COP, IPLV, and NPLV).
  - Specifies testing for both full-load and part-load operations.

AHRI 880: Performance Rating of Air Terminals

- **Purpose**: Standardizes performance metrics for air distribution terminals, including VAV boxes.
- **Key Points:**
  - Provides ratings for airflow, pressure drop, and sound generation.

  – Establishes consistency in comparing products across manufacturers.

AHRI 885: Procedure for Estimating Occupied Space Sound Levels

- **Purpose**: Establishes a methodology for predicting sound levels in occupied spaces from HVAC equipment.
- **Key Points:**
  – Offers guidance on noise attenuation through ductwork and equipment placement.
  – Aims to ensure occupant comfort by minimizing HVAC noise.

AHRI 440: Performance Rating of Room Fan-Coil Units

- **Purpose**: Evaluates fan-coil unit performance for heating and cooling applications.
- **Key Points:**
  – Includes criteria for capacity, airflow, and energy use.
  – Applicable to both 2-pipe and 4-pipe systems.

AHRI 700: Specification for Refrigerants

- **Purpose**: Defines purity requirements for refrigerants used in HVAC systems.
- **Key Points:**
  – Specifies acceptable levels of moisture, acidity, and contaminants.
  – Provides testing methods to ensure refrigerant quality.

AHRI 1060: Performance Rating of Air-to-Air Heat Exchangers for Energy Recovery Ventilation Equipment

- **Purpose**: Covers heat exchangers used in energy recovery ventilators (ERVs).
- **Key Points:**
  – Rates sensible and latent heat recovery efficiencies.
  – Establishes performance under varying airflow and temperature conditions.

AHRI 920: Performance Rating of DX-Dedicated Outdoor Air Systems

- **Purpose**: Focuses on systems designed to handle outdoor air ventilation loads.
- **Key Points:**
  – Defines metrics like Integrated Seasonal Moisture Removal Efficiency (ISMRE) and Integrated Energy Efficiency Ratio (IEER).
  – Ensures compliance with ventilation energy recovery standards.

AHRI 370: Sound Rating of Large Outdoor Refrigerating and Air-Conditioning Equipment

- **Purpose**: Establishes a consistent method for determining sound levels of large outdoor units.
- **Key Points:**

– Provides testing conditions for different environmental setups.
– Helps with noise mitigation strategies for equipment near sensitive zones.

Comparison of AHRI Standards with Other Organizations

Aspect	AHRI standards	ASHRAE standards	CIBSE guides
Focus	Performance testing and certification	Design practices, system integration	Design guidelines and operational standards
Application	Equipment-specific (chillers, coils, etc.)	Comprehensive, including IAQ and controls	Focused on UK/ European practices
Testing methods	Rigorous performance rating procedures	Includes both testing and theoretical design	More qualitative with practical insights
Energy efficiency	Strong focus on metrics like SEER, EER	Broader sustainability goals	Addresses efficiency at a macro level

Importance of AHRI Standards for HVAC Engineers

- **Equipment Selection:**
  – Helps compare performance ratings across manufacturers for chillers, heat pumps, and other HVAC components.
  – Provides validated efficiency and capacity metrics under standardized conditions.
- **Energy Efficiency:**
  – Promotes compliance with regional and international efficiency standards.
  – Encourages the selection of high-performance, environmentally friendly equipment.
- **Sound Control:**
  – AHRI standards on noise provide guidance for creating acoustically comfortable environments.
- **Refrigerant Quality:**
  – Ensures the use of high-quality refrigerants for system reliability and longevity.

Chapter 15
HVAC Software

15.1 Overview

Here's a comprehensive list of software commonly used in HVAC and building services for various tasks such as load calculations, duct and piping design, psychrometrics, and equipment sizing and selection. These tools cater to various stages of HVAC and building services design, from conceptual modelling to detailed engineering and simulation. The choice of software depends on the project's complexity, size, and specific design requirements.

15.2 Load Calculation Software

- **Carrier HAP (Hourly Analysis Program)**
  - Load calculation, energy analysis, and compliance with ASHRAE standards.
- **Trane TRACE 3D Plus**
  - Cooling and heating loads, energy modelling, and life-cycle cost analysis.
- **eQuest**
  - Free energy simulation tool for load calculations and energy optimization.
- **IES VE (Integrated Environmental Solutions Virtual Environment)**
  - Advanced energy modelling for sustainable building design.
- **EnergyPlus**
  - Open-source simulation engine for load calculations and energy efficiency.
- **CHVAC by Elite Software**
  - Simplified HVAC load calculation.

V. Vakiloroaya, *HVAC and Building Services Rules of Thumb*, Springer Tracts in Civil Engineering, https://doi.org/10.1007/978-3-032-10285-0_15

- **TAS Software**
  - Energy modelling and thermal analysis for complex projects.

15.3 Duct Design and Calculation Software

- **DuctSizer by Linric Company**
  - Simplifies duct sizing calculations.
- **Elite Ductsize**
  - Calculates pressure losses and duct sizes based on airflow.
- **MAGiCAD Ventilation**
  - BIM-based software for duct design integrated with Revit.
- **SMACNA Duct Design Tools**
  - Tools based on SMACNA standards for duct design and performance.
- **Revit MEP (Autodesk)**
  - Duct layout and analysis integrated with 3D BIM modelling.

15.4 Piping Design and Calculation Software

Used for water and refrigerant piping in HVAC systems.

- **Pipe Flow Expert**
  - Calculates flow rates, pressure losses, and pipe sizing.
- **Elite Software: Pipe Sizer**
  - Calculates hydronic and steam piping systems.
- **CAREL HumidAir**
  - Specialized for humidification piping calculations.
- **MAGiCAD Piping**
  - Piping design for HVAC systems integrated with Revit.
- **AutoSPRINK**
  - Fire suppression and piping network design.

15.5 Psychrometric Analysis Software

- **PsyCalc**
  - Simplifies psychrometric chart calculations for HVAC design.
- **PsychroSuite**
  - Advanced psychrometric analysis for air handling processes.
- **Carrier Psychrometric Chart Software**
  - Digitized version of traditional psychrometric charts.
- **ASHRAE Psychrometric Analysis Tools**
  - For processes based on ASHRAE guidelines.

15.6 HVAC System and Equipment Sizing Software

- **Mitsubishi Electric Heat Load Calculation and Equipment Selection Tool**
  - For selecting heat pumps and VRF systems.
- **Daikin VRV Xpress**
  - VRF system sizing and selection.
- **Trane Selector**
  - Equipment selection and configuration.
- **HVAC Solution Professional**
  - Comprehensive tool for system design and equipment selection.
- **McQuay Tools**
  - For sizing and selecting chillers, AHUs, and other HVAC equipment.

15.7 Energy Modelling and Simulation Software

- **DesignBuilder**
  - Uses EnergyPlus for advanced building energy modelling.
- **OpenStudio**
  - Free and open-source interface for EnergyPlus simulations.
- **IDA ICE**
  - Energy and comfort simulation for detailed analysis.
- **Green Building Studio (Autodesk)**
  - Cloud-based simulation for energy performance.

15.8 BIM-Based HVAC Design Software

- **Autodesk Revit MEP**
  - BIM-based HVAC design and documentation.
- **MAGiCAD for Revit and AutoCAD**
  - Advanced HVAC design and calculation integrated with BIM.
- **Trimble Nova**
  - HVAC design with a focus on BIM workflows.

15.9 CFD (Computational Fluid Dynamics) Software

- **ANSYS Fluent**
  - Advanced CFD for HVAC airflow and heat transfer analysis.
- **Autodesk CFD**
  - Simulation for airflow and temperature distribution.
- **SimScale**
  - Cloud-based CFD for HVAC and airflow optimization.

15.10 Specialized Tools

- **Coolselector by Danfoss**
  - Refrigeration and cooling system optimization.
- **TRACE Airside by Trane**
  - Airside system design and analysis.
- **IMI Hydronic Engineering Software**
  - Balancing and control in hydronic systems.
- **AHRI Certified Product Selection Tools**
  - Ensures compliance with AHRI standards.

15.11 General CAD and Drafting Tools

- **AutoCAD MEP**
  - Mechanical, electrical, and plumbing design.
- **SketchUp with Extensions**
  - Conceptual HVAC layouts and coordination.

Chapter 16
HVAC Abbreviations and Conversion Factors

16.1 Comprehensive List of HVAC Acronyms and Abbreviations

This comprehensive list provides an overview of common HVAC acronyms and abbreviations that HVAC engineers and professionals use in the building services industry.

General HVAC Terms

- **HVAC**: Heating, Ventilation, and Air Conditioning
- **AHU**: Air Handling Unit
- **RTU**: Rooftop Unit
- **VAV**: Variable Air Volume
- **CAV**: Constant Air Volume
- **FCU**: Fan Coil Unit
- **DX**: Direct Expansion
- **BMS**: Building Management System
- **BAS**: Building Automation System
- **ERV**: Energy Recovery Ventilator
- **HRV**: Heat Recovery Ventilator
- **DOAS**: Dedicated Outdoor Air System
- **VRF**: Variable Refrigerant Flow
- **VRV**: Variable Refrigerant Volume
- **IAQ**: Indoor Air Quality

Refrigeration and Cooling

- **COP**: Coefficient of Performance
- **EER**: Energy Efficiency Ratio
- **SEER**: Seasonal Energy Efficiency Ratio

V. Vakiloroaya, *HVAC and Building Services Rules of Thumb*, Springer Tracts in Civil Engineering, https://doi.org/10.1007/978-3-032-10285-0_16

- **HSPF**: Heating Seasonal Performance Factor
- **CHW**: Chilled Water
- **CHWR**: Chilled Water Return
- **CHWS**: Chilled Water Supply
- **CW**: Condenser Water
- **CWR**: Condenser Water Return
- **CWS**: Condenser Water Supply
- **TXV**: Thermostatic Expansion Valve
- **EXV**: Electronic Expansion Valve

Heating Systems

- **HW**: Hot Water
- **HWR**: Hot Water Return
- **HWS**: Hot Water Supply
- **AFUE**: Annual Fuel Utilization Efficiency
- **FTHW**: Fin Tube Hot Water
- **RTF**: Radiant Tube Furnace

Ductwork and Airflow

- **CFM**: Cubic Feet per Minute
- **L/s**: Litters per Second
- **FPM**: Feet per Minute
- **ESP**: External Static Pressure
- **TSP**: Total Static Pressure
- **NC**: Noise Criterion
- **OD**: Outside Diameter
- **ID**: Inside Diameter

Filters and Ventilation

- **MERV**: Minimum Efficiency Reporting Value
- **HEPA**: High-Efficiency Particulate Air
- **ULPA**: Ultra-Low Penetration Air
- **OAD**: Outside Air Damper
- **VCD**: Volume Control Damper
- **OA**: Outside Air
- **EA**: Exhaust Air
- **RA**: Return Air
- **SA**: Supply Air

Electrical and Controls

- **VFD**: Variable Frequency Drive
- **PLC**: Programmable Logic Controller
- **UPS**: Uninterruptible Power Supply
- **ATS**: Automatic Transfer Switch
- **kW**: Kilowatt

- **kVA**: Kilovolt Ampere
- **PF**: Power Factor
- **RLA**: Rated Load Amperes
- **FLA**: Full Load Amperes

Psychrometrics and Moisture Control

- **DBT**: Dry Bulb Temperature
- **WBT**: Wet Bulb Temperature
- **RH**: Relative Humidity
- **GPP**: Grains per Pound
- **DPT**: Dew Point Temperature
- **SHF**: Sensible Heat Factor
- **SHR**: Sensible Heat Ratio

Codes and Standards

- **ASHRAE**: American Society of Heating, Refrigerating, and Air-Conditioning Engineers
- **IES**: Illuminating Engineering Society
- **NFPA**: National Fire Protection Association
- **IECC**: International Energy Conservation Code
- **IBC**: International Building Code
- **LEED**: Leadership in Energy and Environmental Design
- **EPA**: Environmental Protection Agency

Energy and Efficiency

- **BTU**: British Thermal Unit
- **kWh**: Kilowatt-hour
- **TR**: Ton of Refrigeration (1 TR = 12,000 BTU/h)
- **W/m^2**: Watts per Square Meter
- **EUI**: Energy Use Intensity
- **ESI**: Energy Savings Index

Piping and Hydronic

- **GPM**: Gallons per Minute
- **LPM**: Liters per Minute
- **NPSH**: Net Positive Suction Head
- **PSI**: Pounds per Square Inch
- **kPa**: Kilopascal
- **FTWG**: Feet of Water Gauge
- **HSP**: Heat Source Pump
- **HRP**: Heat Recovery Pump

Combustion and Fuels

- **NG**: Natural Gas
- **LPG**: Liquefied Petroleum Gas

- **NOx**: Nitrogen Oxides
- **CO_2**: Carbon Dioxide
- **CO**: Carbon Monoxide

Miscellaneous

- **ETD**: Entering Temperature Difference
- **LWT**: Leaving Water Temperature
- **FWT**: Flow Water Temperature
- **RWT**: Return Water Temperature
- **U-value**: Thermal Transmittance
- **SDC**: Smoke Detection Control

16.2 Psychrometric Process

Psychrometric involves the study of moist air and its properties, which are critical for understanding and designing HVAC systems. The psychrometric chart is an indispensable tool for HVAC engineers. It allows for the detailed analysis and design of air-conditioning processes, ensuring efficient, comfortable, and cost-effective HVAC systems. Mastery of this chart is fundamental to achieving optimal system performance and meeting indoor environmental quality standards.

Uses of the Psychrometric Chart

- **Understanding Air Properties:**
  - It visually represents the relationships between air temperature, humidity, enthalpy, and other properties.
  - Helps engineers understand how air behaves under different conditions.
- **Process Analysis:**
  - Allows the plotting of air-conditioning processes such as heating, cooling, humidification, and dehumidification.
- **Troubleshooting:**
  - Provides insights into potential issues in HVAC systems, like condensation or improper humidity control.
- **Efficiency Optimization:**
  - Identifies the most energy-efficient methods to achieve desired air conditions.

Applications in HVAC Design

- **Cooling and Heating Load Calculations:**
  - By plotting the initial and desired air conditions, engineers can determine the amount of heat to be added or removed.

- **Moisture Control:**
  - Helps calculate the amount of moisture to be added (humidification) or removed (dehumidification) from the air.
- **Air Mixing:**
  - Used to analyse the properties of mixed air streams, such as outdoor air mixing with recirculated indoor air.
- **Equipment Sizing:**
  - Assists in selecting and sizing HVAC components like fans, coils, humidifiers, and dehumidifiers based on air properties.
- **Energy Recovery Systems:**
  - Evaluates the potential for energy recovery by analysing enthalpy changes.
- **Ventilation Design:**
  - Ensures that indoor air quality standards are met by maintaining proper humidity and temperature levels.

Importance of the Psychrometric Chart in HVAC

- **Visual Representation:**
  - Simplifies complex thermodynamic calculations by providing a graphical view of air properties.
- **Precision in Design:**
  - Ensures accurate sizing of HVAC equipment and systems, which improves performance and reduces energy consumption.
- **Energy Efficiency:**
  - Helps identify energy-saving opportunities by analysing the air-conditioning process.
- **Comfort and Health:**
  - Ensures optimal indoor environmental conditions, improving occupant comfort and preventing issues like mold growth due to improper humidity control.
- **System Integration:**
  - Facilitates the integration of various HVAC components to achieve a cohesive and efficient system.

Key Components of the Psychrometric Chart

Dry-Bulb Temperature (DBT): Horizontal axis; the temperature of air measured with a standard thermometer.

Humidity Ratio: Vertical axis; the amount of water vapour per unit of dry air (g/kg or lb/lb).

Wet-Bulb Temperature (WBT): Indicates the air's temperature when cooled to saturation by evaporation.

Relative Humidity (RH): Curved lines; the percentage of moisture content relative to the air's maximum capacity at a given DBT.

Enthalpy: Diagonal lines; the total heat content (sensible + latent) of the air.

Dew Point Temperature: The temperature at which air becomes saturated and water vapour condenses.

Saturation Line: The curve showing 100% relative humidity.

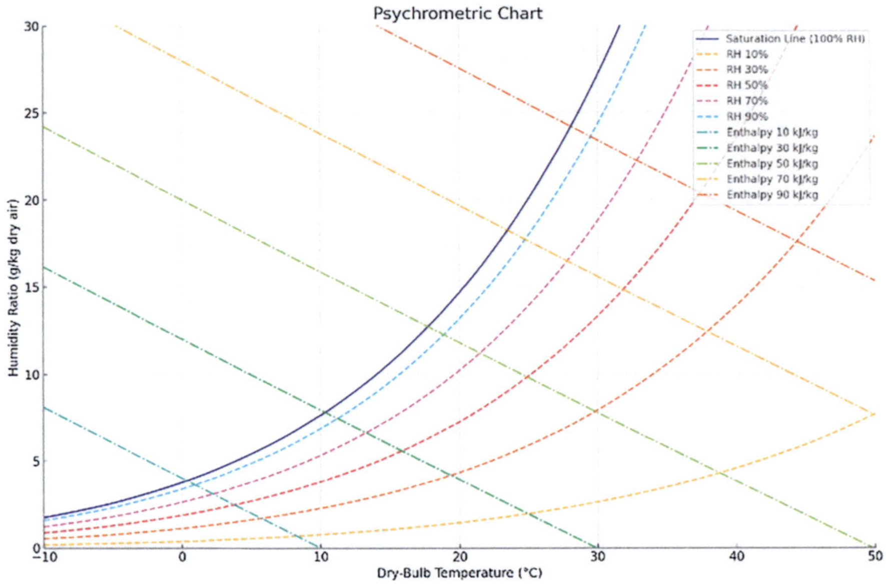

Psychrometric Chart

Dry-Bulb Temperature (°C): Horizontal axis.

Humidity Ratio (g/kg dry air): Vertical axis.

Saturation Line (100% RH): Blue line.

Relative Humidity Lines: Dashed lines for 10%, 30%, 50%, 70%, and 90%.

Enthalpy Lines: Dashed-dotted lines for enthalpy values of 10, 30, 50, 70, and 90 kJ/kg.

Sensible Heating

- **Definition**: Increasing the air temperature without changing its moisture content.
- **Process**: Moves horizontally to the right on the psychrometric chart.
- **Applications**: Space heaters, reheaters, heating coils.

Sensible Cooling

- **Definition**: Decreasing the air temperature without changing its moisture content.
- **Process**: Moves horizontally to the left on the psychrometric chart.
- **Applications**: Cooling coils, air-conditioning systems.

Latent Heating

- **Definition**: Increasing the air's moisture content without changing its temperature.
- **Process**: Moves vertically upward on the psychrometric chart.
- **Applications**: Humidification processes.

Latent Cooling

- **Definition**: Decreasing the air's moisture content without changing its temperature.
- **Process**: Moves vertically downward on the psychrometric chart.
- **Applications**: Dehumidifiers.

Sensible and Latent Heating (Humidification)

- **Definition**: Increasing air temperature and moisture content simultaneously.
- **Process**: Moves diagonally upward to the right.
- **Applications**: Steam humidifiers.

Sensible and Latent Cooling (Dehumidification)

- **Definition**: Decreasing air temperature and moisture content simultaneously.
- **Process**: Moves diagonally downward to the left.
- **Applications**: Air-conditioning coils.

Evaporative Cooling

- **Definition**: Reducing air temperature by adding moisture, with no heat exchange with the environment (adiabatic process).
- **Process**: Moves along a constant enthalpy line.
- **Applications**: Evaporative coolers.

Evaporative Cooling Process in Chart

- **Process Overview:**
  - In evaporative cooling, water absorbs heat from the air to evaporate.
  - The heat comes from the air's sensible heat, causing the dry-bulb temperature to decrease.
  - The air's humidity ratio increases because water vapor is added.
- **Thermodynamic Principles:**
  - The total energy (enthalpy) of the air remains nearly constant.
  - Sensible heat is converted into latent heat as water changes from liquid to vapour.
- **Conditions Favouring Evaporative Cooling:**
  - Hot, dry air conditions provide maximum cooling potential.
  - The process is less effective in humid climates where the air is already saturated with moisture.

Mixing of Air Streams

- **Definition**: Combining two or more air streams with different conditions to produce a mixture with intermediate properties.
- **Process**: Moves to a point between the initial air streams based on mass or volume flow rates.
- **Applications**: Mixing boxes, return air systems.

Heating and Humidification

- **Definition**: Simultaneously increasing air temperature and moisture content.
- **Process**: Moves diagonally upward on the psychrometric chart.
- **Applications**: Steam humidifiers with heating coils.

Cooling and Dehumidification

- **Definition**: Reducing both air temperature and moisture content.
- **Process**: Moves downward diagonally on the psychrometric chart.
- **Applications**: Cooling coils in air-conditioning systems.

Adiabatic Mixing

- **Definition**: Air streams mix without any heat transfer to the surroundings.
- **Process**: Line between the conditions of the mixed air streams.
- **Applications**: Mixing return and outdoor air.

16.3 Conversion Factors

Conversions from SI to English

Quantity	SI unit	English unit	Conversion factor	Formula
Length	meter (m)	inch (in)	1 m = 39.3701°in	in = m × 39.3701
	meter (m)	foot (ft)	1 m = 3.28084°ft	ft = m × 3.28084
	millimetre (mm)	inch (in)	1 mm = 0.0393701°in	in = mm × 0.0393701
	kilometre (km)	mile (mi)	1 km = 0.621371°mi	mi = km × 0.621371
Area	square meter (m^2)	square foot (ft^2)	1 m^2 = 10.7639°ft^2	ft^2 = m^2 × 10.7639
	square meter (m^2)	square inch (in^2)	1 m^2 = 1550°in^2	in^2 = m^2 × 1550
	hectare (ha)	acre	1 ha = 2.47105°acre	acre = ha × 2.47105
Volume	cubic meter (m^3)	cubic foot (ft^3)	1 m^3 = 35.3147°ft^3	ft^3 = m^3 × 35.3147
	litter (L)	gallon (gal)	1 L = 0.264172°gal	gal = L × 0.264172
	litter (L)	cubic inch (in^3)	1 L = 61.0237°in^3	in^3 = L × 61.0237

Quantity	SI unit	English unit	Conversion factor	Formula
Mass	kilogram (kg)	pound (lb)	1 kg = 2.20462°lb	lb = kg × 2.20462
	gram (g)	ounce (oz)	1 g = 0.035274°oz	oz = g × 0.035274
Force	newton (N)	pound-force (lbf)	1 N = 0.224809°lbf	lbf = N × 0.224809
Pressure	pascal (Pa)	pound per square inch (psi)	1 Pa = 0.000145038°psi	psi = Pa × 0.000145038
	kilopascal (kPa)	inches of water (inH_2O)	1 kPa = 4.01865°inH_2O	inH_2O = kPa × 4.01865
Energy	joule (J)	British thermal unit (BTU)	1 J = 0.000947817°BTU	BTU = J × 0.000947817
	kilowatt-hour (kWh)	BTU	1 kWh = 3412.14°BTU	BTU = kWh × 3412.14
Power	watt (W)	horsepower (hp)	1 W = 0.00134102°hp	hp = W × 0.00134102
Temperature	Celsius (°C)	Fahrenheit (°F)	°C = (°F-32)/1.8	°F = (°C × 1.8) + 32
	Kelvin (K)	Rankine (R)	R = K × 1.8	R = K × 1.8
Flow rate	litter per second (L/s)	cubic feet per minute (CFM)	1°L/s = 2.11888°CFM	CFM = L/s × 2.11888

MIX
Papier aus verantwortungsvollen Quellen
Paper from responsible sources
FSC® C105338

If you have any concerns about our products, you can contact us on
ProductSafety@springernature.com

In case Publisher is established outside the EU, the EU authorized representative is:
Springer Nature Customer Service Center GmbH
Europaplatz 3, 69115 Heidelberg, Germany

Printed by Libri Plureos GmbH
in Hamburg, Germany